the design of educational experiments

the design of educational experiments

c. mitchell dayton

Associate Professor of Education
University of Maryland

mcgraw-hill book company

New York
St. Louis
San Francisco
Düsseldorf
London
Mexico
Panama
Sydney
Toronto

This book was set in News Gothic by Trade Composition, Inc., and printed on permanent paper and bound by The Maple Press Company. The designer was J. E. O'Connor; the drawings were done by John Cordes, J. & R. Technical Services, Inc. The editors were Nat LaMar and Antonia Stires. Sally R. Ellyson supervised the production.

the design of educational experiments

Library of Congress Catalog Card Number 72-107285

16174

1 2 3 4 5 6 7 8 9 0 M A M M 7 9 8 7 6 5 4 3 2 1 0

preface

An author undertaking the exposition of advanced statistical procedures for students in the field of education is faced with an enigmatic situation. The bulk of the available advanced procedures was developed two to four decades ago by statisticians applying their talents to design problems in the agricultural and biological sciences. Thus, the traditional array of designs is uniquely suited to these disciplines but not to research in the behavioral sciences. On the other hand, the behavioral sciences have stimulated the development of relatively few designs. In order to expose the student to a variety of techniques for analyzing experiments, there is the temptation to dwell on many procedures which have little real or apparent applicability to the behavioral sciences (e.g., the incomplete-blocks designs). However, to omit these procedures entirely would leave the student unaware of some interesting statistical design principles and unable to comprehend fully some techniques available in popular reference volumes. Also, it is possible that the behavioral scientist has not used some designs popular in other disciplines, not because these designs are not applicable but because he has simply failed to utilize fully the available statistical tools.

This textbook is designed to provide a workable compromise between exhaustive coverage of designs and the practical facts of their utility. Those experimental designs which have proved to be widely usable in education and related research fields are presented in detail along with worked examples of their analysis and interpretation. Less popular and apparently less applicable designs are described in terms of their underlying principles, but the details of analysis and illustration are left to outside references. In general, an attempt has been made to refer to other works which present techniques in a light most congenial to behavioral scientists. Thus, the student may content himself with a pragmatic mastery of popular designs in his own field, or he may press on toward the more recondite regions of experimental design by systematically pursuing the cited references.

Today, failure to achieve exhaustive coverage of the field or experimental design cannot be considered a serious shortcoming in a design book. The available material, if collected into one source, would fill not one but a series of volumes. Thus, any textbook suited for a single coherent course in experimental design must severely restrict its coverage. It is hoped that the selection embodied in this work will prove to be useful in educational design courses.

Among the features contributing to the uniqueness of the present textbook are the following: (1) techniques have been chosen for exposition in terms of their relevance to educational research; (2) no mathematical training beyond ordinary algebra is assumed; (3) explicit coverage is given to cases in which groups of experimental subjects are unequal and disproportionate in size; (4) nested designs and repeated measures designs are treated in detail with educational examples; and (5) emphasis is placed on sets of orthogonal contrasts; higher-order factorial, nested, and other designs are developed in terms of building linear contrasts on the cells of a completely randomized design. This approach prepares the student for utilization of modern linear hypothesis computer programs.

Within each chapter, the discussion of new experimental designs begins with a consideration of general design characteristics and some of the theory underlying the design. Next, hypothetical data are utilized to illustrate the analytical procedures and interpretations which are consistent with the particular design. Finally, summaries of actual examples of the use of the design are presented. Although the worked examples involving hypothetical data do contain the raw data upon which computations are based, it is ordinarily impractical to present such data for the summaries of real research applications. Thus, the analyses of hypothetical data allow the student to follow all computational steps in detail, and the examples of actual research situations provide a realistic framework in which to discuss the interpretation and applicability of the research designs.

Appendix A presents a brief introduction to the elements of matrix algebra. Certain specialized procedures in Chapters 7 and 8 have required the use of matrix notation; although those sections which depend upon matrix notation could be omitted without any serious loss of continuity in terms of an overall understanding of experimental designs, the student who intends to undertake further pursuit of applied statistics after leaving this volume will need to acquire a working knowledge of elementary matrix notation and operations.

I wish to express my appreciation to the students in my advanced statistics course at the University of Maryland who, over the past two years, have read, criticized, and offered suggestions for revision of a draft version of this textbook. Also, the suggestions of an anonymous reviewer who read the manuscript for the publisher were invaluable in improving the clarity and continuity of the text. Finally, I am grateful for permission from several authors and publishers to reproduce tables and other materials from their books and articles. Specifically, Appendix Tables B-6 and B-7 are reproduced with the permission of E. S. Pearson for the Biometrika Trustees; Appendix Table B-8 is reproduced with the permission of Holt, Rinehart, and Winston, Inc.; Appendix Table B-9 with the permission of the American Statistical Association and C. W. Dunnett; and Appendix Table B-10 with the permission of the American Statistical Association and E. B. Page. In addition, the American Educational Research Association, the American Psychological Association (along with G. E. Tagatz and R. C. Anderson), and the Psychometric Society (along with L. S. Feldt) graciously allowed me to quote certain tabular and narrative materials within the textbook.

C. Mitchell Dayton

contents

the design of educational experiments

chapter one

general design principles

introduction

This volume is concerned with statistical designs in which there is one response variable (or dependent variable, or criterion) measured on each experimental subject. These procedures are referred to as *univariate,* as distinguished from *multivariate* techniques in which two or more intercorrelated response variables are included. Note that the term *multivariate analysis* is restricted to cases in which the response variables are interdependent; if two or more response variables are independent, or uncorrelated, the analysis is equivalent to a separate univariate analysis performed with respect to each response variable, ignoring the remaining variables. Certain of the designs considered in this volume may appear, on initial inspection, to belong in the multivariate category. This is true of the analysis of covariance and the repeated measures designs. In the analysis of covariance, a response variable and one or more covariables are measured on each experimental subject. In general, these variables are intercorrelated. However, in the present context, the procedure is classified as univariate since there is only one *response* measure which is subject to the effects of the experimental treatments. The covariables enter into the analysis but do not represent criteria. The case of repeated measures designs is somewhat different. In these designs, each experimental subject receives two or more combinations of treatments and yields a score under each different combination. The analysis is, strictly speaking, of a multivariate nature; however, very restrictive assumptions concerning the nature of the intercorrelations of scores are made by the analytical techniques, and the procedures involved are highly similar to those employed for univariate designs.

treatment variables

All the techniques discussed in this volume are "experimental" designs in the sense that one or more treatment variables, each at two or more levels, are applied to groups of experimental subjects. It is assumed that these treatments, and combinations of these treatments, *may* have differential effects upon the average value of the response measure. Thus, an experimental design is essentially a contrivance to study the differential effects, if any, of the treatment variables upon the response variable. In general, if combinations of two or more treatment variables are simultaneously applied to groups of experimental subjects, it is reasonable to assume that these treatment variables will interact among themselves so that the effect of a combination of two treatment variables is not directly predictable from the individual effects of the separate treatment variables. This fact is fundamental in all the more complex designs; indeed, an understanding of interaction effects is often the most interesting outcome of an educational experiment.

Treatment variables are, in some sense, under the control of the experimenter. This control may be direct, as when the experimenter deliberately manipulates a treatment variable by setting its levels at predetermined values for groups of experimental subjects, or indirect, as when the experimenter makes use of a priori categorizations to group experimental subjects (e.g., according to sex of experimental subject). In the first case we speak of a *manipulated-type* treatment variable; in the second case, of a *selection-type* treatment variable. This distinction, although useful from a general design point of view, has no substantial bearing on the analysis of experiments. That is, it is generally *statistically* irrelevant whether the treatment variables are manipulated or selected, or some mixture of these types is involved. On the other hand, the way in which the levels of the treatment variables are chosen can influence the analysis of the experiment. In view of this, we distinguish two methods of selecting the treatment levels which are to be involved in an experiment. The first, and more common, procedure involves deliberate selection of particular levels and application of these to the experimental subjects. For example, in comparing the relative effectiveness of teaching methods, the experimenter selects the actual methods on a judgmental basis aimed at elucidating contrasts of interest. Thus, one treatment level may be conventional lecture instruction, a second may be closed-circuit television instruction, and a third may be independent study, all based on the same instructional material. Here, the treatment variable may be described as "mode of presenting instructional content." The levels have been chosen by the experimenter to supply specific contrasts of presentation modes. Other possible modes of presentation (e.g., programmed instruction) have been excluded from the experiment by the researcher. A treatment variable whose levels are selected by the general method of this example is called a *fixed-dimension* or a *fixed-effects* treatment variable, since the levels are arbitrarily "fixed" by the experimenter.

A second type of treatment variable, known as a *random-dimension* or a *random-effects* treatment variable, occurs when the treatment levels are randomly selected by the experimenter from a larger set of possible levels. Such a random selection of treatment levels is rarely used with manipulated-type treatment variables. However, in the case of a classification variable which is used to form matched sets of experimental subjects, the matching variable is often of the random-effects type. If the experimental subjects are randomly selected from a population of available subjects and then matched according to, say, their relative rankings on the matching variable, the resulting levels of this selection-type treatment variable are dependent upon the particular random selection of subjects and, hence, constitute a random dimension.

The analysis of an experimental design, whenever more than just one treatment variable is involved in the design, is affected by the types of treatment variables (fixed-effects or random-effects) and their combinations. Thus, throughout this volume, consideration is given to analysis in terms of these two types of treatment variables. Fortunately, no severe complications are introduced by this strategy since the types of treatment variables have an effect only on the later stages of analysis, namely, setting up appropriate F ratios to test specific null hypotheses.

One additional distinction between treatment variables must be recognized in order to fully utilize available analytical procedures. A treatment variable may be either *quantitative* or *qualitative*. A quantitative treatment variable exists when the treatment levels can be scaled in terms of equal units of a characteristic underlying the treatment variable or when, at least, the treatment levels can be rank-ordered with respect to such an underlying characteristic. In the terminology of Stevens (1951, 1968), quantitative treatment variables are based on ordinal, interval, or ratio scales of measurement. On the other hand, if the treatment variable is defined by only a nominal scale of measurement, it is a qualitative treatment variable. That is, the treatment levels are qualitative categorizations rather than points along a quantitative dimension.

For quantitative treatment variables, the experimenter can explore the functional relationship between the treatment dimension and the response variable in the experiment. In fact, for treatment variables which are on at least interval scales, the form of this relationship can be systematically explored in terms of linear, quadratic, and higher-order functions. If a treatment variable is qualitative, no meaning can be attached to the notions of linearity and curvilinearity.

Quantitative treatment variables are rarely utilized in educational research. Treatment levels are often relatively complex aggregates of effects and are only qualitatively distinguishable. This is obviously an accurate description of treatment levels based on instructional methods. No quantitative description of such levels is usually attempted, and often none may be feasible. Live lectures and video-taped lectures represent different modes of presentation of instructional content, but stating this difference in quantitative terms has not been accomplished. The failure of educational researchers to design experiments based on quantitative treatment variables is considered by this author to be a factor which has had a generally unfavorable effect upon progress in understanding educational phenomena. The investigation of complex treatment levels may have some immediate practical utility, but few generalizable results can be found in this way. Since complex treatment levels are difficult, if not impossible, to replicate in future experiments, there is little carry-over from experiment to experiment, and many of the apparently contradictory outcomes from educational experiments can be explained by this fact. In any scientific discipline, experimenters must ultimately focus on the elucidation of functional relationships, and this calls for the investigation of quantitative treatment variables. This does not, of course, rule out studying qualitative selection-type treatment variables, such as sex of experimental subject, within the same experiment.

general types of designs

The purpose of this section is to sketch briefly the types of experimental designs considered in this volume and to introduce a scheme for figuratively representing experimental designs. Experimental designs may be classified primarily in terms of the number of treatment variables they comprise, how these treatment variables are combined for simultaneous application to groups of experimental subjects, and the way in which groups of experimental subjects are assigned to treatment combinations. For representational purposes, capital letters are used for treatment variables, and numerals for the levels of a treatment variable. Also, each group of experimental subjects is denoted G, with subscripts to index the specific group. This scheme will be clarified by the ensuing examples. In the next chapter a notation for individual scores, sums of scores, and so forth, is introduced.

The simplest experimental design is one involving a single treatment variable which is applied at two or more levels. If each level of the treatment is applied to an independent, random group of experimental subjects, the design in Table 1-1 is obtained (the treatment variable is assumed to have four levels). In general, the number of experimental subjects in a group may be the same or different. The one-treatment-variable independent-groups design is used in this textbook as a vehicle for explicating a large number of basic concepts related to the analysis of experiments. For this reason, the somewhat shorter title "com-

table 1-1 *Completely randomized design (CRD)*

	A		
1	2	3	4
G_1	G_2	G_3	G_4

table 1-2 *Two-dimensional factorial design*

		A		
		1	2	3
B	1	G_{11}	G_{12}	G_{13}
	2	G_{21}	G_{22}	G_{23}
	3	G_{31}	G_{32}	G_{33}
	4	G_{41}	G_{42}	G_{43}

pletely randomized design" is used, and the abbreviation CRD will often be utilized for convenience of reference. Chapter 2 is devoted to the completely randomized design.

If two treatment variables are involved in an experimental design *and* if each possible different combination of treatment levels is applied to an independent, random group of experimental subjects, the design is described as a two-dimensional *factorial* design (see Table 1-2, in which treatment variable A has three levels and treatment variable B has four levels). If treatment variable A has p levels and B has q levels, a total of pq groups of experimental subjects is required. Note the use of two subscripts on the G's; the first subscript indicates a level of treatment variable B (a row of the table), and the second subscript indicates a level of treatment variable A (a column of the table). The notion of factorial designs can be easily extended to more than two treatment variables. For three treatment variables A, B, and C with p, q, and r levels, respectively, a total of pqr groups of experimental subjects is needed in the experiment. To represent designs of three or more dimensions, the figure is "collapsed" into two dimensions, as in Table 1-3, for a design with three, two, and four levels. Note that the three subscripts on the G's stand for the levels of C, B, and A, respectively. Higher-dimensional factorial designs can be similarly collapsed for representational purposes. Factorial designs are discussed in Chapter 3.

table 1-3 *Three-dimensional factorial design*

		A					
		1		2		3	
B:		1	2	1	2	1	2
C	1	G_{111}	G_{121}	G_{112}	G_{122}	G_{113}	G_{123}
	2	G_{211}	G_{221}	G_{212}	G_{222}	G_{213}	G_{223}
	3	G_{311}	G_{321}	G_{312}	G_{322}	G_{313}	G_{323}
	4	G_{411}	G_{421}	G_{412}	G_{422}	G_{413}	G_{423}

In so-called "incomplete factorial designs" not all the possible different treatment-level combinations are actually applied to groups of experimental subjects. The principles for building such designs are discussed in Chapter 4.

If one (or more) of the treatment variables in a factorial design is a classification variable used for blocking experimental subjects into relatively homogeneous strata, the design is also described as a *randomized blocks* design. This type of design can be represented in the same way as a factorial design. Chapter 5 is devoted to randomized blocks designs.

The next two categories of designs are based on principles which represent departures from the designs previously discussed. *Nested designs* (Chapter 6), in the simplest case, have one major treatment variable with each of its levels applied to *two or more* separate groups of experimental subjects. In addition, each group of experimental subjects receives a *different* level of a second treatment variable, known as a *nested treatment variable.* This design is represented in Table 1-4 for a case in which there are two groups per major treatment level. The nested treatment variable is often some factor associated with application of the levels of the major treatment variable. For example, in a concept-learning experiment, it may be a matter of practical necessity to have half the experimental subjects at each treatment level supervised by a different research assistant. If three treatment levels are involved, a total of six research assistants will be used. Since a research assistant may introduce his own unique effect on the way the experimental subjects perform in the experiment, the appropriate design is that illustrated in Table 1-4. More complex nested designs occur when additional treatment variables and/or nested variables are introduced. Table 1-5 shows a nested design which has one nested variable added to a two-dimensional factorial design.

Repeated measures designs involve factorial or other combinations of treatment levels, but the same group of experimental subjects is exposed to more than one combination of treatments. Ordinarily, there is complete repetition of a group under all levels of at least one treatment variable. In Table 1-6, the basic design is factorial, but the groups of ex-

table 1-4 *Nested design with one major treatment variable*

	A 1		*A* 2		*A* 3	
B	1 G_1	2 G_2	3 G_3	4 G_4	5 G_5	6 G_6

table 1-5 *Nested design with two major treatment variables A and B and one nested variable C*

		C	A 1	2	3
B	1	1	G_{111}	G_{112}	G_{113}
		2	G_{211}	G_{212}	G_{213}
	2	3	G_{321}	G_{322}	G_{323}
		4	G_{421}	G_{422}	G_{423}

perimental subjects are exposed successively to all the levels of treatment variable A. Thus, a total of 4 (instead of 12) groups of experimental subjects are involved in the experiment. Chapter 7 is devoted to repeated measures designs.

The designs represented here are the major categories discussed in this textbook. Principles underlying a number of additional designs are described in Chapter 9, and references are cited so that further study of these designs may be pursued. Also, in conjunction with any experimental design, it is possible to introduce additional measures which serve as adjustors, or covariables, for the response measure. The basic principles of this *analysis of covariance* are outlined in Chapter 8, and analytical procedures are given for a variety of designs.

logic of hypothesis testing

The student has previously studied concepts involved in the testing of hypotheses by statistical procedures. In this section, the rationale of hypothesis testing is reviewed in a general way, and the terminology adopted in this volume is explained.

A null hypothesis (H_0) is a linear constraint placed on the values of the parameters of one or more populations. Constraints generally take the form of specific equalities (for example, $\pi = .5$ states that the probability of success in a binomial population is .5; and $\mu_1 = \mu_2$ states

table 1-6 *Repeated measures on the A dimension*

		A 1	2	3
B	1	G_1	G_1	G_1
	2	G_2	G_2	G_2
	3	G_3	G_3	G_3
	4	G_4	G_4	G_4

that the means of two populations of scores are equal). In some cases, these constraints can also be conveniently written as differences (for example, $\pi - .5 = 0$, and $\mu_1 - \mu_2 = 0$). During the design phases of an experiment, null hypotheses are set up in terms of relevant aspects of the problem being studied. Then, data are collected which provide estimates of each parameter appearing in a null hypothesis. *Assuming* that the null hypothesis is correct, the probability of the observed data provides a basis for accepting or rejecting the original null hypothesis. If, *assuming* that the null hypothesis is correct, the observed data do not represent an extreme case (i.e., the probability of the observed data is relatively large), then the null hypothesis is accepted.[1] If the observed data represent an unusual case (i.e., the probability of the observed data is relatively small, say .05 or less), then the null hypothesis is rejected as a reasonable statement concerning the population parameters. A statistical test involves determining the probability of the observed data, assuming that the null hypothesis is correct. It is instructive to illustrate these concepts for the simple one-sample z test. This test is suitable for testing the hypothesis that a sample of n observations represents a random sample from a population with a specific mean $\mu = M$, given that the population variance is σ^2. That is, $H_0: \mu = M$, where M is a specific numerical value. We use Y_i to represent the ith score in the sample, and $\bar{Y}$ is the mean computed from the sample. The z test allows us to find the probability that a sample of size n from a population with mean $\mu = M$ will have a mean at least as extreme as $\bar{Y}$. If the population being sampled is normal, then the ratio $(\bar{Y} - \mu)/(\sigma/\sqrt{n})$ can be shown to follow the unit normal, or z, distribution (i.e., the normal distribution with mean of 0 and variance of 1). For relatively large samples, the ratio $(\bar{Y} - \mu)/(\sigma/\sqrt{n})$ will be distributed as z even if the parent population is nonnormal. This result follows from the *central-limit theorem*. The choice of critical values from the z distribution depends upon the alternative hypothesis (i.e., the statement concerning μ which we accept when we reject the null hypothesis). If the alternative hypothesis is *nondirectional* (that is, $\mu \neq M$), critical values are chosen symmetrically in each tail of the z distribution. If z_a is the ath percentile of the z distribution, the critical values can be represented by $z_{\alpha/2}$ and $z_{1-\alpha/2}$, where α is the significance level. For example, with the significance level $\alpha = .05$, we should use $z_{.025} = -1.96$ and $z_{.975} = 1.96$. For the case of a *directional* alternative hypothesis (that is, $\mu > M$ or $\mu < M$) the risk is taken entirely in the corresponding tail of the z distribution. Thus z_α is the critical value when $\mu < M$ is the alternative

[1]The phrase "accept the null hypothesis" is a convenient technical usage, but it should be realized that the meaning of this phrase is that the data *do not allow rejection of the null hypothesis*. Accepting (i.e., failing to reject) a null hypothesis should *not* be understood as statistical evidence that the null hypothesis is true.

hypothesis, and $z_{1-\alpha}$ is the critical value for the alternative hypothesis $\mu > M$.

If the null hypothesis is, in reality, correct, then the proportion α of random samples will yield computed values of z so extreme that the null hypothesis will, nevertheless, be rejected. Rejecting a correct null hypothesis is known as a Type I error, and the significance level α is the risk that a Type I error will occur. On the other hand, if the null hypothesis is not correct, a Type II error occurs when this false null hypothesis is accepted as a correct statement. The risk of a Type II error depends upon the actual value of μ. Intuitively, if μ is near M, Type II errors will occur relatively often; if μ is far from M, these errors will occur less often. The complement of the Type II error is the *power* of the statistical test. That is, if β is the risk of a Type II error, then $1 - \beta$ is power, or the likelihood of rejecting a false null hypothesis.

These concepts are illustrated in Figure 1-1. The distribution labeled 1 is the sampling distribution of the mean, assuming that the null hypothesis is correct (that is, $\mu = M$). The blackened area in the right tail of this distribution represents the extreme cases for which the computed value of z exceeds its critical value and, hence, corresponds to the Type I error risk, assuming the directional alternative hypothesis is $\mu > M$. If the null hypothesis is not correct, then distribution 2 exemplifies the true sampling distribution of the mean (assuming $\mu = M'$ in reality). Thus, all means in the hatched region represent the Type II error risk. Although the value of M' is unknown in practice, it is possible to construct a *power function* from which the power $1 - \beta$ can be found for various values of M'. Referring to Figure 1-1, we see that the right-tail critical value in distribution 1 is $M + z_{1-\alpha}\,(\sigma/\sqrt{n})$, since the standard deviation of the sampling distribution of the mean is $\sigma/\sqrt{n}$. In terms of distribution 2, the equivalent of this critical value is

$$z^* = \frac{M + z_{1-\alpha} - (\sigma/\sqrt{n}) - M'}{\sigma/\sqrt{n}} = z_{1-\alpha} - \frac{\sqrt{n}(M' - M)}{\sigma}$$

For fixed n, this expression can be evaluated for various values of $M' - M$ and the probability of a Type II error determined from tables of

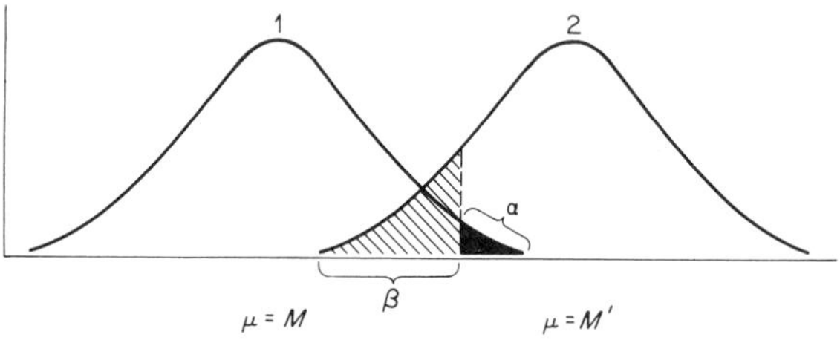

figure 1-1 *Type I and Type II error risks.*

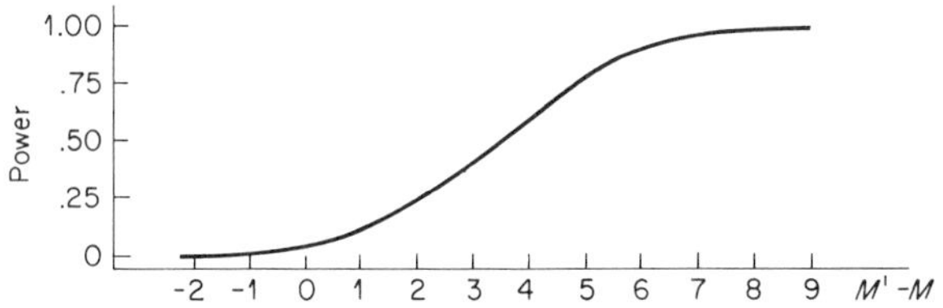

figure 1-2 *Power function for $n = 25$, $\sigma = 10$, $\alpha = .05$, directional z test.*

the unit normal curve. For example, with $n = 25$, $\sigma = 10$, and $\alpha = .05$, the expression for the z score in distribution 2 is $z^* = 1.645 - .5(M' - M)$. Figure 1-2 shows power (that is, $1 - \beta$) as a function of $M' - M$ for this example. The sigmoid shape of the figure is typical of power functions for directional alternative hypotheses. In the case of a nondirectional test, the curve resembles an inverted bell. For larger values of n, the curve slopes upward more rapidly than in the example.

Viewed somewhat differently, the formula $z^* = z_{1-\alpha} - \sqrt{n}(M' - M)/\sigma$ can be used to determine the minimum sample size required to detect a difference $M' - M$ with given Type II error risk. Now $z_{1-\alpha}$, σ, and z^* are fixed; hence, n can be solved in terms of $M' - M$. Figure 1-3 shows a curve of constant power for $\beta = .20$. If σ is known or can be estimated reasonably well before conducting an experiment, constant-power curves such as that in Figure 1-3 can be useful for determining an appropriate sample size.

experimental error

The term *experimental error* is used to refer to the variation in response values observed among experimental subjects who receive the same treatment level, or combination of treatment levels, in an experiment.

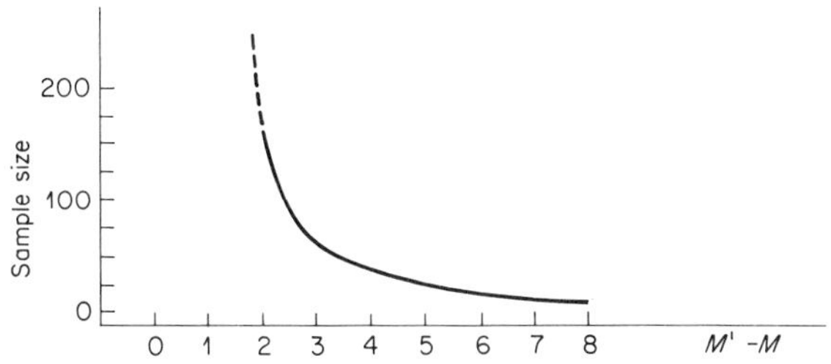

figure 1-3 *Constant-power curve for directional z test, $\sigma = 10$, $\alpha = .05$, $1 - \beta = .80$.*

It is a well-known fact that different human subjects exposed to the same environmental conditions will behave in different ways. Such variation is usually explained in terms of personality differences, perceptual differences, and so forth. In the theory of experimental design, experimental error plays a central role; indeed, a major task facing the experimenter in the behavioral sciences is to control experimental events so that reasonable assumptions can be made about the nature of the experimental error in his experiment. Without some assumptions about experimental error it is obviously impossible to analyze the results of an experiment. It should be noted that what is the rule in the behavioral sciences is virtually outlawed in the more precise physical sciences. That is, a physical scientist ordinarily does not account for experimental error in the analysis of experiments. Rather, he is in the favorable position of being able to control his experimental material and experimental environment to the point where experimental error is of negligible magnitude. If, in fact, it is not negligible, the physical scientist may consider this to be a failure of control, and he will search for more precise methods. The behavioral scientist must, apparently, tolerate such error, and the theory of statistical experimental design provides a tool for designing valid experiments and for testing scientific hypotheses under these circumstances. However, as we shall see, the behavioral scientist pays a price for his ability to analyze data compounded by experimental error. This price is, in effect, the fact that rather restrictive conditions must be met in the experimental design and in the nature of the measurements taken during the experiment.

In this section, the necessary theory relating to experimental error is developed in general terms without reference to specific experimental designs. In the ensuing chapters, this theory is made explicit at the point of discussion of individual designs.

The value of the response measure for individual i, who is exposed to a combination of treatment levels symbolized by t, is Y_{it}. This is the response measure for the ith individual in cell t of the experimental design. The symbol t is adopted for the present as a convenient, neutral manner of identifying a cell or combination of treatment levels. In specific cases, t will be replaced by a set of subscripts identifying the group or cell explicitly. In the analysis of experiments, it is assumed that the response measure Y_{it} is a function of (1) parameters related to the experimental design and (2) experimental error. In symbols,

$$Y_{it} = \mu + \sum_{j=1}^{r} T_j + \epsilon_{it}$$

where μ = an effect (i.e., a component of the Y score attributable to the experimental design) which is common to all experimental subjects in the experiment

$$\sum_{j=1}^{r} T_j = \text{sum of all other effects (including interactions, if any) which contribute to } Y$$

ϵ_{it} = experimental-error component of Y

The effect μ is referred to as the *grand-mean effect* since it corresponds to the overall average Y value achieved by all experimental subjects. This effect is typically separated from the other effects rather than being included in the T_j parameters. The complex effect $\Sigma_{j=1}^{r} T_j$ will be written explicitly for each experimental design at the appropriate place in this volume. The term $\Sigma_{j=1}^{r} T_j$ includes the effects due to levels of treatment variables and interactions of treatment variables. Note that a linear model for Y_{it} is assumed; that is, Y_{it} is conceptualized as an additive sum of effects. In future chapters, it will be shown how null hypotheses can be written by setting restrictions (i.e., linear constraints) on the values of relevant subsets of the T_j parameters. The purpose here, however, is to focus on the component ϵ_{it}, the experimental error. *If* the application of treatment levels were absolutely uniform and *if* there were no intrinsic variability in human behavior, every individual exposed to the same treatment-level combination would achieve an identical Y value, and ϵ_{it} would disappear from the model. Since human behavior is apparently intrinsically variable and since experimenters cannot, in general, precisely control all relevant environmental contingencies, the term ϵ_{it} will, in fact, be nonzero for each experimental subject. We cannot generally, of course, explain the exact sources of the error which is attached to an individual's Y value. If we could, these sources would be taken into account in the experimental design and expressed in the term $\Sigma_{j=1}^{r} T_j$. Thus, ϵ_{it} is due to unknown and unexplained sources.

In order to perform an analysis on the Y values, it is necessary that the experimental error ϵ_{it} conform to a variety of conditions. The first condition is that the average value of ϵ_{it} be 0 within each cell of the experimental design. This condition is only for the purpose of introducing an algebraic simplicity and does not affect the linear model for Y_{it}; this is true since the values of the T_j for a cell can always be adjusted so that the condition is met.

The remaining conditions are more fundamental and impose real restrictions on the experimental error. First, the assignment of individuals within the experiment must be on a random basis. Randomization allows the laws of probability to operate within the framework of the experiment. Since, within a cell, μ and all the T_j are constant quantities, the only within-cell source of variation among Y values is experimental error. If experimental subjects are assigned randomly within the experiment, then probability laws apply to the variation of the experimental error within a cell of the experimental design.

Another condition imposed on experimental error is that the ϵ_{it} values within a cell of the experimental design be distributed normally. Although there are analytical procedures which are free of distributional assumptions in the case of simple experimental designs, the problems of building such procedures for more complex designs are largely unsolved or, at least, intractable computationally even on computers. The reason for the normality condition is clear. If the experimental error is normally distributed, the well-known probability theory connected with normal distributions can be applied to the analysis of the experiment.

A final condition on the experimental error concerns the nature of the within-cell variability for the different cells of the experimental design. It is necessary that the experimental error be constant from cell to cell in the sense that the variability of the ϵ_{it} values is the same, or homogeneous, for all cells in the experimental design. This is the familiar *homogeneity-of-variance* condition. If the experimental error is not constant from cell to cell, this complicates analysis since several different normal distributions must be considered in order to represent the data. Also, it is necessary to "pool" the experimental error from the various cells of the experimental design, and this is complicated under conditions of heterogeneity.

In summary, the conditions imposed on the experimental error are that it be randomly and normally distributed with average within-cell value of 0 and constant variance from cell to cell. How these conditions are utilized to derive appropriate analytical procedures in the case of the completely randomized design is presented in detail in Chapter 2. For the remaining designs considered in this volume, the appropriate derivation is suggested and, in some cases, outlined, but for a rigorous development the student should consult a mathematical statistics textbook which covers experimental designs (e.g., Kendall & Stuart, 1961, or Scheffé, 1959).

In most experimental designs, the magnitude of the experimental error can be estimated in a direct fashion. If two or more experimental subjects are exposed to the same treatment level, or combination of treatment levels, the variation of their Y scores is a direct estimate of the variability of the experimental error. Of course, the larger the number of experimental subjects in a cell of the experimental design, the more precise will be this estimate. A few designs, especially the randomized blocks designs and the repeated measures designs, have cases in which only one score is available per cell. In these cases, no direct estimate of experimental error is possible, although a valid analysis, as respects testing appropriate null hypotheses, may still be possible.

On intuitive grounds, it is apparent that a desirable goal in experimental design is the reduction of experimental error. That is, the experiment is "under control" in direct proportion to the smallness of the variance of experimental error within the cells of the experimental design. In general, the likelihood of detecting false null hypotheses (i.e., power) is related to the magnitude of the experimental error. An experiment with a smaller experimental error is, in general, more powerful than an experiment with more experimental error. There are two distinct possibilities for reducing the effects of error. First, the design of the experiment itself can contribute to this goal. If experimental subjects are blocked into relatively homogeneous groups, for example, these groups can be expected to show less error variability than unselected groups. Also, additional measures can be utilized in order to "adjust" the Y scores and, in effect, to subtract out part of the error term. This latter method is embodied in the analysis of covariance. Second, the conditions of the experiment can be controlled so that extraneous environmental events and variations in the application of the treatment levels do not have unnecessarily large effects on the experimental subjects. Thus, avoiding long time intervals between administration of treatment levels and taking into account the general milieu from which experimental subjects are drawn (e.g., not scheduling data collection near major holidays or social events when utilizing public school children as experimental subjects) will aid in the reduction of experimental error.

Of course, reduction of experimental error is not the only way in which a powerful experiment can be achieved. The most universal device for achieving power is to increase the total size of the experiment, that is, include more experimental subjects per cell of the design. In some circumstances, in fact, a researcher may have little choice except to rely on size as a method for achieving power. The most efficient strategy from a research-design point of view is, however, to plan an experiment with the greatest reduction in experimental error and then to choose the number of experimental subjects which gives whatever degree of power is desired. Due to practical limitations, a compromise strategy is to use all the available experimental subjects in the most efficient design possible. In any case, statistical and design approaches to reducing experimental error are important considerations.

expected values

In the case of the simplest experimental design, the completely randomized design, there is no difficulty in arriving at the appropriate test for the hypothesis of equal effects for the various treatment levels. This is the familiar ratio between the *mean square for treatments* and *mean square for error* (or *mean square within groups*). However, for more complex designs there may be a variety of hypotheses to test, and the appropriate hypothesis-testing procedure may be less than obvious, or the "obvious" choice may be, in fact, incorrect. In this volume we adopt a general approach which greatly simplifies the problem of setting up appropriate hypothesis-testing procedures. This approach involves the notion of *expected values* of response variables and expected values of certain functions of such response variables. In this section, the theory of expected values is developed in the abstract. In Chapter 2 this theory is applied to the completely randomized design, and the general approach to setting up hypothesis-testing procedures is developed.

Consider a single response measure Y. The expected value of the variable Y, or of a variable defined as a function of Y, is "the average long-run value of the variable" (Edwards, 1964). More precisely, the expected value is the mean of the sampling distribution of the variable (or function of the variable). Algebraically we write $E(Y)$ for the "expected value of the response variable Y" and $E[F(Y)]$ for the "expected value of the function $F(Y)$ of the response variable Y." The symbol E in the expressions $E(Y)$ and $E[F(Y)]$ is a mathematical operator. In the paragraphs below we present some elementary rules for manipulation of the expected-value operator and show some results for common statistical quantities.

We let Y_i be a specific value of the discrete response variable Y. It is assumed that Y is a discrete variable and that a total of N Y elements exist. If Y is continuous, only minor changes are necessary, and these will be considered shortly. Since a given value of Y may occur more than once, let F_i be the frequency of the ith Y value and let $F_i/N = P(Y_i)$ be the relative frequency, or proportion, of Y_i in the total set of Y elements. Of course $\Sigma_{i=1}^{k} F_i = N$ and $\Sigma_{i=1}^{k} P(Y_i) = 1$ if there are k different values for Y. $P(Y_i)$ is also the probability associated with Y_i in the sense that if we randomly select a single Y, our chances of selecting Y_i are F_i in N [for example, for the Y values 1, 3, 3, 4, 6, 6, 7, 9, the chances of randomly selecting a value of 3 are 2 in 8; or the probability associated with $Y_i = 3$ is $P(Y_i) = .25$]. The Y_i and their associated probabilities $P(Y_i)$ constitute the distribution of Y.

If we draw a single Y, the expected value for the response variable is simply the sum of the cross products of the Y_i and the $P(Y_i)$. That is, for k different possible values of Y, $E(Y) = \Sigma_{i=1}^{k} Y_i[P(Y_i)]$. Replacing $P(Y_i)$ by F_i/N gives $E(Y) = \Sigma_{i=1}^{k} F_iY_i/N$. This is the familiar formula for the mean computed from a grouped-data frequency distribution. If each Y_i occurs only once (or if Y_i is repeated F_i separate times), then $P(Y_i) = 1/N$ for all i, and $E(Y) = \Sigma_{i=1}^{N} Y_i/N$, which is the mean expressed in raw data form. Thus, the expected value for a single randomly drawn Y is simply the mean of the Y's, or μ_y. This result is not surprising since the sampling distribution of single Y's is the distribution of the variable itself, and μ_y is the mean of the distribution of Y values.

If Y is continuous, it is necessary to reinterpret $P(Y_i)$, since the relative frequency of a given value is meaningless when an infinity of different values is possible. For continuous variables, $P(Y)$ is defined as the *probability density function*. To find the probability that Y_i is between two numerical values a and b, the probability density function must be integrated over the interval from a to b. Symbolically, $P(a < Y_i < b) = \int_a^b P(Y)\, dY$. By analogy with the discrete case, $E(Y) = \int_{-\infty}^{+\infty} P(Y)(Y\, dY)$ if Y extends from $+\infty$ to $-\infty$. If the distribution of Y is bounded, the limits of integration are adjusted accordingly. Since readers of this textbook may not be fluent in calculus and since all necessary results are unchanged if only the discrete case is considered, we shall not pursue the extension to continuous variables. Students familiar with integral calculus can, however, easily carry out the required generalization.

Three rules which apply to the expected-value operator are useful in deriving additional results. First, the expected value of a constant is the constant itself [i.e., if c is a constant quantity, $E(c) = c$]. Second, constant terms can be factored out. Thus, if c is a constant quantity and Y is a response variable, $E(cY) = cE(Y) = c\mu_y$. Third, the expected-value operator can be distributed over a linear function of a variable [for example, $E(Y^2 + 2Y) = E(Y^2) + 2E(Y)$]. These three rules and the result previously found for $E(Y)$ allow a number of useful results.

The expected value of a squared deviation from the mean is

$$E(Y - \mu_y)^2 = \sum_{i=1}^{k} [P(Y_i)(Y_i - \mu_y)^2] = \sum_{i=1}^{k} \frac{F_i(Y_i - \mu_y)^2}{N}$$

which is, by definition, the variance of the Y's. Hence, $E(Y - \mu_y)^2 = \sigma_y^2$; also,

$$\sigma_y^2 = E(Y^2 - 2\mu_y Y + \mu_y^2) = E(Y^2) - 2\mu_y E(Y) + \mu_y^2 = E(Y^2) - \mu_y^2$$

From this we see that the expected value of a squared score is $E(Y^2) = \sigma_y^2 + \mu_y^2$. If n different randomly selected Y^2 scores from the same distribution are summed, we have

$$E\left(\sum_{i=1}^{n} Y_i^2\right) = \sum_{i=1}^{n} E(Y_i^2) = n\sigma_y^2 + n\mu_y^2 = n(\sigma_y^2 + \mu_y^2)$$

Some well-known relationships can be easily derived by using expected values. If n Y scores are randomly selected, their mean is $\overline{Y} = \Sigma_{i=1}^{n} Y_i/n$. The expected value of $\overline{Y}$ is

$$E(\overline{Y}) = E\left(\sum_{i=1}^{n} \frac{Y_i}{n}\right) = \frac{1}{n} E\left(\sum_{i=1}^{n} Y_i\right) = \frac{1}{n} E(Y_1 + Y_2 + \cdots + Y_n)$$

$$= \frac{1}{n} (\mu_y + \mu_y + \cdots + \mu_y) = \frac{1}{n} (n\mu_y = \mu_y)$$

We have thus shown that the mean of the sampling distribution of the sample mean is μ_y. The variance of this sampling distribution can also be found.

$$\sigma_{\bar{y}}^2 = E(\bar{Y} - \mu_y)^2 = E(\bar{Y}^2) - \mu_y^2$$

But

$$E(\bar{Y}^2) = E\left(\sum_{i=1}^{n} \frac{Y_i}{n}\right)^2 = \frac{1}{n^2}E\left(\sum_{i=1}^{n} Y_i\right)^2 = \frac{1}{n^2}E(Y_1 + Y_2 + \cdots + Y_n)^2$$

If the polynomial is expanded, there are n Y^2 terms and $n(n-1)$ terms which are cross products of Y's (for example, Y_1Y_2, Y_1Y_3, and so forth). Each of the Y^2 terms has expectation $\sigma_y^2 + \mu_y^2$, and each cross-product term has expected value μ_y^2. Hence

$$E(\bar{Y}^2) = \frac{n(\sigma_y^2 + \mu_y^2) + n(n-1)\mu_y^2}{n^2} = \frac{\sigma_y^2 + n\mu_y^2}{n}$$

Using this result, we obtain

$$\sigma_{\bar{y}}^2 = \frac{1}{n}(\sigma_y^2 + n\mu_y^2) - \mu_y^2 = \frac{\sigma_y^2}{n} + \mu_y^2 - \mu_y^2 = \frac{\sigma_y^2}{n}$$

In the reference cited previously for Edwards (1964), a number of additional elementary results are given which may be of interest to some readers.

degrees of freedom

The concept of degrees of freedom plays an important role in the statistical analysis of data. The proper evaluation of any statistic derived from a test of significance depends upon being able to determine appropriate degrees of freedom; and, from a purely practical point of view, the user of a statistical test will have to calculate degrees of freedom in order to enter tables of percentiles of test statistics (e.g., in order to determine the significance of a computed value of t from a single sample, the researcher must utilize the theoretical t distribution with $n - 1$ degrees of freedom). But, from a more fundamental point of view, the researcher employing analysis-of-variance designs must understand the role of degrees of freedom in the analysis and interpretation of these designs.

The degrees of freedom for any statistic are equal to the number of independent scores (or observations, quantities, etc.) entering into the computation of the statistic. If there is a total of n scores and, for a specific statistic, the degrees of freedom are $n - k$, this implies that there are k independent constraints placed on the scores during the computation of the statistic. For example, the mean of n scores has n degrees of freedom since there is no constraint placed on the scores during computation of the mean. However, the variance of n scores has only $n - 1$ degrees of freedom since the constraint $\Sigma_{i=1}^{n} Y_i = n\overline{Y}$ is placed on the scores during the computation of the variance [i.e., if the numerator of the sample variance is defined as $\Sigma_{i=1}^{n} (Y_i - \overline{Y})^2$, the value of $\overline{Y}$ represents a constraint placed on the values of the scores]. In other words, since $\overline{Y}$ appears in the computation of the sample variance, the scores entering into the computation are not completely free to vary, and $\Sigma_{i=1}^{n} (Y_i - \overline{Y}) = 0$ must be true for the scores. This lack of freedom can be seen to represent a single constraint if we consider the fact that, given $\overline{Y}$, we are free to assign arbitrarily only $n - 1$ values. For example, with $n = 4$ and $\overline{Y}$ given as 10, if we choose *any* three values for the first three scores, the value of the fourth score is completely determined (e.g., the three arbitrary values might be 8, 10, and 13; then the fourth score *must* be 9 so that $\Sigma_{i=1}^{4} Y_i = 40$ and $\overline{Y} = 10$).

Let us apply the above conception of degrees of freedom to some common test statistics to show that our results agree with the usual formulas for degrees of freedom given for these tests. Consider, first, the one-sample z test which is based on the test statistic $z = (\overline{Y} - \mu)/(\sigma/\sqrt{n})$. Both μ and σ are presumed to be known from sources other than the scores entering into the analysis; thus, they place no constraint on the values of the scores. Also, the computation of $\overline{Y}$ is free of constraints; therefore, the distribution of z is independent of any consideration of degrees of freedom (sometimes this is expressed by saying that z is based on infinite degrees of freedom). Turning to the one-sample t statistic, we have $t = (\overline{Y} - \mu)/(S/\sqrt{n})$. Although μ is known from outside sources, the value of S (the sample variance) is computed with one constraint as noted above, and the distribution of t does depend upon degrees of freedom; specifically, the degrees of freedom appropriate to this case are $n - 1$. As a final example, consider the two-sample t test. The computed test statistic is $t = (\overline{Y}_1 - \overline{Y}_2)/S_d$, where S_d is the standard error of the difference between pairs of sample means; that is,

$$S_d = \sqrt{\frac{(n_1 - 1)S_1^2 + (n_2 - 1)S_2^2}{n_1 + n_2 - 2}}$$

The total number of observations is $n_1 + n_2$; since two variances must be estimated from the data (S_1^2 and S_2^2), this corresponds to placing two constraints on the scores, and the degrees of freedom are $n_1 + n_2 - 2$.

chapter two

the completely randomized design

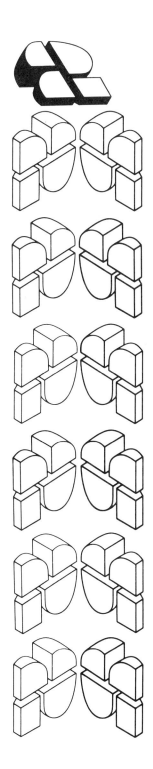

introduction

The reader of this volume has, in all likelihood, had previous instruction in the completely randomized design, or one-way analysis of variance. In this chapter, the CRD is examined in detail and with some rigor. We begin by setting up the model for a score and then proceed to show that an appropriate null hypothesis can be established and tested under this model. The method, based on expected values of random variables, is very general and is applicable to all remaining designs considered in this textbook. The remainder of this chapter is concerned with related material. Testing the homogeneity-of-variance assumption is discussed, as well as strategies to employ when the condition cannot be met. The logic of specific contrasts, beyond the global hypothesis first tested in the CRD, is developed, and a variety of useful procedures are illustrated. Finally, the question of the power of the CRD is entertained, and a method is described by which the power of the CRD may be used as a kind of bench mark against which the power of error-reducing designs may be compared.

the model and expected values for the CRD

The research design underlying the CRD is one in which two or more independent, random groups of experimental subjects are being contrasted in terms of mean values of a single response measure. In order for such contrasts to be scientifically meaningful, the groups, or samples, must have histories which reflect differential treatment. This latter condition is to be interpreted rather broadly; e.g., when a sample of males is contrasted with a sample of females, the sexes of the experimental subjects represent the levels of the treatment variable. Nevertheless, certain groups may confound two or more differential factors, and contrasts of such groups are ambiguous. If, for example, a randomly selected group of high school girls were given initial instruction in the elements of probability theory via programmed instruction and a second, randomly selected group of high school boys were given equivalent content via closed-circuit television instruction, a contrast of the average posttest achievement of these two groups would give ambiguous results with respect to either sex of experimental subjects or instructional medium. There is no way, in this example, of obtaining separate estimates of the effects of these two treatment variables. If the boys achieve more, on the average, than the girls, this may be an effect of the sex of the experimental subjects, and the mode of instruction may have no effect. Or the difference may be entirely due to instructional mode, with sex having no differential effect. A third possibility, somewhat more complex, is that the *combination* of sex and mode results in differences that are not directly the effect of either treatment variable when considered separately. The confounding of the sex variable with the mode of instruction may appear to pose no threat in practice since it is difficult to believe that a researcher would be so naïve as to design such an experiment. However, more subtle confounding of factors may intrude into a design and go undetected for a time. Indeed, examples of this sort are not difficult to find in the research literature. A not uncommon example in research contrasting levels of instructional variables is the failure to control the total amount of instructional time given to different groups. Thus, time may be confounded with the effects of the instructional variable.

The discussion above should alert the student to the criticality of design elements and control when the CRD, or other analysis-of-variance design, is contemplated as a paradigm for an experiment. To mechanically analyze data because they apparently conform to the general structure of a design is an inexcusable lapse of scientific attitudes. Design and analysis of experiments are inextricably bound together.

The data in the CRD can be represented as Y scores in the format of Table 2-1. We let the first subscript on Y indicate the position of the score in the group, and the second subscript the group. In the jth group, in a total of p groups, there are, all together, n_j scores which sum to $Y_{.j}$ and have a mean equal to $Y_{.j}/n_j = \bar{Y}_{.j}$.

In terms of the completely randomized design, the model for a score is especially simple. If linear components are assumed, each Y score is affected only by the treatment group of which it is a member and by experimental error. As mentioned earlier, it is traditional to separate a grand-mean effect from the treatment effect. Thus, $Y_{ij} = \mu + T_j + \epsilon_{ij}$, where T_j is the effect of treatment level j. The grand mean μ is the average effect of being in the experiment and is estimated by the

table 2-1 *Representation of data in the* CRD

	TREATMENT LEVELS 1	2	3	$\cdots$	p	
	Y_{11}	Y_{12}	Y_{13}	$\cdots$	Y_{1p}	
	Y_{21}	Y_{22}	Y_{23}	$\cdots$	Y_{2p}	
	$\cdots$	$\cdots$	$\cdots$	$\cdots$	$\cdots$	
	Y_{n_11}	Y_{n_22}	Y_{n_33}	$\cdots$	Y_{n_pp}	
Sums	$Y_{.1}$	$Y_{.2}$	$Y_{.3}$	$\cdots$	$Y_{.p}$	Grand sum $Y_{..}$
Means	$\bar{Y}_{.1}$	$\bar{Y}_{.2}$	$\bar{Y}_{.3}$	$\cdots$	$\bar{Y}_{.p}$	Grand mean $\bar{Y}_{..}$

mean of all the Y scores, $\bar{Y}_{..}$. Hence, the T_j are really deviation effects (or differential effects). That is, a score in group 1 is affected by both μ and T_1, as well as by experimental error. However, μ is common to all treatment levels, and only the T_j serve to differentiate the treatment levels. Viewing this slightly differently, we may rewrite the model as $Y_{ij} - \mu = T_j + \epsilon_{ij}$. Here the deviation $Y_{ij} - \mu$ is explained in terms of the treatment level and experimental error. The analysis of data which fit the CRD depends upon the particular hypothesis of interest. The most general and initially relevant hypothesis posits that the terms T_j are each equal to 0; that is, $T_1 = T_2 = \cdots = T_p = 0$. This hypothesis of equal effects, or null hypothesis, is equivalent to stating that the model for Y_{ij} can be rewritten as $Y_{ij} = \mu + \epsilon_{ij}$. In this formulation, the treatment levels have no differential effects (i.e., there is no T_j term in the model), and a score deviates from the grand mean only because of experimental error. In essence, then, we are interested in comparing two models. If the null hypothesis $T_1 = T_2 = \cdots = T_p = 0$ is correct, then $Y_{ij} = \mu + \epsilon_{ij}$. Otherwise, the more complete model $Y_{ij} = \mu + T_j + \epsilon_{ij}$ is needed. The analysis of variance provides an analytical tool for deciding between these two models. The parallelism between the null hypothesis and the appropriate model for Y_{ij} should be carefully noted since the ensuing developments hinge upon contrasting the appropriateness of these two models and, hence, allow a decision concerning the tenability of the null hypothesis $T_1 = T_2 = \cdots = T_p = 0$.

The general approach of the analysis of variance involves obtaining, for each null hypothesis, two separate estimates of the population variance of the response measures. Within each treatment group (i.e., cell of the design) the only source of variability among the scores is experimental error. A fundamental assumption in the analysis is that the variance of this error is the same for all treatment groups (i.e., homogeneity of variance is assumed). Hence, there is just one population variance, σ^2, to estimate, and this is the common error variance of the populations represented by the p samples. In terms of the model, the mean of the jth population is $\mu + T_j$. For simplicity, let $\mu_j = \mu + T_j$ be this population mean. The grand mean μ is the average of the μ_j. Note that the sum of the T_j, $\Sigma_{j=1}^{p} T_j$, is 0 and the mean T_j is 0.

For the completely randomized design, it is quite simple to find two estimates of σ^2, one of which is always unbiased and one of which is unbiased only if the null hypothesis $T_1 = T_2 = \cdots = T_p = 0$ is correct. Prior to introducing the mathematics of these estimation procedures, let us consider their general logic. Under the assumption of homogeneity of variance, the treatment groups represent random samples from populations with equal variances. The equality of these population variances is independent of any assumption concerning the means of the populations. Hence, a single estimate of the common population variance σ^2 can be obtained by pooling the variance estimates from the treatment groups. This pooling is carried out by adding the sums of squares [i.e., the variance numerators of the form $\Sigma_{i=1}^{nj} (Y_{ij} - \overline{Y}_{.j})^2$] for the p groups and then dividing by the appropriate degrees of freedom (that is, $\Sigma_{j=1}^{p} n_j - p$, since 1 degree of freedom is lost for each treatment group). The estimate resulting from this pooling of the sample variances is the *mean square for error* MS_{error} (or *within-groups mean square*).

The second estimate of σ^2 is obtained from the means of the treatment groups. The variance of these means is $\Sigma_{j=1}^{p} (\overline{Y}_{.j} - \overline{Y}_{..})^2/(p - 1)$; and *if* the null hypothesis is correct, this variance is an unbiased estimate of the variance of the sampling distribution of the mean for means drawn from the parent population (which have identical means and variances and are normal distributions under our assumptions). If the treatment groups are the same size, say n, the variance of this sampling distribution is σ^2/n. Hence, $n\, \Sigma_{j=1}^{p} (\overline{Y}_{.j} - \overline{Y}_{..})^2/(p - 1)$ is an unbiased estimate of σ^2. For treatment groups of unequal size, an unbiased estimate is given by $\Sigma_{j=1}^{p} n_j(\overline{Y}_{.j} - \overline{Y}_{..})^2/(p - 1)$. This is the *mean square for treatment* MS_{treat} (or *among-groups mean square*). If the null hypothesis is not correct, the mean square for treatment is no longer an unbiased estimate of σ^2. In fact, in this case the mean square for treatment will estimate a quantity *larger* than σ^2 since sample means from populations with different means will show greater variability than sample means from a single population (or from populations with equal means).

These results can be put on a more precise footing by utilizing the concept of expected values. In the ensuing derivations it is assumed, for simplicity, that the treatment groups each contain the same number of subjects. Hence, $n_1 = n_2 = \cdots = n_p = n$, and the total experiment contains pn subjects. The numerator of the mean square for error is, then,

$$\sum_{=1}^{p}\left[\sum_{i=1}^{n} (Y_{ij} - \overline{Y}_{.j})^2\right]$$

and is referred to as the *sum of squares for error* (SS_{error}). This sum of squares is composed of p terms, each of which is a sum of squares for one of the treatment groups. The expected value of any one of these separate sums of squares is

$$E\left[\sum_{i=1}^{n} (Y_{ij} - \overline{Y}_{.j})^2\right] = E\left[\left(\sum_{i=1}^{n} Y_{ij}^2\right) - n\overline{Y}_{.j}^2\right] = E\left[\sum_{i=1}^{n} Y_{ij}^2\right] - E(n\overline{Y}_{.j}^2)$$

From the derivations in Chapter 1, the left-hand term has expectation $n\sigma^2 + n\mu_j^2$; also, $E(n\overline{Y}_{.j}^2) = \sigma^2 + n\mu_j^2$. Thus,

$$E\left[\sum_{i=1}^{n} (Y_{ij} - \overline{Y}_{.j})^2\right] = n\sigma^2 + n\mu_j^2 - \sigma^2 - n\mu_j^2 = n\sigma^2 - \sigma^2 = (n - 1)\sigma^2$$

This is the expected value of the sum of squares for just one treatment group. For all p groups we have

$$E(\text{SS}_{\text{error}}) = E\sum_{j=1}^{p}\left[\sum_{i=1}^{n}(Y_{ij} - \overline{Y}_{.j})^2\right] = \sum_{j=1}^{p}(n-1)\sigma^2 = (pn - p)\sigma^2$$

Since $\text{MS}_{\text{error}} = \text{SS}_{\text{error}}/(pn - p)$, the result is

$$E(\text{MS}_{\text{error}}) = \frac{(pn - p)\sigma^2}{pn - p} = \sigma^2$$

Hence, we have formally shown that the mean square for error has as its expected value the common population variance σ^2. Note that this result was obtained without making any assumption concerning the equality or lack of equality of the population means. However, homogeneous variances were assumed in the derivation.

Now we turn our attention to the mean square for treatment MS_{treat}. Its numerator is the sum of squares for treatment SS_{treat}. The expected value of this sum of squares is $E[\Sigma_{j=1}^{p} n(\overline{Y}_{.j} - \overline{Y}_{..})^2]$. The sum of squares for treatment can be written in the equivalent form

$$\text{SS}_{\text{treat}} = n\sum_{j=1}^{p}\overline{Y}_{.j}^2 - pn\overline{Y}_{..}^2$$

Thus,

$$E(\text{SS}_{\text{treat}}) = E\left(n\sum_{j=1}^{p}\overline{Y}_{.j}^2\right) - E(pn\overline{Y}_{..}^2)$$

The left-hand term has the expected value

$$E\left(n\sum_{j=1}^{p}\overline{Y}_{.j}^2\right) = \sum_{j=1}^{p}(\sigma^2 + n\mu_j^2) = p\sigma^2 + n\sum_{j=1}^{p}\mu_j^2$$

The right-hand term is

$$E(pn\overline{Y}_{..}^2) = \sigma^2 + pn\mu^2$$

Combining these results, we obtain

$$E(\text{SS}_{\text{treat}}) = p\sigma^2 + n\sum_{j=1}^{p}\mu_j^2 - \sigma^2 - pn\mu^2 = (p-1)\sigma^2 + \left(n\sum_{j=1}^{p}\mu_j^2 - pn\mu^2\right)$$

Rewriting the term in parentheses, we have

$$E(\text{SS}_{\text{treat}}) = (p - 1)\sigma^2 + n\sum_{j=1}^{p} (\mu_j - \mu)^2$$

Since $\text{MS}_{\text{treat}} = \text{SS}_{\text{treat}}/(p - 1)$, the final expression is

$$E(\text{MS}_{\text{treat}}) = \sigma^2 + \frac{n\sum_{j=1}^{p} (\mu_j - \mu)^2}{p - 1} = \sigma^2 + \frac{n\sum_{j=1}^{p} T_j^2}{p - 1}$$

If the null hypothesis is correct, the T_j are all equal to zero, and the right-hand term in the equation for the expected value is zero. Hence, when the null hypothesis is correct, $E(\text{MS}_{\text{treat}}) = \sigma^2$. On the other hand, if the null hypothesis is not correct, the right-hand term will be a positive quantity and MS_{treat} will estimate a variance larger than σ^2. For treatment groups of differing sizes, only a minor change results. The expected value for the mean square for error is the same as above, but the mean square for treatment has the expected value

$$E(\text{MS}_{\text{treat}}) = \sigma^2 + \frac{\sum_{j=1}^{p} n_j T_j^2}{p - 1}$$

For the equal n case (and in general) the term $\Sigma_{j=1}^{p} T_j^2/(p - 1)$ will be replaced by the variance-like symbol σ_t^2. Thus, the expectation becomes $\sigma^2 + n\sigma_t^2$. Although σ_t^2 is not, for the fixed-effects model, a population variance in the usual sense, this notation is more compact than the cumbersome summation and will be used in writing expectations.

From distributional theory in mathematical statistics, it is known that the ratio of two independent variance estimates is an F variate if the two estimates are based on random samples from a single population.[1] Hence, *if* the null hypothesis is correct, the ratio $\text{MS}_{\text{treat}}/\text{MS}_{\text{error}}$ is distributed as F with $p - 1$ and $\Sigma_{j=1}^{p} n_j - p$ degrees of freedom. The expected value of the ratio $\text{MS}_{\text{treat}}/\text{MS}_{\text{error}}$ will be, of course, 1. If the null hypothesis is not correct, the numerator MS_{treat} will be, on the average, larger than MS_{error}, and the expected value of the ratio will be larger than 1. Thus, the F distributions (Appendix Table B-5) provide critical values for deciding on the tenability of the null hypothesis. If an obtained ratio $\text{MS}_{\text{treat}}/\text{MS}_{\text{error}}$ is smaller than, say, the 95th percentile of the appropriate F distribution, the null hypothesis can reasonably be accepted. If the ratio is larger than the 95th percentile of the appropriate F distribu-

[1]A concise derivation of this condition can be found in Chapter 1 of Mann (1949).

tion, it can be decided that it is not reasonable that MS_{treat} and MS_{error} both estimate the same variance and, hence, that the null hypothesis is untenable. Note that only a relatively large value of $\text{MS}_{\text{treat}}/\text{MS}_{\text{error}}$ implies rejection of the null hypothesis because when the null hypothesis is incorrect, the MS_{treat} will estimate a quantity larger than σ^2. In other words, the test of the null hypothesis involves critical values in just the right tail of the appropriate F distribution.

The method for testing the creditability of the null hypothesis for the completely randomized design can, in principle, be extended to all more complex experimental designs. That is, an F ratio is set up in such a manner that its expected value is 1 only if the null hypothesis in question is correct and is larger than 1 otherwise. Such a ratio is easy to set up if the expected value for each mean square is known. Therefore, we adopt the convention of presenting expected values for mean squares for each design discussed in this textbook. In general, these expected values will be displayed in the form of Table 2-2.

In the derivations of expected values for the mean square for error and mean square for treatment we have tacitly assumed that the treatment levels represent fixed effects. That is, the p levels of the treatment were directly set by the experimenter rather than representing p randomly selected levels from some larger set of P levels. When the fixed-effects model is appropriate, the sample effects $\overline{Y}_{.j} - \overline{Y}_{..}$ estimate the corresponding T_j, and there is one estimate for each different T_j. Under the random-effects model, only p out of a possible total of P of the T_j values can be estimated from the sample means. In general, P is very large relative to p. Indeed, if the treatment variable is continuous, there are infinitely many possible levels of the treatment variable, and P is infinite. In this case, the term $\Sigma_{j=1}^{p} n_j\, T_j^2$ is no longer appropriate as part of the expected value for the mean square for treatment. However, only a minor change is introduced. The distribution of possible treatment levels has a mean of 0; the T_j values also have a variance, and we shall denote this σ_t^2 (in this case, σ_t^2 is a true variance). For the case of equal sample sizes, the expected value for MS_{treat} becomes $\sigma^2 + n\sigma_t^2$. It is apparent that the same ratio $\text{MS}_{\text{treat}}/\text{MS}_{\text{error}}$ is appropri-

table 2-2 *Expected values of mean squares for the* CRD *(fixed effects and equal sample sizes)*

source	df	E(MS)
Treatment	$p-1$	$\sigma^2 + n\sum_{j=1}^{p} T_j^2/(p-1) = \sigma^2 + n\sigma_t^2$
Error	$pn - p$	σ^2

table 2-3 *Expected values of mean squares for the CRD (random effects and equal sample sizes)*

source	df	E(MS)
Treatment	$p-1$	$\sigma^2 + n\sigma_\tau^2$
Error	$pn-p$	σ^2

ate for testing the null hypothesis and that the random-effects model introduces no change in the testing procedure. In more complex experimental designs, random effects will result in changed hypothesis-testing procedures; for this reason, expected values for mean squares will be presented in terms of both random effects and fixed effects. Table 2-3 presents these results for the CRD.

It should be noted that the results for the CRD are little changed if equal or unequal sample sizes comprise the treatment groups. Minor computational simplifications are introduced when data are collected on equal-sized samples, but no fundamental difference exists. For most of the more complex experimental designs, unequal sample sizes do result in complexities, and these will be explicitly discussed.

computational procedures for the CRD

From the point of view of actually carrying out the computations required to analyze the CRD, the formulas presented for SS_{treat} and SS_{error} are not convenient. The usual computational approach involves first finding the *total sum of squares*, which is based on deviations of the scores from the grand mean $\bar{Y}_{..}$. Thus,

$$SS_{total} = \sum_{j=1}^{p} \sum_{i=1}^{n_j} (Y_{ij} - \bar{Y}_{..})^2$$

It can be shown that the treatment and error sums of squares sum to SS_{total}. That is, $SS_{treat} + SS_{error} = SS_{total}$. We shall demonstrate this for the case of equal sample sizes and leave the extension to samples of unequal size as an exercise. The sum $SS_{treat} + SS_{error}$ is

$$n\sum_{j=1}^{p} (\bar{Y}_{.j} - \bar{Y}_{..})^2 + \sum_{j=1}^{p} \left[\sum_{i=1}^{n} (Y_{ij} - \bar{Y}_{.j})^2 \right]$$

Expanding the square,

$$\left(n\sum_{j=1}^{p}\overline{Y}_{.j}^{2}-2n\overline{Y}_{.j}\overline{Y}_{..}+np\overline{Y}_{..}^{2}\right)$$
$$+\left(\sum_{j=1}^{p}\sum_{i=1}^{n}Y_{ij}^{2}-2\sum_{j=1}^{p}\sum_{i=1}^{n}Y_{ij}\overline{Y}_{.j}+n\sum_{j=1}^{p}\overline{Y}_{.j}^{2}\right)$$
$$=\left(n\sum_{j=1}^{p}\overline{Y}_{.j}^{2}-2np\overline{Y}_{..}^{2}+np\overline{Y}_{..}^{2}\right)$$
$$+\left(\sum_{j=1}^{p}\sum_{i=1}^{n}Y_{ij}^{2}-2n\sum_{j=1}^{p}\overline{Y}_{.j}^{2}+n\sum_{j=1}^{p}\overline{Y}_{.j}^{2}\right)$$
$$=\sum_{j=1}^{p}\sum_{i=1}^{n}Y_{ij}^{2}-np\overline{Y}_{..}^{2}=\sum_{j=1}^{p}\sum_{i=1}^{n}(Y_{ij}-\overline{Y}_{..})^{2}$$

Thus, $SS_{treat} + SS_{error} = SS_{total}$. This fact is computationally useful since the easiest terms to compute are SS_{treat} and SS_{total}. Then, SS_{error} can be found by subtraction as $SS_{error} = SS_{total} - SS_{treat}$. Also, the following formulas involve less labor than those above:

$$SS_{total}=\sum_{j=1}^{p}\sum_{i=1}^{n_j}Y_{ij}^{2}-\frac{Y_{..}^{2}}{\sum_{j=1}^{p}n_j}$$

$$SS_{treat}=\sum_{j=1}^{p}\frac{Y_{.j}^{2}}{n_j}-\frac{Y_{..}^{2}}{\sum_{j=1}^{p}n_j}$$

$$SS_{error}=SS_{total}-SS_{treat}$$

Once these sums of squares are computed, the treatment and error sums of squares are converted to mean squares by dividing by their degrees of freedom, and the F ratio is set up. The results are usually displayed in an analysis-of-variance summary table (Table 2-4). For simplicity, we let $n_t = \Sigma_{j=1}^{p} n_j$.

table 2-4 *Analysis-of-variance summary table*

source	df	sum of squares	mean square	F
Treatment	$p-1$	SS_{treat}	MS_{treat}	MS_{treat}/MS_{error}
Error	n_t-p	SS_{error}	MS_{error}	
Total	n_t-1	SS_{total}		

table 2-5 *Hypothetical data for three groups*

	GROUP 1	2	3	totals
	70	51	88	
	89	50	80	
	79	54	57	
	80	75	79	
	82	52	74	
	71	63	65	
	56	89	82	
	58	64	80	
	54	73	64	
	72	79	75	
Sum of scores	711	650	744	2,105
Sum of squared scores	51,827	43,902	56,180	151,909
Group mean	71.1	65.0	74.4	$70.17 = \overline{Y}_{..}$
Sum of squares	1,274.90	1,652.00	826.40	$3{,}753.30 = \mathrm{SS}_{\mathrm{error}}$
Variance	141.66	183.56	91.82	
Standard deviation	11.90	13.55	9.58	

To illustrate the computations involved in a one-way analysis of variance, consider the hypothetical data in Table 2-5 (these data were generated by drawing three random samples of size 10 from a table of rectangularly distributed random numbers). The sum and sum of squared scores are reported for each sample under the corresponding column of the table. The grand sum $Y_{..}$ is 2,105, and the grand sum of squared scores, $\Sigma_{j=1}^{p} \Sigma_{i=1}^{nj} Y_{ij}^{2}$, is 151,909. Thus, the total sum of squares is

$$\mathrm{SS}_{\mathrm{total}} = 151{,}909 - \frac{2{,}105^2}{30} = 4{,}208.17$$

The treatment sum of squares is

$$\mathrm{SS}_{\mathrm{treat}} = \frac{711^2}{10} + \frac{650^2}{10} + \frac{744^2}{10} - \frac{2{,}105^2}{30} = 454.87$$

Then, by subtraction,

$$\mathrm{SS}_{\mathrm{error}} = \mathrm{SS}_{\mathrm{total}} - \mathrm{SS}_{\mathrm{treat}} = 4{,}208.17 - 454.87 = 3{,}753.30$$

table 2-6 *Analysis-of-variance summary table for hypothetical data*

source	df	sum of squares	mean square	F
Treatment	2	454.87	227.44	1.64
Error	27	3,753.30	139.01	
Total	29	4,208.17		

These results can be summarized in an analysis-of-variance summary table (Table 2-6) and the mean squares and F ratio computed. The resulting F ratio is 1.64, which, with 2 and 27 degrees of freedom, is a nonsignificant value (using $\alpha = .05$). Thus, we have no reason to doubt the hypothesis that the three sets of numbers were sampled from populations with equal means (i.e., that $\mu_1 = \mu_2 = \mu_3$ is true). It is interesting to consider the interpretation which would have been necessary if the null hypothesis had been rejected. Because this was a contrived data situation, it was known that the data were, in fact, random samples from a single population with known properties (a rectangular distribution of two-digit numbers with the range of permissible sample elements restricted to integers from 50 to 90). Thus, rejection of the null hypothesis could only be interpreted as an instance of Type I error. In the remainder of this book, we shall deal with more realistic data examples.

tests for homogeneity of variance

The sum of squares for error is obtained by pooling the sums of squares from the treatment groups constituting the CRD. Fundamental to this pooling procedure is the condition that the variance of each treatment group estimates the same common population variance σ^2. If this condition is not met by the data, the entire analysis is suspect from a mathematical point of view. The homogeneity condition is $\sigma_1{}^2 = \sigma_2{}^2 = \cdots = \sigma_p{}^2 = \sigma^2$, where $\sigma_j{}^2$ is the variance of the population from which the jth sample was drawn. By assuming that the sample variances $S_1{}^2, \ldots, S_p{}^2$ are each independent estimates of σ^2, the sampling distribution of sets of p sample variances can be derived. This distribution is complex, and, in general, approximation procedures are used in lieu of direct evaluation. In this section, two widely used procedures and one newer technique for testing the homogeneity-of-variance, or equivariance, assumption are described. The first test, known as Bartlett's test, can be used with samples of the same or of different sizes. The second test, Hartley's $F_{\max}$ test, is computationally much simpler than Bartlett's test but is designed for use only with samples of equal sizes. The final procedure, Levene's test, has wide applicability but has not been thoroughly studied from a mathematical point of view.

Given p samples of sizes $n_1, n_2, \ldots, n_p$, Bartlett (1954) has shown that the statistic

$$\frac{2.303}{C}\left[\left(\sum_{j=1}^{p} n_j - p\right)\log(\mathrm{MS}_{\mathrm{error}}) - \sum_{j=1}^{p}(n_j - 1)\log(S_j^2)\right]$$

where

$$C = 1 + \frac{1}{3(p-1)}\left[\sum_{j=1}^{p}\frac{1}{n_j - 1} - \frac{1}{\sum_{j=1}^{p}(n_j - 1)}\right]$$

is distributed as chi-square with $p - 1$ degrees of freedom when these samples arise by random sampling from a single normal population. Thus, if the samples constituting the CRD are hypothesized as coming from populations with equal variances, Bartlett's statistic provides the basis for testing this hypothesis. For an α-level-significance test, the $(1 - \alpha)$th percentile of the chi-square distribution with $p - 1$ degrees of freedom provides the requisite critical value. The computations necessary to find Bartlett's statistic in an actual application are best laid out in tabular form. Table 2-7 shows this tabular scheme in a general form, and Table 2-8 illustrates the computations for the data from Table 2-5. The value of Bartlett's statistic is 1.01; compared with the 2 degrees of freedom chi-square distribution, this value is nonsignificant, and we conclude that the population variances are homogeneous.

It is known that Bartlett's statistic is affected by population nonnormality as well as by heterogeneity of variance. Thus, when a significant value occurs, there is some ambiguity concerning the source of this event. This fact, although it introduces some uncertainty concerning the exact interpretation of Bartlett's statistic, does not invalidate its use in applied situations. Since a normality, as well as an equi-variance, assumption underlies the analysis of variance, a significant value of

table 2-7 *Computational plan for Bartlett's statistic*

sample	$n - 1$	$1/(n-1)$	S^2	log S^2	$(n - 1)$ log S^2
1	$n_1 - 1$	$1/(n_1 - 1)$	S_1^2	log S_1^2	$(n_1 - 1)$ log S_1^2
2	$n_2 - 1$	$1/(n_2 - 1)$	S_2^2	log S_2^2	$(n_2 - 1)$ log S_2^2
⋯	⋯	⋯	⋯	⋯	⋯
p	$n_p - 1$	$1/(n_p - 1)$	S_p^2	log S_p^2	$(n_p - 1)$ log S_p^2
	$\sum_{j=1}^{p}(n_j - 1)$	$\sum_{j=1}^{p} 1/(n_j - 1)$			$\sum_{j=1}^{p}(n_j - 1)\log S_j^2$

table 2-8 *Example of computation of Bartlett's statistic*

sample	$n-1$	$1/(n-1)$	S^2	$\log S^2$	$(n-1)\log S^2$
1	9	.1111	141.66	2.1512	19.3608
2	9	.1111	183.56	2.2638	20.3742
3	9	.1111	91.82	1.9629	17.6661
	27	.3333			57.4011

$$\chi^2 = \frac{2.303}{C}[27(2.1430) - 57.4011] = \frac{2.303}{C}(.4599)$$

$$C = 1 + \frac{1}{6}\left(.3333 - \frac{1}{27}\right) = 1.05$$

$$\chi^2 = \frac{2.303}{1.05}(.4599) = 1.01$$

Bartlett's statistic indicates a departure from model assumptions. Fortunately, nonnormality and heterogeneity of variance imply similar types of strategies. That is, if either nonnormality or heterogeneity of variance is characteristic of a set of populations, the experimenter can (1) show that the design is one that is not seriously affected by such departures from assumptions or (2) apply a transformation to the response measures. The rationales for these strategies are discussed at the end of this section.

For the special case in which the p treatment groups are of the same size, the simpler Hartley's $F_{\max}$ test may be used rather than Bartlett's test. The $F_{\max}$ statistic is simply the ratio of the largest to the smallest of the sample variances. If $S^2_{\max}$ is the largest of the $S_j{}^2$ and $S_{\min}$ is the smallest, then $F_{\max} = S^2_{\max}/S^2_{\min}$. Hartley has worked out and tabled (Pearson & Hartley, 1958) the distribution of $F_{\max}$ for samples from constant size 2 through 60 and for 2 through 12 treatment groups. Appendix Table B-6 presents .05 and .01 critical values of the $F_{\max}$ statistic. For the data in the example (Table 2-5), $S^2_{\max} = 183.56$, $S^2_{\min} = 91.82$, and $F_{\max} = 2.00$. With $\alpha = .05$, this value is nonsignificant.

The final procedure for testing the equi-variance assumption, Levene's test, is different from the other two in that it does not base its computation or model directly on the sample variances. Rather, the procedure involves performing an ordinary one-way analysis of variance on the absolute deviations of scores from the treatment-group means. Within each treatment group of the analysis-of-variance design, the scores are subtracted from the mean for that group; the *absolute values* of these deviation scores become the entries for the computation of a one-way analysis of variance. The hypothesis being tested is that the mean absolute deviation from the mean is constant across the groups. Studies of the Levene test (Glass, 1966) indicate that it is relatively insensitive to nonnormality of the underlying distributions (at least for cases in which the treatment groups are of equal sizes) and that it compares well in power with Bartlett's test. Thus, with further investigation, this test may prove to have wide applicability since there is no necessary restriction to equal-sized samples, the computations are relatively simple, and standard statistical tables are utilized (i.e., tables of the F distribution).

There is a good deal of evidence that the analysis of variance is virtually unaffected by violations of normality and homogeneity of variance *if* the samples entering into the analysis are of the same, or approximately the same, size. Results reported by Box (1954) and Norton (cited in Lindquist, 1953) directly support this contention for analysis-of-variance designs, and a study by Boneau (1960) gives similar evidence for the two-sample case (t test). Thus, unless the equivariance hypothesis is of interest in its own right, the experimenter who uses samples of equal size can proceed directly to testing the hypothesis concerning treatment effects. On the other hand, for unequal sample sizes, the experimenter should first test for homogeneity of variance, using Bartlett's statistic (or, perhaps, the Levene test). If this condition proves tenable, the analysis of variance may be carried out. If heterogeneous variance (or nonnormality) is indicated, the transformations suggested below should be utilized.

Before turning to a consideration of specific transformations which may prove useful in practice, two preliminaries are relevant. First, as mentioned previously, the test based on Bartlett's statistic may reflect either nonnormality or heterogeneity of variance. In practice, if the Bartlett statistic exceeds the chi-square critical value, the experimenter can apply the recommended transformations since either violation may be cured by this route. If, in addition, a test specifically for the normality assumption is desired, the usual chi-square goodness-of-fit procedures should be used.[2] Once again, if normality is not sustained, a transformation of the data is the indicated remedy.

The second preliminary revolves around the defensibility of any transformation procedure. It is probably true that for any set of samples, there will be some fortuitous transformation which exaggerates group differences in such a way that significant treatment effects on the transformed response measure will occur. Thus, the experimenter

[2]Briefly, this involves, for each sample, forming a grouped-data frequency distribution of the scores in the sample. By using the sample mean and standard deviation, the expected frequency of cases per category can be computed through use of tables of the unit normal distribution. For k categories, if F_j is the observed frequency and F'_j is the expected frequency for the jth category, then $\Sigma^k{}_{j=1}\,[(F_j - F'_j)^2/F'_j]$ is distributed as χ^2 with $k - 3$ degrees of freedom. This test is applied independently to each of the samples. Unless relatively large samples are used (say 50 or more), the test has little power.

faces the possibility not only that a transformation stabilizes the variances but also that it gives rather misleading rejections of null hypotheses. (The opposite, of course, may also occur.) This possibility is not to be taken lightly, but a satisfactory strategy is available. In some cases, a transformation may, from previous experiments or from theoretical considerations, have a history of being especially useful for specific types of data. Such a "standard" transformation does not appear to pose any real problem. Also, the original choice of unit for the response measure may have been arbitrary. A theoretically defensible transformation may stabilize variances without danger of being fortuitous (e.g., original measures in terms of "elapsed time" may be transformed to "speed of response"). The one basically indefensible strategy is to deliberately use transformations to enhance the magnitude of the obtained F ratio. Modern statistical programs for electronic computers offer the capacity of repeating an analysis several times, using a new transformation each time. The temptation toward indiscriminate use of such capabilities can only lead to misleading experimental results.

If heterogeneity of variance is found for a set of treatment groups, Table 2-9 can serve as a guide for selecting an appropriate variance-stabilizing transformation. Any standard collection of mathematical tables may be consulted to carry out the square root and logarithmic transformations. Appendix Table B-12 provides values for $\sin^{-1}$ transformations. Most modern statistical computer programs (especially those for larger computers such as the IBM 7094 or Univac 1108) have internal provision for selection and application of these, as well as many additional, transformations.

A transformation is applied by replacing each original Y score by its transformed equivalent Y'. Prior to an analysis of variance on the Y' values, the homogeneity-of-variance condition must be tested on these transformed scores by means of the Bartlett statistic. A rough check on the adequacy of a transformation can be made by simply comparing the ranges of the transformed scores for the samples (since the suggested

table 2-9 *Data transformation guide*

data characteristics	recommended transformation
$\bar{Y}$ approximately equal to S^2 for all samples (data are counts)	$Y' = \sqrt{Y} + \sqrt{Y+1}$
$\bar{Y}$ approximately proportional to S for all samples	$Y' = \log(Y+1)$
$\bar{Y}^2$ approximately equal to S^2 for all samples (data are proportions or percentages)	$Y' = 2\sin^{-1}\sqrt{Y}$

transformations are monotonic under the recommended data circumstances, only the largest and smallest Y score in each sample must be transformed in order to calculate these ranges). Generally, a transformation which tends to equalize the sample ranges will also tend to equalize the sample variances.

In summary, if a given response measure is believed likely to yield heterogeneous variances, it behooves the experimenter to design his study with samples of equal sizes since the analysis of variance can tolerate heterogeneity of variance and nonnormality under these conditions. Otherwise, Bartlett's statistic should be used to test the equivariance assumption. If this assumption proves untenable, the prudent choice of a data transformation is suggested. It should be noted that any of the procedures discussed in this section may be applied to scores in the cells of a multidimensional experimental design as well as to data in the CRD.

contrast procedures for qualitative treatment variables

In the event that the null hypothesis $T_1 = T_2 = \cdots = T_p$ is rejected in a completely randomized design, the experimenter must continue his analysis of the data in order to isolate specific inequalities among the treatment effects. For the case of just two treatment levels (that is, $p = 2$), there is, of course, no problem since the relative magnitudes of the sample means immediately allow us to decide which population has the larger mean. However, whenever more than two treatment levels are involved, rejection of the null hypothesis suggests only that there are some inequalities among the p treatment effects. The location and direction of these inequalities must be found by further analysis. Note that for a total of p effects, there are $\frac{1}{2}p(p-1)$ pairwise-contrast effects of the form $T_i - T_j$. For example, if $p = 4$, the pairwise contrasts are $T_1 - T_2$, $T_1 - T_3$, $T_1 - T_4$, $T_2 - T_3$, $T_2 - T_4$, and $T_3 - T_4$. Each of these contrasts may be set up as a null hypothesis. That is, $T_1 = T_2$, $T_1 = T_3$, and so forth. However, not all these contrasts are independent of one another. Consider, for $p = 3$, the set of three contrasts $T_1 - T_2$, $T_1 - T_3$, and $T_2 - T_3$. If $T_1 = T_2$ and if $\frac{1}{2}(T_1 + T_2) = T_3$, then, by implication, $T_1 = T_2 = T_3$. There are thus only two *independent* contrasts among the three effects. In general, for p effects, only $p - 1$ independent contrasts may be set up. Since there are $p - 1$ degrees of freedom for treatments in the analysis of variance, we say that there is one independent contrast for each degree of freedom.

In this section, we discuss procedures which are applicable when the treatment levels are arbitrarily set levels of a qualitative treatment variable. First, we consider testing the set of $\frac{1}{2}p(p-1)$ pairwise contrasts. As noted above, not all these contrasts are independent. Second, testing just a set of $p - 1$ independent contrasts is discussed. It may seem unnecessary, in practice, to have available both these types of testing procedures for contrasts. Indeed, the testing of the $\frac{1}{2}p(p-1)$ pairwise contrasts can lead to internally contradictory conclusions which cannot occur if only independent contrasts are tested. For example, in the first instance with $p = 3$, it is quite possible to conclude that $T_1 = T_2$, $T_2 = T_3$, but $T_1 \neq T_3$. This result is clearly intransitive and presents a logical contradiction. No such contradiction can occur for the independent contrasts $T_1 - T_2$ and $\frac{1}{2}(T_1 + T_2) - T_3$. On the other hand, testing all pairwise contrasts is a straightforward attack on the problem of locating significant differences and is often used in educational research of a pilot, or preliminary, nature, where specific predictions concerning theoretical outcomes are not possible.

Let us begin by considering a perplexing problem which arises whenever any type of contrast procedure is utilized. If the null hypothesis for the CRD, $T_1 = T_2 = \cdots = T_p$, is tested by the analysis of variance and if the .05 significance level is chosen for this test, then there are 5 chances in 100 (or 1 chance in 20) that we shall erroneously reject a *true* null hypothesis. This familiar interpretation of the significance level as the risk of a Type I error is a basic fact in any hypothesis-testing situation, and when setting the significance level we are, in effect, saying that this is the amount of risk we are willing to tolerate when we interpret the outcome of an experiment. However, whenever two or more separate hypothesis decisions are made, the total risk of *at least one* false rejection is $1 - (1 - \alpha)^m$, where m is the total number of significance tests and α is the level of significance for each test.[3] For example, if three hypotheses are tested, each at $\alpha = .05$, the total Type I risk is $1 - .95^3 = .14$; also, for 10 hypothesis tests, the risk is .40. The implications of this for testing a set of contrasts are clear. If $\alpha = .05$, say, is used for testing each of the contrasts, the total Type I risk can be enormous. With only five treatment levels, there is a total of 10 pairwise contrasts, and the total Type I risk is .40. An alternative strategy might be to test each contrast at a relatively stringent level ($\alpha = .001$, for example). Then the total Type I risk would remain within reasonable limits even for a large number of tests. There are, however, two interrelated objections to this strategy. First, use of sufficiently stringent significance levels would result in tests of low power; thus, the risk of Type II errors (i.e., accepting false null hypotheses) would be inflated. Secondly, if $\alpha = .05$ is used in the analysis of variance of the CRD, it

[3]This formula is based on the binomial distribution. The probability in a total of m binomial trials that all results are alike is $(1 - \alpha)^m$, where $1 - \alpha$ is the probability of the result on one trial. In the present case, $1 - \alpha$ is the probability of accepting a true null hypothesis. Hence, $(1 - \alpha)^m$ is the probability of accepting m out of m true null hypotheses. Conversely, the complement of this quantity, or $1 - (1 - \alpha)^m$, is the probability of rejecting one or more of these true null hypotheses.

seems reasonable to test the individual pairwise contrasts at a comparable significance level; otherwise, none of the pairwise contrasts might be judged significant, yet this contradicts the null-hypothesis rejection from the CRD.

Within the context of a single experiment, there is no entirely satisfactory resolution to the problem raised above. Operationally, however, the experimenter can feel more confidence in his hypothesis decision if the contrasts which he tests are preplanned (or a priori) and directional alternatives are set up. For example, if the experimenter predicts a specific rank order among the treatment effects and if the analysis of the pairwise contrasts conforms to this a priori ordering, the experimenter will have few qualms concerning the total Type I risk. However, in many experimental contexts, such a priori decisions cannot be reasonably established. The major exception occurs in experiments which evolve from a background of theoretical reasoning. A well-defined theory (or theories) will yield specific predictions concerning the outcomes of a suitably designed experiment. Unfortunately, at present, a relatively small proportion of educational research is based on theory. Although the experimenter may have "hunches" (i.e., guesses) concerning the outcomes of an experiment, these can seldom provide a satisfactory foundation for the specification of a priori contrasts.

In order to properly test specific contrasts among sample means, a fundamental distinction must be maintained between preplanned (or a priori) comparisons and postmortem (or a posteriori) comparisons. Preplanned contrasts arise, presumably, from the nature of the theoretical and empirical background out of and around which the research is being conducted. They exist prior to the collection of data; this is a critical point since once the sample data are known, the experiment represents an entirely new probability situation. Postmortem comparisons by their very nature capitalize on chance occurrences within the experimental situation. A not uncommon postmortem procedure is to contrast the two most extreme, out of p, sample means. Which treatment levels result in the most extreme outcomes is not known until the data are collected; it is critical in the process of assigning a probability to the corresponding contrast to take into account the fact that the extremes were located after collection and inspection of the empirical data. There are statistically sound procedures for carrying out postmortem comparisons, and they will be dealt with in this chapter. The point being emphasized here is that the researcher must select his analytical procedures partly in terms of the *origins* of the contrasts which he is testing.

In the remainder of this section, procedures for analyzing sets of specific contrasts are developed and illustrated. The first three techniques, multiple t tests, Duncan's multiple range test, and the Newman-

Keuls test, test the set of all possible pairwise contrasts and are applicable to the postmortem testing of data. Orthogonal contrasts test only sets of independent contrasts and are typically utilized when a priori contrasts have been defined. However, sets of orthogonal contrasts may also be constructed after inspection of the data, and a modification of the usual testing procedure is presented which is appropriate in this circumstance. The final procedure, Dunnett's test, is applicable only in the situation in which one of the treatment groups represents a control condition; in Dunnett's test, all other treatment groups are compared (one at a time) with the control group, and the overall level of significance for the set of tests is controlled by the experimenter.

The two-sample t test can be used to test each possible pairwise comparison. In this application, the mean square for error is used to estimate the population variance, and the degrees of freedom for the test are the sum of the degrees of freedom from the two samples entering into the contrast. If, for example, $p = 3$, t tests could be used to test $T_1 = T_2$, $T_1 = T_3$, and $T_2 = T_3$. In detail, the first test would be

$$t = \frac{\bar{Y}_{.1} - \bar{Y}_{.2}}{\sqrt{\text{MS}_{\text{error}}\,(1/n_1 + 1/n_2)}}$$

The difference between the sample means, $\bar{Y}_{.1} - \bar{Y}_{.2}$, is our best sample estimate of the population contrast $T_1 - T_2$. The denominator is the estimated standard error of the difference between pairs of sample means (i.e., it is the estimate of the standard deviation of the sampling distribution of differences between the means of pairs of samples). The computed value of t can be evaluated against the t distribution with $n_1 + n_2 - 2$ degrees of freedom. To illustrate the complete procedure, consider the data in Table 2-10, which display significant

table 2-10 *Exemplary table of means and analysis-of-variance summary table*

treatment group	number	mean
1	10	11.41
2	10	15.60
3	10	19.47

source	df	sum of squares	mean square	F
Treatment	2	324.99	162.50	17.05
Error	27	257.22	9.53	
Total	29	582.21		

results (at $\alpha = .01$) for three treatment groups. The sample means are 11.41, 15.60, and 19.47; the mean square for error is 9.53. For the hypothesis $T_1 = T_2$, the estimated standard error of the difference is $\sqrt{\text{MS}_{\text{error}}(1/n_1 + 1/n_2)} = 1.38$. Then $t = -4.19/1.38 = -3.04$. With $n_1 + n_2 - 2$ degrees of freedom, this computed value of t is significant at conventional levels. The t values for the remaining two contrasts $T_1 - T_3$ and $T_2 - T_3$ are -5.84 and -2.80, respectively. These values are also significant (at least at the .05 level). Note that if all samples are of the same size, the estimated standard error of the difference need be computed only once.

Although once quite popular, multiple t tests are less often used today for testing the set of pairwise contrasts. This reduced usage is largely attributable to the fact of compounded Type I error for the total set of tests. It should be noted that the use of multiple t tests based on the usual critical values for the t distributions results in significance levels which are really indeterminate within the framework of the test. For example, when t is computed from the two most extreme means, the fact that these are the two most extreme among a set of p means is not utilized in the test. It is apparent that *by chance alone*, a greater range of observed sample means will occur for larger values of p (for example, for sampling from *one* population, a pair of sample means would show less variability than, say, five sample means). However, the tabled t distributions are relevant only for the case in which a total of two samples have been selected and are being compared.

A number of alternative procedures for testing the total set of pairwise contrasts have been developed. Each of these procedures is based on some strategy for dealing with the Type I error problem. Among the more widely used of these is Duncan's multiple range test. Duncan (1955) derived this test by considering the set of sample means in rank order of magnitude. Then critical values are selected so that the Type I error for all contrasts involving adjacent pairs of means is α (say .05 or .01). Contrasts between pairs of means which span k rank positions are tested with a Type I error of $1 - (1 - \alpha)^{k-1}$. Duncan refers to these total Type I error risks as "protection levels." Duncan's test represents a compromise procedure with respect to the problem of Type I error. For sample means which are relatively near each other in magnitude (i.e., rankwise are close together), the Type I error is maintained at a relatively small level, whereas for means differing by larger amounts, more risk of the false rejection of a null hypothesis for a contrast is taken.

Duncan's test is designed for use with samples of equal size. This common sample size is n. If the samples differ in size, but not markedly, the harmonic mean $\tilde{n} = p/(\Sigma_{j=1}^{p} 1/n_j)$ may be used in place of n in the test. The standard error of a mean is defined as $S_m = \sqrt{\text{MS}_{\text{error}}/n}$ (or

$S_m = \sqrt{\text{MS}_{\text{error}}/\tilde{n}}$ for unequal sample sizes). Tables of "studentized ranges" (Appendix Table B-8) are available which present standard errors of standardized differences between pairs of means spanning varying numbers of rank positions in an ordered set of p sample means. The product of the appropriate studentized range and the standard error of a mean provides a sample estimate of the standard error of the difference between pairs of means separated by a specific number of rank positions. These products are referred to as *shortest significant ranges*. Thus, to utilize Duncan's multiple range test, these steps are followed:

1. Using the mean square for error from the analysis-of-variance summary table, find the standard error of a mean: $S_m = \sqrt{\text{MS}_{\text{error}}/n}$ (or, for unequal sample sizes, use $\tilde{n}$ in lieu of n).
2. Enter the row of Appendix Table B-8 corresponding to n (or $\tilde{n}$ rounded to the nearest integer), and find the studentized ranges for rank differences from 1 through $p - 1$.
3. Compute the set of shortest significant ranges by multiplying each studentized range by S_m.
4. Rank-order the sample means, and systematically compute all pairwise contrasts.
5. Hypothesis decisions are made by comparing each pairwise contrast with the appropriate shortest significant range. Any pairwise contrast which exceeds the appropriate shortest significant range is adjudged significant (provided, however, that it is not contained within the range of a pair of means already adjudged nonsignificant).

The Duncan test is carried out by testing the largest mean differences first and progressing to adjacent pairs. This order of testing may reduce the total number of comparisons required since any pair of means contained within a pair already adjudged nonsignificant must also be nonsignificant.

We return again to the data and the analysis-of-variance summary in Table 2-10 to illustrate the application of the Duncan test. The standard error of a mean is $S_m = \sqrt{9.53/10} = \sqrt{.953} = .976$. From Appendix Table B-8 the studentized ranges at the .05 level of significance are 3.20 and 3.34 (found from the row for 9 degrees of freedom for rank differences of 1 and 2, respectively). Thus, the shortest significant ranges are 3.12 [that is, (3.20)(.976)] for adjacent pairs of means and 3.26 [that is, (3.34)(.976)] for pairs of means differing by two rank positions. The sample means, in rank order, are $\bar{Y}_{.3} = 19.47$, $\bar{Y}_{.2} = 15.60$, and $\bar{Y}_{.1} = 11.41$. The pairwise contrasts are $3 - 2 = 3.87$, $3 - 1 = 8.06$, and $2 - 1 = 4.19$. Both of the adjacent pairs (that is, $3 - 2$ and $2 - 1$) exceed the critical value of 3.12, and the $3 - 1$ contrast is likewise significant.

The final procedure covered here for testing all pairwise contrasts is known as the Newman-Keuls technique. In this procedure, the significance level for all pairwise contrasts is always maintained at a constant; that is, if $\alpha = .05$ is selected as the significance level, each pairwise contrast, regardless of the rank separation in standing, is tested at $\alpha = .05$. The Newman-Keuls technique utilizes a studentized range statistic which is similar to that entering into Duncan's test; however, Duncan's values have been selected in terms of protection levels which vary depending upon the rank separation of pairs of means. The values which must be used in the Newman-Keuls test are based on a constant level of significance. Appendix Table B-7 contains the relevant critical studentized ranges for the Newman-Keuls test. The procedure is identical (except for the choice of studentized ranges) with the Duncan test. That is, the standard error of a mean is computed by using the mean square for error from the analysis of variance (again, $\tilde{n}$ may be used in lieu of n for samples of unequal, but not too disparate, sizes). Shortest significant ranges are computed by multiplying the standard error of a mean by the respective studentized ranges (from Appendix Table B-7), and the pairwise contrasts are compared against these shortest significant ranges. In general (except for adjacent means), however, the Newman-Keuls shortest significant ranges will be larger than those from the Duncan test; therefore, the Newman-Keuls procedure is the more conservative of the two.

In order to choose between the available procedures for testing a set of postmortem, pairwise contrasts, the researcher must adopt some stance relative to the problem of the risk of Type I errors in a series of significance tests. The most conservative approach would involve setting the overall α level for the entire set of tests at α; although this procedure has not been discussed, the interested reader will find an appropriate test (the Tukey test) in Myers (1966, pp. 334–335). Among the procedures which have been discussed, the use of multiple t tests is probably the least noteworthy since pairs of means which differ by several rank positions have true α levels which may be enormous compared with the nominal risk of a Type I error. On the other hand, use of the Newman-Keuls procedure results in all pairwise contrasts being tested at the same α level, and this is directly under the control of the researcher since the studentized-range statistic takes into account the rank separation of the pair of means entering into a contrast. The Duncan test represents a middle-of-the-road procedure since the level of significance is greater for pairwise contrasts involving means separated by larger numbers of rank positions. However, many users find this an objectionable trait in the Duncan procedure and prefer to utilize a test with a constant risk of Type I errors. A useful summary of multiple-comparison procedures as well as a strategy for deciding between contrast procedures is presented by Hopkins and Chadbourn (1967). Also included in their schema are the techniques of orthogonal contrasts and Dunnett's test, which are discussed in the next two sections of this chapter.

orthogonal contrasts

For p treatment levels, it is always possible to design a set of $p - 1$ linear contrasts which are mutually orthogonal (i.e., statistically independent). In this section, we develop procedures for testing hypotheses based on such sets of orthogonal contrasts. Often, a single set of orthogonal contrasts will be sufficient to test all a priori hypotheses which are of interest to the researcher. However, it may occur that overlapping, nonorthogonal sets of contrasts are necessary in specific settings. Consider, again, the case of a CRD with three treatment levels. Further, assume that the first treatment group received a control treatment (e.g., a placebo treatment), while treatment groups 2 and 3 were variations of an experimental treatment dimension. A meaningful and orthogonal set of contrasts would be $T_1 - \frac{1}{2}(T_2 + T_3)$ and $T_2 - T_3$. The first contrast compares the average of the experimental treatment effects with the control effect. Empirically, the question being answered can be stated: Is, on the average, any treatment better than no treatment at all? Whether or not this contrast proves significant, the second contrast compares the two levels of the experimental treatment to see whether differential effects are produced.

Now let us change the meaning of the treatment levels. Suppose the three treatments are differing amounts of treatment along one dimension in which, at best, only a rank-ordering of levels is possible[4] (e.g., different intensities of individualized, in-school vocational counseling). It might then be of interest to contrast successive differences of the form $T_1 - T_2$, $T_2 - T_3$. These contrasts are not independent, yet in an applied situation we may still wish to test them.

For samples of *equal size*, the following conditions define a set of orthogonal contrasts:

1. The sum of the contrast coefficients for each contrast must be 0.
2. The sum of cross products of coefficients for every pair of contrasts must be 0.

In general, the ith contrast of sample means (based on equal sample sizes) may be represented as $C_i = \Sigma_{j=1}^{p} c_{ij}\bar{Y}_{.j}$. The above

[4]This restriction to an ordinal treatment dimension is assumed here since if the levels were scaled in terms of equal units, the procedures of orthogonal polynomials (discussed in the next section) would be appropriate.

conditions require that $\Sigma_{j=1}^{p} c_{ij} = 0$ for all i, and that for all i and k, $\Sigma_{j=1}^{p} c_{ij}c_{kj} = 0$, where the coefficients are for the ith and kth contrasts. For example, with three treatment levels, the contrasts $T_1 - \frac{1}{2}(T_2 + T_3)$ and $T_2 - T_3$ are orthogonal. To demonstrate this, we rewrite the contrasts with specific coefficients:

$$1T_1 - .5T_2 - .5T_3$$

$$0T_1 + 1T_2 - 1T_3$$

The sum of the coefficients for each contrast is 0 (that is, $1 - .5 - .5 = 0$ and $0 + 1 - 1 = 0$). Also, the sum of the cross products of the coefficients is $(1)(0) + (-.5)(1) + (-.5)(-1) = 0 - .5 + .5 = 0$. For more than three treatment levels, cross products must be computed for all possible pairs of contrasts [in general, there will be $(p - 1)(p - 2)/2$ pairs of contrasts].

A unique, and fundamentally important, feature of sets of orthogonal contrasts is that they form the basis for a complete *partitioning* of the treatment sum of squares from the analysis of variance. As will be demonstrated in later chapters, more complex analysis-of-variance designs may be constructed by beginning with a CRD and then building orthogonal contrasts on the cells. For example, a 2×2 factorial design has four treatment groups or cells. These four cells may be considered to be the levels of a one-way analysis of variance, or CRD. The sum of squares for rows, the sum of squares for columns, and the interaction term may be defined in terms of orthogonal contrasts on the four cells. Thus, a fundamental unity underlies all analysis-of-variance techniques.[5]

For a set of p sample means, sets of $p - 1$ orthogonal contrasts can be constructed; thus, the orthogonal contrasts "consume" the degrees of freedom available in the set of sample means. There are, in general, many different sets of orthogonal contrasts which may be built on the same set of sample means. Ordinarily, the research design and the researcher's interests will lead to the choice of one or, at least, a small number of orthogonal sets. Each set will absorb the $p - 1$ degrees of freedom and will completely partition the treatment sum of squares. If all samples are of the same size n, then a sum of squares for the ith contrast is defined as

$$\text{SS}_{C_i} = \frac{n\left(\sum_{j=1}^{p} c_{ij}\overline{Y}_{.j}\right)^2}{\sum_{j=1}^{p} c_{ij}^2}$$

[5]Indeed, even greater generality can be achieved through the so-called "general linear hypothesis." Although this is beyond the scope of this volume, readers versed in matrix algebra may wish to consult Scheffé, *The analysis of variance*, 1959.

For $p - 1$ orthogonal contrasts, $\Sigma_{i=1}^{p-1} \text{SS}_{C_i} = \text{SS}_{\text{treat}}$, where SS_{treat} is the treatment sum of squares based on the p cells of the one-way analysis of variance. The data and the analysis-of-variance summary from Table 2-10 may be used to illustrate the additive property of sums of squares based on sets of orthogonal contrasts.

Consider, first, the pair of contrasts $\bar{Y}_{.1} - \frac{1}{2}(\bar{Y}_{.2} + \bar{Y}_{.3})$ and $\bar{Y}_{.2} - \bar{Y}_{.3}$. As shown previously, this pair of contrasts is orthogonal. The sum of squares for the first contrast is

$$\text{SS}_{C_1} = \frac{10[11.41 - \frac{1}{2}(15.60 + 19.47)]^2}{1^2 + .5^2 + .5^2} = 250.10$$

For the second contrast,

$$\text{SS}_{C_2} = \frac{10(15.60 - 19.47)^2}{1^2 + 1^2} = 74.88$$

The sum of squares for treatment from Table 2-10 is 324.99, and $\text{SS}_{C_1} + \text{SS}_{C_2} = 324.98$. Each of the sums of squares based on a contrast may be converted to a mean square and tested against the within-cells-error term. Since each contrast has 1 degree of freedom, the mean squares are identical in value with the sums of squares (when p is larger than 3, it is sometimes desirable or necessary to pool several of the sums of squares; the number of contrasts which are pooled together determines the degrees of freedom of the resulting sum of squares, and it may be tested against within-cells error, also; examples of this procedure will be seen in subsequent chapters which deal with more complex designs). Thus, $\text{MS}_{C_1} = 250.10$ and $\text{MS}_{C_2} = 74.88$. The F ratios, each of which has 1 and 27 degrees of freedom, are $250.10/9.53 = 26.24$ and $74.88/9.53 = 7.86$. Both of these values reach significance at conventional levels. It should be noted that, of the treatment sum of squares, 77 percent is accounted for by the first contrast [that is, $100(250.10/324.99) = 77$], and 23 percent is accounted for by the second contrast.

Any other pair of orthogonal contrasts defined on the three sample means would also consume the treatment sum of squares. To demonstrate this, consider the contrasts $\frac{1}{2}(\bar{Y}_{.1} + \bar{Y}_{.2}) - \bar{Y}_{.3}$ and $\bar{Y}_{.1} - \bar{Y}_{.2}$. These are clearly orthogonal. The sums of squares are

$$\text{SS}_{C_1} = \frac{10[\frac{1}{2}(11.41 + 15.60) - 19.47]^2}{.5^2 + .5^2 + 1^2} = \frac{355.81}{1.5} = 237.20$$

and

$$\text{SS}_{C_2} = \frac{10(11.41 - 15.60)^2}{1^2 + 1^2} = \frac{175.56}{2} = 87.78$$

Thus, $SS_{C_1} = SS_{C_2} = 237.20 + 87.78 = 324.98$, which is within a rounding error of $SS_{\text{treat}} = 324.99$. For these contrasts, the first consumes 73 percent of the treatment variability, and the second accounts for 27 percent.

The discussion of orthogonal contrasts has, until this point, been restricted to the case in which all samples are the same size. In practice this is often possible, but many times, due to experimental mortality, loss of data, etc., unequal size groups may result. Thus, it is necessary to become familiar with the procedures for use with samples of any sizes. If we return to our original conditions for defining a contrast and for orthogonality, certain changes are necessary. A contrast for samples of varying sizes is defined as

$$C_i = \sum_{j=1}^{p} n_j c_{ij} \bar{Y}_{.j} \qquad \text{where} \quad \sum_{j=1}^{p} n_j c_{ij} = 0$$

Note that the sample sizes n_j enter into the expressions as weights. The sum of squares for a contrast becomes

$$SS_{C_i} = \frac{\left(\sum_{j=1}^{p} n_j c_{ij} \bar{Y}_{.j}\right)^2}{\sum_{j=1}^{p} n_j c_{ij}^2}$$

For two contrasts to be orthogonal, the weighted sum of cross products $\sum_{j=1}^{p} n_j c_{ij} c_{kj}$ must be 0. It is apparent that the construction of orthogonal sets is much more tedious when the sample sizes are not the same. However, for $p = 3$ treatment groups, computation of orthogonal-contrast coefficients is reasonably easy. Consider three groups with the following sizes and means:

	TREATMENT GROUP 1	2	3
Sample size	5	8	12
Mean	$\bar{Y}_{.1}$	$\bar{Y}_{.2}$	$\bar{Y}_{.3}$

Suppose that the two contrasts of interest involve comparing group 1 with group 2 and comparing the average of groups 1 and 2 with group 3 (for the case of equal sample sizes, this would be the contrasts with coefficients 1, −1, 0 and $\frac{1}{2}$, $\frac{1}{2}$, −1). By definition, $5c_{11} + 8c_{12} + 12c_{13} = 0$; since group 3 is omitted from the first contrast, we set $c_{13} = 0$. Then

$5c_{11} = -8c_{12}$. A convenient solution is $c_{11} = 8$ and $c_{12} = -5$. Note that $c_{11} = n_2$, while $c_{12} = -n_1$. In terms of means, the contrast is $8\overline{Y}_{.1} - 5\overline{Y}_{.2}$.

For the second contrast, we have $5c_{21} + 8c_{22} + 12c_{23} = 0$. However, for orthogonality to be met, we must impose the additional restriction $5c_{11}c_{12} + 8c_{12}c_{22} + 12c_{13}c_{23} = 0$. Since $c_{11} = 8$, $c_{12} = -5$, and $c_{13} = 0$, we have $40c_{21} = 40c_{22}$; thus, $c_{21} = c_{22}$. Call this common value $c_{2.}$. Substituting in the original contrast, we have $5c_{2.} + 8c_{2.} + 12c_{23} = 0$. Then $13c_{2.} = -12c_{23}$. A convenient solution is $c_{2.} = c_{21} = c_{22} = 12$ and $c_{23} = -13$. Note that $c_{2.} = n_3$ and $c_{23} = -(n_1 + n_2)$. In terms of means, the second contrast is $12(\overline{Y}_{.1} + \overline{Y}_{.2}) - 13\overline{Y}_{.3}$. Thus, in this example the pair of orthogonal-contrast coefficients is 8, -5, 0 and 12, 12, -13.

As mentioned earlier, sets of orthogonal contrasts can, of course, be set up after the data from an experiment have been collected and inspected by the researcher. In this circumstance, the use of conventional significance values from the F table is inappropriate. However, a simple alternative is available which generally results in a conservative test for postmortem contrasts. The procedure, which is due to Scheffé, merely substitutes the critical value $(p - 1)F'$, where F' is the critical value of F for $p - 1$ and $np - p$ (where this is the degrees of freedom for error) degrees of freedom. Otherwise, the sums of squares for the contrasts and the formation of F ratios proceed in the same manner as when a priori contrasts are involved.

The formulas presented in this section for contrasts have been in terms of treatment means. Equivalent expressions can be set up for computing with the treatment sums. The advantage of the equations based on treatment sums is that somewhat greater computational accuracy may be achieved, since these sums are often whole numbers whereas the means are decimal quantities which must be rounded off. However, if the means are expressed to several decimal places (say four or five), the two approaches will yield virtually identical results. For treatment groups of the same size n, a contrast based on sums is defined as

$$C_i = \sum_{j=1}^{p} c_{ij}Y_{.j} \qquad \text{where} \quad \sum_{j=1}^{p} c_{ij} = 0$$

and the associated sum of squares is

$$SS_{C_i} = \frac{C_i^2}{n\left(\sum_{j=1}^{p} c_{ij}^2\right)}$$

Since $\overline{Y}_{.j} = Y_{.j}/n$, the student can easily verify that these formulas are identical with those based on means. Similarly, if the groups vary in size, a contrast based on sums is

$$C_i = \sum_{j=1}^{p} c_{ij} Y_{.j} \qquad \text{where} \quad \sum_{j=1}^{p} n_j c_{ij} = 0$$

and the associated sum of squares is

$$SS_{C_i} = \frac{C_i^2}{\sum_{j=1}^{p} n_j c_{ij}^2}$$

The conditions for orthogonality of two contrasts based on sums are the same as for means (that is, $\Sigma_{j=1}^{p} c_{ij} c_{kj} = 0$ for groups of the same size, and $\Sigma_{j=1}^{p} n_j c_{ij} c_{kj} = 0$ for groups of varying sizes).

Dunnett's test

In some experiments, one of the experimental groups may be a control condition, and the purpose of the study is to evaluate the differential effects of two or more other treatment conditions in comparison with the control. Dunnett's test was developed for just this purpose and allows the researcher to fix the overall level of significance for the set of tests at .05, .01, or any other desired value. For p levels of the treatment, there will be $p - 1$ comparisons possible between a single control group and the other treatment groups. For convenience, assume that the control group is the pth level of the treatment. Then the test statistic is $t = (\bar{Y}_{.j} - \bar{Y}_{.p})/S_m$, where $S_m = \sqrt{2\text{MS}_{\text{error}}/n}$, and MS_{error} is taken from the analysis-of-variance summary table. Normally, n is the number of observations per treatment group. If the groups vary in size, the harmonic mean $\tilde{n}$, as defined earlier, can be substituted for n in the formula. Appendix Table B-9 presents critical values for the t values from Dunnett's test (note that these special tables are required to evaluate Dunnett's t's; ordinary t tables are inappropriate). Note that each of the $p - 1$ t values is compared with the same tabular critical value. Dunnett's test is illustrated in connection with the third exemplary application of CRDs at the end of this chapter.

contrast procedures for quantitative treatment variables

Quantitative treatment variables exist when the treatment levels themselves represent points along a continuum which is at least ordinal. If, in fact, the continuum can be scaled in terms of units of equal size, it is possible to use the procedures of orthogonal contrasts to investigate systematically a variety of functional relationships between the treatment dimension and the response variable. In this context, the contrasts are referred to as *orthogonal polynomials*, and the procedures are known, collectively, as *trend analysis*. If the treatment continuum is orderable but equal units cannot be defined, then Page's L test can be applied if the researcher can specify an ordered alternative to the null hypothesis.

orthogonal polynomials

The techniques of orthogonal contrasts can be extended to yield sums of squares corresponding to specific degrees of polynomial regression between the treatment variable and the response variable. For a CRD with p levels of the treatment variable, it is possible to investigate regression up to a $(p-1)$st-degree polynomial. Thus, for five levels, a fourth-degree polynomial can be fitted to the data. The use of regression procedures in an analysis-of-variance context closely parallels ordinary correlation and regression applications. However, in the application of orthogonal polynomials, the regression line is fitted to the *means* of the response variable. It is, remember, the variability among the sample means which accounts for the magnitude of the treatment sum of squares in the analysis of variance; furthermore, a set of orthogonal contrasts consumes or accounts for this treatment sum of squares. Thus, if these contrasts are constructed so as to correspond to specific types of polynomial regression, they also will account for the treatment sum of squares and will represent the variability among the treatment means. The concept of the regression of response means on treatment levels can best be understood in terms of a diagram (Figure 2-1). In the figure, only points corresponding to treatment means have been plotted (the error sum of squares represents the scatter of observations around these sample means and is used, of course, in testing the significance of the regression sums of squares). The horizontal axis is labeled in terms of treatment levels; without loss of generality, the first five integers are used to represent the five treatment levels (i.e., only an interval scale of measurement is assumed to underlie the treatment dimension). For simplicity, only first-degree (i.e., linear) and second-degree (i.e., quadratic) regression curves are shown on the figure. In reality, up to a fourth-degree polynomial could be fitted to the data. It should be noted that a $(k-1)$st-degree polynomial can always be constructed which will be a *perfect* fit to k points in space (of course, a lower-order polynomial *may* be a perfect fit in special circumstances). For example, a straight line (which is a first-degree polynomial) can always be fitted to two points in space, and a conic section (which is a second-degree polynomial or a quadratic function) can always be fitted

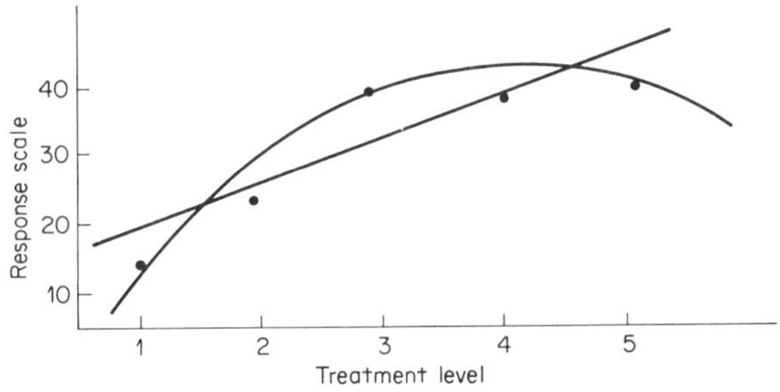

figure 2-1 *Linear and quadratic regression of sample means on treatment levels.*

to three points in space. Thus, for the example in Figure 2-1, the fourth-degree polynomial would pass through all five points representing the means.

The general polynomial of degree $p - 1$ can be defined as

$$Y' = a_0 + a_1X + a_2X^2 + a_3X^3 + \cdots + a_{p-1}X^{p-1} = \sum_{j=0}^{p-1} a_jX^j$$

In the analysis-of-variance situation, X represents a specific level of the treatment variable, and Y' is the predicted value of the response-variable mean. There are all together p unknown coefficients a_j. It would be possible to apply the usual least-squares techniques to find estimators for these coefficients. In fact, the procedures of trend analysis, utilizing orthogonal contrasts, are equivalent to the least-squares solution but stay completely within the context of the analysis of variance. In practice, it is often of no particular interest to learn the actual numerical values of the regression coefficients, and the use of orthogonal polynomials allows one to sidestep their computation while obtaining sums of squares which may be used to test the significance of the contribution of a given degree of polynomial regression.

Prior to discussion of the actual procedures of trend analysis, it is necessary to be clear about the nature of the hypothesis tests which are of interest. If a linear component proves to be significant, then it is certain that any higher-degree polynomial can fit the data at least as well as the straight line and, thus, must also be significant (e.g., a quadratic function cannot be a worse fit to the means than a straight line, since the straight line itself is a degenerate form of a quadratic with $a_2 = 0$). What one really wants to know is whether or not the higher-degree polynomial is a *significantly* better fit to the means than the lower-order polynomial. Thus, the sums of squares which are of interest are those representing independent contributions for the polynomials. For example, a sum of squares for quadratic regression should represent only that portion of the quadratic regression not already contained in the linear regression sum of squares, and so forth for higher-order polynomials. The techniques of trend analysis as developed here yield independent sums of squares. Therefore, each test of significance for a particular degree of polynomial is concerned only with the variability accounted for by that degree of polynomial but *not* accounted for by lower-degree polynomials. In effect, at each stage of the analysis we are testing the hypothesis that the relevant population regression coefficient is equal to 0. For example, when a linear component is tested, the hypothesis is $H_0: A_1 = 0$ (assuming $A_2 = 0$, $A_3 = 0$, etc.), where A_j is the population regression coefficient corresponding to a_j. For the quadratic component, the hypothesis is $H_0: A_2 = 0$ (assuming $A_3 = 0$, $A_4 = 0$, etc.), and this test is independent of the value of A_1 (that is, it is immaterial whether $A_1 = 0$ or $A_1 \neq 0$).

Sums of squares corresponding to specific degrees of polynomial regression are especially easy to find if the treatment levels represent *equally spaced* points along the treatment continuum. In this circumstance, the width of the spacing may be taken as a unit, and, for p levels of the treatment, the first p positive integers may be used to represent the levels (this procedure is not necessary but does simplify the arithmetic somewhat). By letting X_i represent the magnitude of the ith treatment level and $\bar{X}$ the mean of the p treatment levels, coefficients for polynomials through the fifth degree (that is, quintics) may be found from the following expressions:

$$c_{1i} = X_i - \bar{X}$$

$$c_{2i} = c_{1i}^2 - \frac{p^2 - 1}{12}$$

$$c_{3i} = c_{1i}^3 - c_{1i}\frac{3p^2 - 7}{20}$$

$$c_{4i} = c_{1i}^4 - c_{1i}^2\frac{3p^2 - 13}{14} + \frac{3(p^2 - 1)(p^2 - 9)}{560}$$

$$c_{5i} = c_{1i}^5 - c_{1i}^3\frac{5(p^2 - 7)}{18} + c_{1i}\frac{15p^4 - 230p^2 + 407}{1008}$$

It is rarely of interest to investigate regression even up to a quintic, since the form of the relationship becomes exceedingly complex. In general, linear, quadratic, and cubic components are of primary interest. Often, all higher-order components are lumped together and tested (and, for simplicity, their pooled sum of squares is found by subtraction from the treatment sum of squares). To avoid the use of the formulas to find polynomial coefficients, many of the more widely used cases are tabulated in Appendix Table B-11.

The actual computation of sums of squares proceeds exactly in the same manner as for orthogonal contrasts in general. Thus, a sum of squares for linear regression is derived from

$$\text{SS}_{\text{lin}} = n\frac{\left(\sum_{j=1}^{p} c_{1j}\bar{Y}_{.j}\right)^2}{\sum_{j=1}^{p} c_{1j}^2}$$

and so forth, assuming equal sample sizes. Returning to the data in Table 2-10, let us assume that the three treatment levels represent equally spaced amounts of a treatment. Using either the equations for polynomial-contrast coefficients or Appendix Table B-11 shows that the contrast corresponding to a linear trend among the sample means is $\bar{Y}_{.3} - \bar{Y}_{.1}$ and to a quadratic trend is $\bar{Y}_{.1} - 2\bar{Y}_{.2} + \bar{Y}_{.3}$. Thus,

$$\text{SS}_{\text{lin}} = \frac{10(19.47 - 11.41)^2}{1^2 + 1^2} = \frac{10(8.06^2)}{2} = 324.80$$

and

$$\text{SS}_{\text{quad}} = \frac{10[11.41 - 2(15.60) + 19.47]^2}{1^2 + 2^2 + 1^2} = \frac{10(-.32)^2}{6} = .17$$

Note that $324.80 + .17 = 324.97$, which is, within rounding, the value for the treatment sum of squares. To test the significance of each of these components, an F ratio is formed with the error mean square entering into the denominator. Since $\text{MS}_{\text{lin}} = \text{SS}_{\text{lin}}/1$ and $\text{MS}_{\text{quad}} = \text{SS}_{\text{quad}}/1$, these ratios are $F_{\text{lin}} = 324.80/9.53 = 34.08$ and $F_{\text{quad}} = .17/9.53 = .02$. Thus, the linear trend is highly significant, and the quadratic falls far short of significance. Note that the linear component accounts for 99.9 percent of the treatment variability, whereas the remaining .1 percent is explained by quadratic regression.

When the sample sizes are not the same and/or unequally spaced levels of the treatment are used in the experimental design, the standard formulas and tables of polynomial-contrast coefficients cannot be used. In these circumstances, the most expedient solution is the utilization of high-speed digital-computer programs which can effortlessly carry out the otherwise tedious mathematical computations. A general solution requires the use of matrix algebra, and useful computing schemes are presented by Peng (1967).

Page's L test

If the treatment levels can be ordered but cannot be scaled in terms of equal units of an underlying dimension, a procedure developed by Page (1963) has a great deal of potential usefulness. Page's test is "nonparametric" in the sense that measurement scale and distributional properties (e.g., normality) commonly assumed in the development of classical statistical tests (such as the analysis of variance) are not assumed by Page. Thus, Page's L test can be used with data which are only ranks. Indeed, if the response measures are stronger than ranks (e.g., ordinary physical measures), they must be rank-ordered prior to the use of Page's L test.

For a one-way analysis of variance with p treatment levels, the null hypothesis is $H_0: T_1 = T_2 = \cdots = T_p$. If this hypothesis is rejected, the decision is that *at least one* of the equalities does not hold. In Page's L test, a specific alternative hypothesis is investigated; that is, the investigator must specify a rank order among the treatment levels. This rank order may be based on an assumed continuum underlying the treatment dimension or on any other rational basis which is relevant to the research situation. If the treatment levels are objectively rankable, the specific alternative hypothesis can be set up to test for a "linear" trend or for any other trend of interest to the researcher (e.g., the rank-order analogy of linear trend would use the ordered alternative hypothesis $H_1: T_1 > T_2 > \cdots > T_p$ or $H_1: T_1 < T_2 < \cdots < T_p$; for an analogy to quadratic trend, one might use, for six treatment levels, $H_1: T_4 > T_3 > T_5 > T_6 > T_2 > T_1$, which suggests a hyperbolic sequence of treatment means). Of course, when the treatment dimension is only rank-orderable, no specific meaning can be given to linear, quadratic, and higher-order components since the spacings between treatment levels are arbitrary. Nevertheless, this approach may be of interest in certain research situations.

Page's L test is restricted to cases in which the samples are of equal size and, furthermore, subjects are matched across samples. Although a complete discussion of matching is included in Chapter 5 of this volume, the process is conceptually simple and is widely used in educational research. Essentially, *prior* to assignment to treatment groups, homogeneous subgroups of p subjects are formed on the basis of a relevant matching variable (e.g., intelligence scores or previous scholastic performance); ordinarily, this is done by simply ranking all subjects on the matching variable and choosing subgroups of p subjects, starting with those occupying the p highest ranks. Somewhat more complex strategies must be followed if matching on two or more outside variables is desired. The major precaution in matching (and one not uncommonly violated in practice) is that the matching must be done prior to assignment to treatment groups; pseudo-matching of subjects after the application of treatments is a fallacious procedure which capitalizes on chance. Returning to Page's L test, we see that the data take the form of n subjects assigned to p treatment groups, with matched p-tuples of subjects across the groups. Table 2-11 presents data from a fictitious experiment reported by Page. The details of the test involve, first, setting up the specific ordered alternative to the null hypothesis (since this test is nonparametric, the null hypothesis and alternative hypothesis cannot, strictly speaking, be written in terms of population effects; rather, a general location parameter, say λ, would be substituted for T in the hypotheses). Page (1963, p. 218) characterizes his fictitious situation as follows:

> The hypothesis under investigation is that speed of learning depends partly upon the similarity of practice material used in pre-training sessions. Let us say that Group 1 employed practice material most similar to that of the criterion learning; Group 2 next most similar; Group 3 least similar; and Group 4 had no pre-training.

The null hypothesis is $H_0: \lambda_1 = \lambda_2 = \lambda_3 = \lambda_4$. In view of the purpose of the fictitious experiment, the ordered alternative hypothesis of interest is $H_1: \lambda_1 > \lambda_2 > \lambda_3 > \lambda_4$. From Table 2-11, it is apparent that the data, as reported, are ranks across the rows of the data table; that is, subjects within matched groups have been ranked from 1 through p in terms of their performance on the criterion task. If the data to be analyzed are numerical scores, they must be converted to ranks in this

table 2-11 *Ranking matrix from fictitious experiment*

matched group	1	2	3	4	
I	2	1	3	4	
II	1	3	4	2	
III	1	3	2	4	
IV	1	4	2	3	
V	3	1	2	4	
VI	1	2	4	3	
$\sum_{i=1}^{n} Y_{ij}$	9	14	17	20	$\sum_{i=1}^{n}\sum_{j=1}^{p} Y_{ij} = 60$
$X_j \sum_{i=1}^{n} Y_{ij}$	9	28	51	80	$\sum_{j=1}^{p} (X_j \sum_{i=1}^{n} Y_{ij}) = 168 = L$
(Ranking) X_j	1	2	3	4	

Source: Page (1963).

fashion prior to the computation of the L statistic. X_j is used to represent the hypothesized rank order of the jth level of the treatment. Then the test statistic L is simply the sum of the cross products of the X_j and the sum of ranks for the treatments. That is, $L = \Sigma_{j=1}^{p} (X_j \Sigma_{i=1}^{n} Y_{ij})$, where Y_{ij} is the rank for the ith individual in the jth treatment group. The significance of L must be determined by reference to special tables (Appendix Table B-10). These tables extend from 3 through 10 treatment groups and up to 50 subjects per treatment group. For larger data tables, an approximate chi-square statistic is presented by Page (1963, p. 224).

For the fictitious data, $L = 168$; referring to Appendix Table B-10, this value is significant beyond the .01 level, but not at the .001 level. Thus, there is evidence that the a priori ordering of effects, embodied in the alternative hypothesis, does characterize the populations being studied. There has been some investigation of the relative power of Page's L test and the ordinary analysis of variance performed on the same data (Boersma, DeJonge, & Stellwagen, 1964). These comparisons are somewhat shaky since the null hypothesis and the alternative hypothesis are different for the two tests; they may be, however, of interest to the prospective user of Page's L test.

relative power of experimental designs

In this chapter, the completely randomized design, or one-way analysis of variance, has been used as a vehicle to introduce a number of topics essential to the understanding and utilization of the full range of available experimental designs. Certain of the designs considered in later chapters (e.g., randomized blocks designs, repeated measures designs, and the analysis of covariance) have the advantage of *potentially* yielding a smaller experimental-error term than the CRD. That is, by either the use of homogeneous subgroups of subjects or statistical adjustment through regression techniques, these designs account for some of the variability among subjects which, in the CRD, is included in the error sum of squares. These procedures, if successful, are obviously highly desirable since they will lead to a greater rate of rejection of false null hypotheses. Since these designs call for extra data collection (e.g., for a covariate or for a matching variable) or for special experimental arrangements (e.g., individual treatment order randomization, as in the repeated measures designs), it is worthwhile to investigate the degree to which they do, in fact, increase the sensitivity of an experiment relative to a CRD. The CRD is the simplest design from a conceptual point of view and is also the easiest to implement in practice. Its power can be increased by the simple stratagem of increasing the size of samples entering into the experiment. Thus, a natural approach to a discussion of the relative power of experimental designs is in terms of the sample sizes which are required to reject a specific null hypothesis when, in fact, that null hypothesis is false and some alternative hypothesis characterizes the populations being studied. Unfortunately, this approach leads to a number of complexities; there is generally no logical way of specifying an alternative hypothesis (or even a simple class of them) which is of interest if the null hypothesis is false. Furthermore, the sample size in many experimental situations may not be particularly flexible. Time and other limitations may dictate an experiment of a certain size; the problem is to decide whether or not a design more sophisticated than the CRD is worthwhile from a practical point of view. For these reasons, and because of its great simplicity and ease of interpretation, the relative power (or relative efficiency) of designs is defined in this volume as a ratio of error variances. If one design has an error mean square MS_1, based on degrees of freedom df_1, and a second design has an error mean square MS_2, based on degrees of freedom df_2, then the relative efficiency of the first design, compared with the second design, is defined as

$$RE = \frac{(df_1 + 1)(df_2 + 3)MS_2}{(df_2 + 1)(df_1 + 3)MS_1}$$

This ratio follows from Fisher's (1960) definition of the "amount of information" which a variance estimator carries concerning the population variance. Except for the weighting terms which involve degrees of freedom, RE is simply a ratio of error mean squares. In many practical settings, the only relevant use of RE would be to make a *post hoc* evaluation of the relative benefit from having applied an error-reducing design to a research situation. Thus, if the design actually utilized were randomized blocks, the data could be reanalyzed as a CRD, and a comparison of error mean squares made in terms of relative efficiency.

exemplary applications of completely randomized designs

This section contains summaries of some research applications of CRDs. The studies presented have been selected from research journals or other reports of original research efforts. Because of space limitations, most journals reporting educational research present only summary statistics and results from tests of significance; raw data upon which analysis is based are ordinarily not available. However, even if such data were available, sample sizes are often quite large, and the reporting of raw data in this volume would be very uneconomical. Nevertheless, the data reported in our summaries usually omit only relatively simple computational steps such as finding means, variances, and sums of squares. The essentials of the analysis are retained, and interpretations consistent with the data can be evaluated.

example 1 Johnson (1968) reported a study in which four methods of training education students in observing pupil classroom behavior were compared with respect to the accuracy with which relevant details of video-taped classroom scenes could be recalled. Each treatment group consisted of 21 education students who had been randomly selected from summer school enrollees in a large state university. The first group received training which consisted in viewing instructional video tapes of classroom scenes and, at the same time, completing items in a programmed booklet which posed questions about the video-taped scenes and provided feedback concerning the correctness of student responses to these questions. The second group viewed the same video tapes as the first group, but were provided with booklets which were not programmed and did not provide feedback information. The third group listened to a tape recording of a lecture concerned with observation of pupil behavior, and the fourth group received no special training. The treatment session was 35 minutes in length, and immediately following treatment *all* subjects viewed a new video tape and completed a 21-item multiple-choice examination which posed questions concerning the pupil behavior viewed on the video tape. The criterion measure was the number of items answered correctly on the 21-item test.

The basic design of the experiment is a CRD with 21 observations per treatment group; also, the treatment dimension represents a fixed factor. Since subjects were randomly selected for treatment-group membership, the experiment can be legitimately approached by the methods of statistical inference. The author of the study carried out two analytical procedures. First, the data from the four groups were treated by analysis-of-variance procedures. Then, Duncan's multiple range test was applied to the group means since a significant F resulted from the analysis of variance. The analysis-of-variance summary table is:

source	*df*	sum of squares	mean square	F	p
Treatments	3	143.65	47.88	8.36	<.01
Error	80	458.39	5.73		
Total	83	602.04			

The null hypothesis $H_0: T_1 = T_2 = T_3 = T_4$ can be rejected at the .01 level of significance on the basis of the obtained F ratio (the author does not report a homogeneity-of-variance test, although this is not critical since the groups were all the same size; however, from the standard deviations, which were reported, variances can be computed, and $F_{\max} = 9.61/2.89 = 3.33$, which, with 3 and 20 degrees of freedom, is significant at the .05 level). Since a significant F ratio from the analysis of variance was obtained, the author proceeded to apply Duncan's test to locate specific pairwise contrasts which represent significant differences. The standard error of a mean is $S_m = \sqrt{5.73/21} = .5224$; from Appendix Table B-8, the studentized ranges with $n = 21$ are (for means separated by 1, 2, and 3 rank positions, respectively) 2.95, 3.10, and 3.18. Thus, the shortest significant ranges are 1.54, 1.62, and 1.66 (for the .05 level of significance).[6] The means, in rank order, were $\bar{Y}_{.1} = 18.33$, $\bar{Y}_{.3} = 16.66$, $\bar{Y}_{.2} = 15.81$, and $\bar{Y}_{.4} = 14.76$. The contrasts for adjacent means are $1 - 3 = 1.67$, $3 - 2 = .85$, and $2 - 4 = 1.05$; for means separated by two rank positions, the contrasts are $1 - 2 = 2.52$ and $3 - 4 = 1.90$; and for means separated by three rank positions, the one contrast is $1 - 4 = 3.57$. Note that only the contrasts $3 - 2$ and $2 - 4$ fail to reach significance at the .05 level.

From these outcomes, the author concludes that training with questioning and immediate feedback is superior to the other three types of training and that some training is better than no training at all. It should be noted that these conclusions can be statistically generalized only to the population of education students from which random samples were chosen.

Although the author of this study did not report orthogonal contrasts among the sample means, it is interesting to note that the following set of contrasts are especially relevant to the research questions posed in the experiment:

	GROUP 1 video tape + programmed booklet	GROUP 2 video tape only	GROUP 3 tape-recorded lecture	GROUP 4 control	C_i	SS_{C_i}
Mean:	18.33	15.81	16.66	14.76		
Contrast						
1	+1	+1	+1	−3	6.52	74.39
2	+1	−1	0	0	2.52	66.68
3	+1	+1	−2	0	.82	2.35

The first comparison involves contrasting the control group with the three groups receiving some form of training. The sum of squares

[6]The author apparently made a computational error since the shortest significant ranges are reported as 1.47, 1.55, and 1.60. Seemingly, he erroneously used $n = 80$.

(and mean square, since each contrast is based on 1 degree of freedom) is 74.39; utilizing the mean square for error from the analysis of the CRD gives the F ratio as $74.39/5.73 = 12.98$, which is significant beyond the .01 level of significance. The second contrast involves the two groups receiving training which utilized video tapes. This sum of squares is 66.68, and the F ratio is $66.68/5.73 = 11.64$; this value is also significant beyond the .01 level. Finally, the third contrast compares the two groups trained with video tapes with the group trained via the tape-recorded lecture. This sum of squares is 2.35, and the F ratio is less than 1 (and, obviously, nonsignificant). It should be noted that the sums of squares associated with these three contrasts total 143.42, which is somewhat less than the sum of squares for treatment from the analysis-of-variance summary table. This inaccuracy was introduced since group means were reported only to two decimal places in the research article.

example 2 Allen and McDonald (1966) report an experiment involving learner-selection of instructional materials. The goal of the study was to determine whether self-selection had an enhancing effect upon achievement of a specific body of subject matter. The experimental task involved learning the rules of an original game modeled along the lines of chess and checkers. Students in one educational psychology course at a university were randomly divided among three treatment conditions. Group 1, comprising 23 students, learned the game by means of a specially prepared linear programmed booklet. Group 2, comprising 21 students, studied the same programmed booklet, but each student had the option of selecting the order in which he studied the four sections of the booklet. A third group, comprising 21 students, learned the game in a one-to-one situation with an instructor who responded to all questions posed by the student but did not initiate instruction. The authors of the research report indicate no reason for the slightly disparate numbers in the three groups; a likely explanation is that certain subjects in groups 2 and 3 failed to complete the experimental task or did not follow instructions. At the completion of one experimental session, each subject was given a 27-item criterion test related to the hypothetical game.

The authors report mean *error* scores from the test, variances of these error scores, and within-group sums of squares per group. The design is basically a CRD with unequal numbers of observations per cell. The randomization requirements of such a design have been met, and the treatment dimension is fixed. Although the authors only report an F ratio of 11.53 (which is significant at the .01 level with 2 and 62 degrees of freedom), the data reported allow reconstruction of an analysis-of-variance summary table:

source	df	sum of squares	mean square	F	p
Treatment	2	180.78	90.39	11.53†	<.01
Error	62	471.17	7.60		
Total	64	651.95			

†90.39/7.60 = 11.89, which is somewhat greater than the value reported by the authors of the research article. Thus, some rounding error has been introduced into the reproduction of the sum of squares for treatment.

Based on the significant F value, the authors' conclusion was that self-selection of instructional materials does have an effect upon achievement. The mean error scores for the three groups were 6.08, 5.83, and 9.52. Multiple t tests were then utilized to isolate specific pairwise contrasts which reach significance. These were as follows: group 1 versus group 2: $t = .44$, which is nonsignificant; group 1 versus group 3: $t = 3.96$, which is significant at the .01 level; and group 2 versus group 3: $t = 4.00$, which is significant at the .01 level. Since the means reported are *error* scores, it appears that groups 1 and 2 did significantly better than group 3, but that groups 1 and 2 are not, themselves, different. These outcomes conflict with the notion that self-selection of instructional materials is more effective than following an imposed sequence of material. It should, however, be understood that these results can be generalized only to the original class of 65 educational psychology students who took part in the experiment, since no randomization external to the experiment was conducted.

The use of multiple t tests allows for an inflation of the overall level of significance for the set of three tests. Had the authors wished to control the level at .01, the Newman-Keuls procedure would have been appropriate. Since the sample sizes vary, their harmonic mean

$$\tilde{n} = \frac{3}{1/23 + 1/21 + 1/21} = \frac{3}{.13872} = 21.6263$$

is used in lieu of n in the computational formulas. The standard error of a mean is

$$S_m = \sqrt{\frac{\text{MS}_{\text{error}}}{\tilde{n}}} = \sqrt{\frac{7.60}{21.6263}} = \sqrt{.3514} = .5928$$

Interpolating in Appendix Table B-7 for 22 degrees of freedom, we find the studentized ranges for means separated by one and two rank positions to be, respectively, 3.99 and 4.59. Multiplying these by the standard error gives the shortest significant ranges as 2.37 and 2.72. The adjacent mean contrasts are $\bar{Y}_{.3} - \bar{Y}_{.1} = 9.52 - 6.08 = 3.44$, which exceeds the shortest significant range at the .01 level, and $\bar{Y}_{.1} - \bar{Y}_{.2} = 6.08 - 5.83 = .25$, which is nonsignificant. The contrast for means separated by two rank positions is $\bar{Y}_{.3} - \bar{Y}_{.2} = 9.52 - 5.83 = 3.69$, which is significant. Thus, the conclusions from the Newman-Keuls tests are the same as those from the multiple t tests.

example 3 Scannell and Marshall (1966) describe an experiment aimed at assessing the effect of certain composition errors upon the grading of essay examination papers. Three types of composition errors were considered: punctuation, spelling, and grammar. Sample answers to a twelfth-grade American history question were specially prepared; all together, five forms were produced: Form G contained grammatical errors; Form S contained spelling errors; Form P contained punctuation errors; Form C contained all three types of errors; and Form N, the control form, contained no composition errors. The graders—the subjects of this experiment—were a group of 165 undergraduate education students at a state university. From the total group, five groups of 33 students each were arbitrarily chosen (the procedure as described by the authors is not strictly random but did involve arbitrarily assigning a given answer form to a student; the failure to randomize is, however, a weakness in this study, and especially so since randomization could have been easily accomplished). Each group of students individually graded an answer of a specific form, using an A through F scale. These grades were converted to scores, allowing nine points for an A, seven points for a B, and so forth.[7] Dunnett's test was applied to the results in order to assess the contrast of each group with the control group. The results from Dunnett's test can be summarized as follows:

group	number	mean grade	$\bar{Y}_{.j} - \bar{Y}_{.p}$	t	p
Form N	33	6.15			
Form G	33	5.39	−.76	−2.06	$>.05$
Form P	33	5.70	−.45	−1.22	$>.05$
Form S	33	5.21	−.94	−2.55	$<.05$
Form C	33	5.06	−1.09	−2.95	$<.05$

[7]The actual study involved a second stage of data collection (4 weeks after the initial administration); since analysis-of-variance procedures were not applied during the second stage, it is omitted here.

The mean square for error, as reported, was 2.253. With $n = 33$, then

$$S_m = \sqrt{\frac{2\text{MS}_{\text{error}}}{n}} = \sqrt{\frac{2(2.253)}{33}} = \sqrt{.1362} = .369$$

The t values, as shown, were computed by using S_m and the contrasts of means with the control group. From Appendix Table B-9 with 160 degrees of freedom for error and five treatment groups, the critical value of t is approximately -2.17 (by interpolation) for a *directional* test at the .05 level of significance (directional tests were appropriate since the authors were predicting that the inclusion of composition errors would lower the assigned grades). The comparisons involving groups grading forms S and C exceeded this critical value and are significant. The authors conclude that prospective teachers are affected by composition errors when they grade essay examinations. The most influence occurs when a variety of errors occur (Form C). These results can, however, only be generalized to the group of 165 students who took part in the experiment since no external randomization was employed.

chapter three

factorial designs

introduction

The completely randomized design is suitable for investigating differential treatment effects along a single dimension of treatment variation. Factorial designs involve the simultaneous application of two or more different treatments within a single experiment. The factorial arrangement of treatments has two major advantages: (1) since the same groups of subjects are utilized in estimating effects from the two or more treatment dimensions, there is an economy of both time and personnel; and (2) the simultaneous application of treatments makes possible the elucidation of interactive effects among the treatment dimensions themselves (that is, one or some combination of treatments may have an accelerating effect or a moderating effect upon the action of some other treatment).

In a factorial design, there are as many groups of experimental subjects as there are possible combinations of the two or more treatment dimensions being studied. Thus, if two treatment dimensions are involved and the first occurs at three levels and the second at four levels, a total of 12 groups of subjects must be used in the experiment. There is a complete "crossing" of the levels of the treatment dimensions. In order to obtain a valid experiment, subjects must be assigned to the treatment groups by a random process. Furthermore, as in all experimental designs, the external generalizability of results is limited by the base of randomization utilized in obtaining the original pool of experimental subjects.

There is a wide variety of educational research situations to which factorial designs are particularly suited. In fact, it is difficult to think of research areas in education in which student responses cannot reasonably be hypothesized to be a function of a veritable host of environmental influences. In teaching reading to elementary school students, for example, the performance of students is likely to be affected by the rate at which new material is introduced, the frequency of testing for progress, the sequence in which new material is introduced, the degree of individualization of instruction, the total amount of time devoted to teaching reading during a day or a week, and so forth. In order to carry out meaningful research in the area of teaching reading, the investigator would wish to vary, not one, but several of these factors during a single experiment. It is highly likely that certain factors would be interactive, and the only road to discovering this is through factorial arrangements of treatments.

In discussing factorial designs, specific designs are labeled in terms of the number of levels of the treatments involved. Thus, the simplest possible design would have just two treatments, each occurring at only two levels, and it would be called a 2×2 *factorial design*. A more complex design might have four different treatment dimensions, the first having three levels, the second two levels, the third five levels, and the fourth six levels; this design would be labeled a $3 \times 2 \times 5 \times 6$ *factorial design*. The implied multiplication used in this naming scheme also yields the number of cells, or independent random groups of subjects, involved in the experiment. Thus, a 2×2 design has four cells, or groups of subjects; a $3 \times 2 \times 5 \times 6$ design requires 180 cells. With only 10 subjects per cell, this latter design would require a total of 1,800 experimental subjects! It is apparent that really large-scale factorial designs become rather impractical unless the experimenter has substantial resources for research.

This chapter is devoted to procedures for analyzing factorial designs and, also, to discussions of the interpretation of such designs. Since the 2×2 factorial design is the simplest, it is the first design presented. Next, additional designs with all treatments occurring at only two levels are covered; however, the number of different treatment dimensions is three or more. The next topic is concerned with general designs having no restrictions on the number of levels for the factors. Finally, procedures are presented for the case in which the experimental groups receiving treatment combinations are of different sizes for different treatment combinations.

the 2 × 2 factorial design

If we assume that each treatment group, or cell, has the same number n of subjects, the data from a 2×2 factorial design can be represented as in Table 3-1.

Note that the response measures in the cells are indexed in terms of row, column, and cell position. The restriction to designs with equal numbers of observations per cell will be maintained throughout this chapter until the section which deals with the computational problems that arise when unequal cell sizes are employed. Actually, equality of cell size is not necessary to a straightforward solution. If the numbers of cases are *proportional* by rows (or columns), no computational problem arises. For example, if both cells at level 1 of the A factor contain 10 observations and both cells at level 2 contain 15 observations, the rows (and columns) are proportional, and the usual computational approach is applicable.

Since there will be occasion to refer to specific cells of factorial designs, a standard notation is adopted. The symbol (ij) is used to refer to the cells defined by the intersection of the ith row and jth column in the data table. Thus, for the 2×2 factorial design, the upper left cell is (11), the upper right cell is (12), the lower left cell is (21), and the lower right

table 3-1 *Representation of data for a 2 × 2 factorial design*

		TREATMENT A 1	2
TREATMENT B	1	Y_{111}	Y_{121}
		Y_{112}	Y_{122}
		Y_{113}	Y_{123}
		.	.
		.	.
		.	.
		Y_{11n}	Y_{12n}
	2	Y_{211}	Y_{221}
		Y_{212}	Y_{222}
		Y_{213}	Y_{223}
		.	.
		.	.
		.	.
		Y_{21n}	Y_{22n}

cell is (22). This same scheme will be utilized in higher-dimensional designs, always starting with the dimension having the highest alphabetic coding [e.g., in a four-dimensional design, the dimensions would be coded A, B, C, and D; the first cell would be (1111), and so forth].

In order to construct the structural model for a score in the 2 × 2 factorial design, we must define the relevant sources of possible variation among response measures. Clearly, there may be an effect due to being at a specific level of A, another effect due to being at a specific level of B, and an effect due to random error. In addition, however, a specific combination of levels from the two treatment dimensions may also produce a unique effect. Thus, the total model is $Y_{ijk} = \mu + \alpha_j + \beta_i + \alpha\beta_{ji} + \epsilon_{ijk}$. Once again, a grand-mean effect μ is separated from the treatment effects. The term α_j is the effect of being at the jth level of the A treatment, β_i is the effect of being at the ith level of the B dimension, $\alpha\beta_{ji}$ is the effect of being in cell (ij), and ϵ_{ijk} is the random-error component unique to subject k in cell (ij). The interaction term $\alpha\beta_{ji}$ is, in effect, a component representing the non-additivity of the main effects for the treatment dimensions. If the A and B dimensions were strictly additive in their effects on the scores, then $\alpha\beta_{ji}$ would always be 0 and could be omitted from the model. Often, combinations of treatments do produce unique effects, unpredictable from the simple effects themselves, and the interaction term is necessary in the model. Of course, the significance of the contribution of an interaction effect can be tested within the analysis-of-variance framework, just as main effects can be tested.

Sums of squares and mean squares can be computed which correspond to the main effects, interaction effect and experimental error. If $\overline{Y}_{...}$ is the grand mean of all the scores in the data table, ignoring row and column classifications, then each score deviates from this grand mean: $Y_{ijk} - \overline{Y}_{...}$. This deviation may be considered to be composed of four components, each component representing one of the sources of variation from the model for a score. By letting $\overline{Y}_{.j.}$ be the mean of all scores at level j of the A dimension, $\overline{Y}_{i..}$ be the mean of all scores at level i of the B dimension, and $\overline{Y}_{ij.}$ be the mean of the scores in cell (ij), the deviation of a score from the grand mean can be rewritten as

$$(Y_{ijk} - \overline{Y}_{ij.}) + (\overline{Y}_{.j.} - \overline{Y}_{...}) + (\overline{Y}_{i..} - \overline{Y}_{...}) + (\overline{Y}_{ij.} - \overline{Y}_{.j.} - \overline{Y}_{i..} + \overline{Y}_{...})$$

The first term corresponds to the deviation of a score from its cell mean, and this component estimates experimental error; the second term corresponds to the deviation of a column mean from the grand mean, and this component estimates the effect due to the A dimension; the third term corresponds to the deviation of a row mean from the grand mean, and this component estimates the effect due to the B dimension; and the final term corresponds to the deviation of the cell mean from the value predictable on the basis of the relevant row and column means, and this component estimates the interaction of the A and B dimensions. These concepts can be clarified by reference to Table 3-2. For simplicity, only two scores have been entered per cell; the cell means and marginal means have been labeled. If we arbitrarily pick the first score in cell (12), the deviation from the grand mean, $Y_{ijk} - \overline{Y}_{...}$,

table 3-2 *Sample* 2 × 2 *factorial design*

		DIMENSION A 1	DIMENSION A 2	
DIMENSION B	1	8 6 $\overline{Y}_{11.} = 7.0$	11 5 $\overline{Y}_{12.} = 8.0$	$\overline{Y}_{1..} = 7.5$
	2	11 13 $\overline{Y}_{21.} = 12.0$	4 8 $\overline{Y}_{22.} = 6.0$	$\overline{Y}_{2..} = 9.0$
		$\overline{Y}_{.1.} = 9.5$	$\overline{Y}_{.2.} = 7.0$	$\overline{Y}_{...} = 8.25$

is $11 - 8.25 = 2.75$. The four components which constitute this deviation are

$$Y_{ijk} - \bar{Y}_{ij.} = 11 - 8.0 = 3.0$$

$$\bar{Y}_{.j.} - \bar{Y}_{...} = 7.0 - 8.25 = -1.25$$

$$\bar{Y}_{i..} - \bar{Y}_{...} = 7.5 - 8.25 = -.75$$

$$\bar{Y}_{ij.} - \bar{Y}_{.j.} - \bar{Y}_{i..} + \bar{Y}_{...} = 8.0 - 7.0 - 7.5 + 8.25 = 1.75$$

Note that $3.00 + (-1.25) + (-.75) + 1.75 = 2.75$, which is the same as the original deviation of score Y_{121} from the grand mean; also both positive and negative effects contributed to the magnitude of this score. Interpretively, these components are 3.0 units of experimental error; -1.25 units attributable to the effect of being at level 2 of the A dimension; $-.75$ units attributable to the effect of being at level 1 of the B dimension; and 1.75 units which are due to interaction, or due to being in cell (12). The interaction component, 1.75 units, can be given additional interpretation. Using just the grand mean and marginal means, we can attempt to reproduce the cell mean by a purely additive process. The effect of level 2 of the A dimension is -1.25; the effect of level 1 of the B dimension is $-.75$. Thus, our predicted value for the mean of cell (12) would be $8.25 + (-1.25) + (-.75) = 6.25$. However, the actual cell mean is 8.0, which is 1.75 units larger than the predictable value. This 1.75 units is unique to cell (12) and is unpredictable from the marginals and grand mean; it is, therefore, due to the interaction of the A and B dimensions.

Turning from our consideration of deviations per se, sums of squares can be based on each of the four components which have been isolated. A sum of squares for error (or within-cells sum of squares) is found strictly from within-cell variability:

$$SS_{error} = \sum_{i=1}^{2} \sum_{j=1}^{2} \left[\sum_{k=1}^{n} (Y_{ijk} - \bar{Y}_{ij.})^2 \right]$$

Note that the term in brackets is simply the sum of squared deviations from a cell mean; the double summation operation performed on this bracketed term merely pools these sums of squared deviations across the four cells of the 2×2 table. Similarly, the sums of squares for the main effects are

$$SS_A = 2n\left[\sum_{j=1}^{2} (\bar{Y}_{.j.} - \bar{Y}_{...})^2 \right]$$

$$SS_B = 2n\left[\sum_{i=1}^{2} (\bar{Y}_{i..} - \bar{Y}_{...})^2 \right]$$

The constant multipliers $2n$ occur because the marginal sums are pooled over two cells each containing n observations. Finally, the interaction sum of squares is

$$SS_{AB} = n\left[\sum_{i=1}^{2}\sum_{j=1}^{2}(\bar{Y}_{ij.} - \bar{Y}_{.j.} - \bar{Y}_{i..} + \bar{Y}_{...})^2\right]$$

From a computational point of view, these formulas are rather inconvenient; the following computational approach is recommended.

$$SS_A = \sum_{j=1}^{2}\frac{Y_{.j.}^2}{2n} - \frac{(Y_{...})^2}{4n}$$

$$SS_B = \sum_{i=1}^{2}\frac{Y_{i..}^2}{2n} - \frac{(Y_{...})^2}{4n}$$

$$SS_{AB} = \sum_{i=1}^{2}\sum_{j=1}^{2}\frac{Y_{ij.}^2}{n} - \frac{(Y_{...})^2}{4n} - SS_A - SS_B$$

$$SS_{\text{total}} = \sum_{i=1}^{2}\sum_{j=1}^{2}\sum_{k=1}^{n} Y_{ijk}^2 - \frac{(Y_{...})^2}{4n}$$

$$SS_{\text{error}} = SS_{\text{total}} - SS_A - SS_B - SS_{AB}$$

Each sum of squares can be converted to a corresponding variance, or mean square, by dividing by appropriate degrees of freedom. In the 2×2 factorial design, the main effects and interaction each have 1 degree of freedom, and the degrees of freedom for the error term are $4(n - 1)$. Thus, the results can be summarized as in Table 3-3.

When a 2×2 factorial design is applied, the null hypotheses of interest are concerned with row, column, and interaction effects. Thus, separate hypotheses must be established; utilizing the parametric notation from the structural model for a score, these hypotheses are

$$H_{0_1}: \alpha_1 = \alpha_2 = 0$$

$$H_{0_2}: \beta_1 = \beta_2 = 0$$

$$H_{0_3}: \alpha\beta_{11} = \alpha\beta_{12} = \alpha\beta_{21} = \alpha\beta_{22} = 0$$

That is, H_{0_1} states that the effects of the two levels of the A factor are equal; if such is the case, these effects may be set equal to 0 and absorbed into the grand-mean effect. Similarly, H_{0_2} posits the equality of the effects of the levels of the B dimension, and H_{0_3} states that there is no interaction of the A and B dimensions (that is, in effect, that the cell

table 3-3 *Analysis-of-variance summary table for a 2 × 2 factorial design*

source	*df*	sum of squares	mean square
Columns (A)	1	SS_A	$MS_A = SS_A/1$
Rows (B)	1	SS_B	$MS_B = SS_B/1$
Interaction (AB)	1	SS_{AB}	$MS_{AB} = SS_{AB}/1$
Error	$4(n-1)$	SS_{error}	$MS_{error} = SS_{error}/4(n-1)$
Total	$4n-1$	SS_{total}	

means can be perfectly reproduced from the marginal means and the grand mean). The three hypotheses are completely independent, and any one or more of them can be correct or incorrect without any effect on the validity of the remaining hypotheses. This follows from the fact that the partitioning of the total sum of squares yielded completely independent terms. The independence of these hypotheses is not always clearly understood and bears illustration. Figure 3-1 shows three possible *population* relationships between the dimensions of a 2 × 2 factorial design. The points plotted are population means corresponding to the cells of the design; lines connecting the points for the levels of B are merely for ease in reading the figures. In Figure 3-1a, there is a clear differential between the levels of B (that is, level 2 of B produces a higher mean than level 1 of B); however, averaged over the levels of B, there is no effect due to A. Also, there is no interaction of A and B since cell means can be perfectly reproduced from the marginal means. In Figure 3-1b, the opposite condition is shown. There is a clear differential between the levels of the A treatment dimension (with level 1 producing the higher mean score), but the levels of B (averaged over A) have no differential effect, and, again, there is no interaction of the A and B dimensions. Figure 3-1c illustrates a strong interaction effect with no differential effects for the main dimensions. Note that levels 1 and 2 of the A dimension average out to the same value, as do the levels

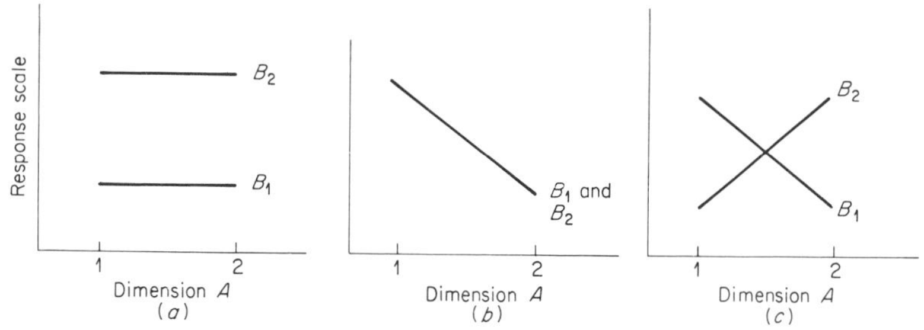

figure 3-1 *Illustrative population cell means in 2 × 2 factorial designs.*

of the B dimension. There is, however, a clear interaction between the A and B treatments, with the combination of level 1 of A and level 1 of B producing an enlarged effect, the combination of level 1 of A and level 2 of B producing a suppressed effect, and so forth. From these figures it is apparent that any one of the null hypotheses can be incorrect while the remaining two are correct. Similarly, the student should convince himself that any two of the hypotheses can be incorrect while the remaining null hypothesis is correct.

In order to test each of the three relevant null hypotheses for a 2×2 factorial design, recognition must be taken of the nature of the treatment variables; that is, test statistics must be set up taking into account whether the treatment levels represent fixed components, random components, or a combination of fixed and random components. By following the general lines of the derivations for the completely randomized design (Chapter 2), the expectations for mean squares shown in Table 3-4 can be found.[1] Although the distinction between fixed effects and random effects did not alter the hypothesis-testing procedure appropriate for the CRD, this is not the case, in general, for factorial designs. Here, it is necessary to distinguish three cases: the *fixed-effects model,* in which the levels of both the A and the B dimension are fixed by the experimenter; the *random-effects model,* in which the levels of both the A and the B dimension are selected randomly from a universe of possible levels; and the *mixed model,* in which either A is fixed and B is random or B is fixed and A is random.

From Table 3-4 it is apparent that the error mean square is the appropriate denominator in F ratios for both the A and the B dimension only in the case of the fixed-effects model. However, if both the A and

table 3-4 *Expected mean squares for the 2 × 2 factorial design*

MEAN SQUARE	FIXED-EFFECTS MODEL (both dimensions fixed)	RANDOM-EFFECTS MODEL (both dimensions random)	MIXED MODEL (A random and B fixed)†
MS_A	$\sigma^2 + 2n\sigma_\alpha^2$	$\sigma^2 + 2n\sigma_\alpha^2 + n\sigma_{\alpha\beta}^2$	$\sigma^2 + 2n\sigma_\alpha^2$
MS_B	$\sigma^2 + 2n\sigma_\beta^2$	$\sigma^2 + 2n\sigma_\beta^2 + n\sigma_{\alpha\beta}^2$	$\sigma^2 + 2n\sigma_\beta^2 + n\sigma_{\alpha\beta}^2$
MS_{AB}	$\sigma^2 + n\sigma_{\alpha\beta}^2$	$\sigma^2 + n\sigma_{\alpha\beta}^2$	$\sigma^2 + n\sigma_{\alpha\beta}^2$
MS_{error}	σ^2	σ^2	σ^2

†If A is fixed and B is random, reverse the role of α and β in the expectations.

[1]For designs more complex than the CRD, no attempt will be made in this volume to detail the steps involved in actually finding these expected values. The interested student should consult Edwards (1964).

table 3-5 *Data from a fictitious 2 × 2 factorial design*

		TREATMENT A open-book	closed-book	sums
		37.5	38.2	
		39.2	37.3	
	Review	40.3	39.1	363.7
		32.8	31.4	
		35.5	32.4	
TREATMENT B				
		33.4	28.2	
		32.1	30.4	
	No review	35.2	27.6	301.4
		31.8	29.2	
		28.3	25.2	
	Sums	346.1	319.0	665.1

the B factors represent random dimensions, both main effects must be tested against the interaction mean square. Note that the ratios MS_A/MS_{AB} and MS_B/MS_{AB} each have expectation 1 if the corresponding null hypothesis is correct; if the null hypothesis is incorrect, the expectation is larger than 1 since the numerator will contain a variance term (that is, $2n\sigma_\alpha^2$ or $2n\sigma_\beta^2$) which reflects the nonzero treatment effects. Also, the appropriate F ratio for use in testing the interaction effect is $MS_{AB}/MS_{\text{error}}$. The random-effects model illustrates the fact that it is not always the error mean square which is used in testing experimental effects in factorial or other designs. Turning to the mixed model, if A is the random dimension, it is interesting to note that it is the mean square for the B dimension which contains the extra interaction term. Thus, the effect for A, the random dimension, is tested against the within-cells-error mean square. However, the B dimension, which is fixed, is tested against the interaction (i.e., the F ratio for the B effect is MS_B/MS_{AB}). The interaction mean square itself is still tested against within-cells error.

If, after choice of the appropriate F ratio, one or both of the hypotheses based on main effects are rejected, the researcher has no problem in deciding which level of the factor is producing the larger or smaller effect on response measures. The marginal means (that is, $\bar{Y}_{.1.}$ and $\bar{Y}_{.2.}$ for the A dimension, and $\bar{Y}_{1..}$ and $\bar{Y}_{2..}$ for the B dimension) will rank-order the treatment effects. If the interaction proves significant, interpretation is eased if a graph of the cell means is plotted after the fashion of Figure 3-1. If the levels of B are plotted against A, the absence of interaction will be reflected by the parallelism of the line segments. Deviations from parallelism indicate interaction, and the figure makes specific interpretation of the interaction relatively straightforward.

In order to illustrate the complete computations and the interpretation of a 2×2 factorial design, consider the data in Table 3-5. The data, which are fictitious, represent criterion scores on a 50-item multiple-choice examination. The hypothetical research situation was:

> It is traditional practice in education to review instructional units prior to administering examinations on these units. Also, some instructors prefer open-book examinations, while others use only closed-book examinations. This experiment contrasted groups which did with those which did not have a review session prior to taking a unit examination and groups which were allowed use of notes, textbooks, etc., with those taking the examination under closed-book conditions. The design was a 2×2 factorial as below:

		TREATMENT A	
		open-book	closed-book
TREATMENT B	Review	(11)	(12)
	No review	(21)	(22)

> The examination was a 50-item multiple-choice test which was fairly highly speeded (40-minute time limit). The instructional unit was comprised of material on inequalities presented to intact classroom groups of ninth-grade algebra students. The unit of analysis was classroom means.

From the description, the procedure can be inferred to involve random assignment of classroom groups to the cells of the experimental design (in experiments with intact groups, the individual student cannot be the unit of analysis since students are not randomly assigned to cells; however, classroom means can be used in the analysis if the unit of randomization is the classroom). Each classroom was subjected to one of the four combinations of treatments (i.e., five classrooms received a review and took the examination under open-book conditions, five classrooms received a review and took the examination under closed-book conditions, and so forth).

The relevant null hypotheses for the experiment are

$$H_{01}: \alpha_{\text{open}} = \alpha_{\text{closed}}$$

$$H_{02}: \beta_{\text{rev}} = \beta_{\text{no rev}}$$

$$H_{03}: \alpha\beta_{\text{open,rev}} = \alpha\beta_{\text{open,no rev}} = \alpha\beta_{\text{closed,rev}} = \alpha\beta_{\text{closed,no rev}}$$

However, prior to testing these hypotheses, the homogeneity-of-variance assumption should be checked. Since the cell sizes are constant, there is no real danger from lack of homogeneity; in the present case, however, the equality of variances would be of interest in its own right since it is possible that the treatments have an effect on variability as well as on means of the response measure. The procedures of Chapter 2 for testing the equi-variance assumption can be applied to any analysis-of-variance design if the tests are applied to *cell* variances. Table 3-6 presents the necessary information from the cells (i.e., the cell sums, sums of squared scores, etc.). Selecting the largest and smallest cell variances, we have $S^2_{\text{max}} = 12.4370$ and $S^2_{\text{min}} = 3.7920$. Thus, $F_{\text{max}} = 12.4370/3.7920 = 3.28$. For four groups and five observations per group, the computed F_{max} does not exceed the critical value from Appendix Table B-6 (note that 20.6 is needed for rejection at the .05 level of significance). Thus, there is no reason to suspect that the homogeneity-of-variance condition does not hold for these data.

table 3-6 *Cell summary for a fictitious* 2 × 2 *factorial design*

	open-book	closed-book
Review	$\sum_{k=1}^{5} Y_{11k} = 185.3$ $\sum_{k=1}^{5} Y_{11k}^2 = 6{,}903.07$ $S_{11}^2 = 8.9630$ $\bar{Y}_{11.} = 37.06$	$\sum_{k=1}^{5} Y_{12k} = 178.4$ $\sum_{k=1}^{5} Y_{12k}^2 = 6{,}415.06$ $S_{12}^2 = 12.4370$ $\bar{Y}_{12.} = 35.68$
No review	$\sum_{k=1}^{5} Y_{21k} = 160.8$ $\sum_{k=1}^{5} Y_{21k}^2 = 5{,}197.14$ $S_{21}^2 = 6.4530$ $\bar{Y}_{21.} = 32.16$	$\sum_{k=1}^{5} Y_{22k} = 140.6$ $\sum_{k=1}^{5} Y_{22k}^2 = 3{,}968.84$ $S_{22}^2 = 3.7920$ $\bar{Y}_{22.} = 28.12$

Turning to the main hypotheses of interest, the necessary sums of squares can be found by using the computational formulas and the figures from Tables 3-5 and 3-6.

$$SS_A = \frac{346.1^2}{10} + \frac{319.0^2}{10} - \frac{665.1^2}{20} = 22{,}154.621 - 22{,}117.900 = 36.721$$

$$SS_B = \frac{363.7^2}{10} + \frac{301.4^2}{10} - \frac{665.1^2}{20} = 22{,}311.965 - 22{,}117.900 = 194.065$$

$$SS_{AB} = \frac{185.3^2}{5} + \frac{160.8^2}{5} + \frac{178.4^2}{5} + \frac{140.6^2}{5} - \frac{665.1^2}{20} - SS_A - SS_B$$

$$= 22{,}357.530 - 22{,}117.900 - 36.721 - 194.065 = 8.844$$

$$SS_{\text{total}} = 22{,}484.11 - 22{,}117.900 = 366.210$$

By subtraction,

$$SS_{\text{error}} = 366.210 - 36.721 - 194.065 - 8.844 = 126.580$$

As a check on the accuracy of the computations, SS_{error} can also be computed directly from within-cells information. Thus,

$$SS_{\text{error}} = \sum_{i=1}^{2}\sum_{j=1}^{2}\left[\sum_{k=1}^{n}(Y_{ijk} - \bar{Y}_{ij.})^2\right] = \sum_{i=1}^{2}\sum_{j=1}^{2}(n-1)S_{ij}^2$$

$$= 4(8.9630 + 12.4370 + 6.4530 + 3.7920) = 126.5800$$

which checks with the figure computed by the subtractive process. These results, along with degrees of freedom and mean squares, are displayed in Table 3-7.

From the description of the experiment, it is apparent that both the A and the B dimension represent fixed effects; thus, the appropriate term against which to test all effects is MS_{error}. On this basis, the appropriate F ratios for each of the null hypotheses are $F = 36.72/7.91 =$

table 3-7 *Analysis-of-variance summary table for fictitious data*

source	*df*	sum of squares	mean square
A	1	36.72	36.72
B	1	194.06	194.06
AB	1	8.84	8.84
Error	16	126.58	7.91
Total	19	366.21	

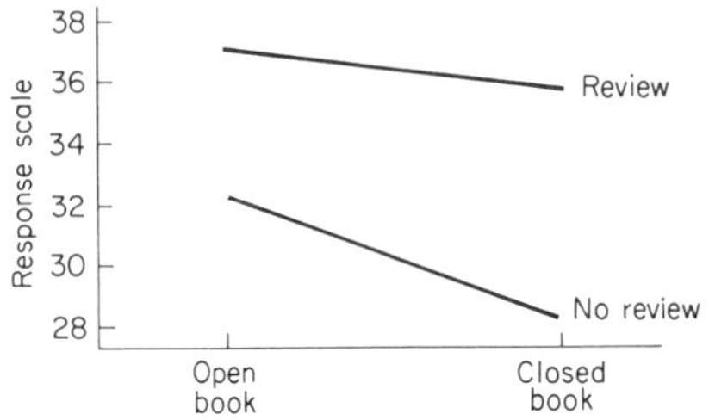

figure 3-2 *Interaction graph for fictitious experiment.*

4.64 (significant at the .05 level); $F = 194.06/7.91 = 24.53$ (significant at the .01 level); $F = 8.84/7.91 = 1.12$ (nonsignificant). Thus, both main effects, open-book versus closed-book and review versus no review, are significant, while the interaction is negligible. Inspection of the marginal means (Table 3-5) reveals that open-book is superior to closed-book as a mode of taking examinations and that review produces more achievement than no review. However, these two dimensions are additive and do not result in a significant interaction. The graph of the levels of B against the levels of A is shown in Figure 3-2. *In the sample,* the no-review group does show a larger differential between the open-book condition and the closed-book condition; this difference is, however, unreliable and cannot be said to characterize the procedures for the general population.

the 2 × 2 factorial and orthogonal contrasts

In the previous section, the analysis of the 2×2 factorial design was predicated on a traditional partitioning of the total sum of squares into components attributable to main effects, interaction, and error. In this section, we take a different tack to arrive at the same result. This approach, which is in terms of defining appropriate orthogonal contrasts on the cells of the 2×2 factorial design, represents an especially general and powerful method for analyzing experimental designs.

We begin by rearranging the data in the form of a one-dimensional design; for reference, however, the 2×2 design cell labels are also retained (Table 3-8). Since the CRD has four levels, we can form sets of three orthogonal contrasts to absorb the degrees of freedom in the design. We must, however, choose these contrasts to conform to the contrasts of interest from the point of view of the 2×2 factorial design. A contrast which compares the two levels of the A treatment dimension is (in terms of CRD level codes) $1 + 3 - 2 - 4$. That is, we sum scores in cells (11) and (21) and compare them with the sum from cells (12) and (22). In terms of cell means, this contrast becomes $\overline{Y}_1 - \overline{Y}_2 + \overline{Y}_3 - \overline{Y}_4$.

table 3-8 *One-dimensional representation of a 2 × 2 factorial design*

	TREATMENT LEVELS			
	1	2	3	4
Cell	(11)	(12)	(21)	(22)
Means	$\bar{Y}_1$	$\bar{Y}_2$	$\bar{Y}_3$	$\bar{Y}_4$

Note that the implied coefficients for the contrast are either +1 or −1 and that the sum of these coefficients is 0. For the B effect, the relevant contrast is $\bar{Y}_1 + \bar{Y}_2 - \bar{Y}_3 - \bar{Y}_4$ [that is, cells (11) and (12) compared with cells (21) and (22)]. Again, all coefficients are either +1 or −1 and their sum is 0. Also, the contrast for the B effect is orthogonal to that for the A effect since $(1)(1) + (-1)(1) + (1)(-1) + (-1)(-1) = 1 - 1 - 1 + 1 = 0$. The only remaining effect for a contrast that is needed is the interaction of A and B. If the B effect is strictly additive to the A effect, then we expect the difference $\bar{Y}_1 - \bar{Y}_3$ to be the same as the difference $\bar{Y}_2 - \bar{Y}_4$. For example, in Figure 3-1*a*, this is equivalent to stating that the distances between the two lines, at their end points, are equal. Clearly, this corresponds to our notion of a lack of interaction. Thus, a relevant contrast can be formed by comparing these two differences:

$$\bar{Y}_1 - \bar{Y}_3 - (\bar{Y}_2 - \bar{Y}_4) = \bar{Y}_1 - \bar{Y}_2 - \bar{Y}_3 + \bar{Y}_4$$

All coefficients are either +1 or −1 and their sum is 0; also, the student should check that this contrast is orthogonal to those for the A and B effects by forming the sums of cross products of the coefficients.

The contrast for the interaction of the A and B treatment dimensions could have been found by a more mechanical procedure. For any two main effects, the contrast representing their interaction can be found by multiplying together, coefficient by coefficient, the contrasts for the individual effects. To show this, we rewrite the original contrasts in column form and explicitly show the coefficients:

A effect: $(+1)\bar{Y}_1 + (-1)\bar{Y}_2 + (+1)\bar{Y}_3 + (-1)\bar{Y}_4$

B effect: $(+1)\bar{Y}_1 + (+1)\bar{Y}_2 + (-1)\bar{Y}_3 + (-1)\bar{Y}_4$

AB effect: $(+1)\bar{Y}_1 + (-1)\bar{Y}_2 + (-1)\bar{Y}_3 + (+1)\bar{Y}_4$

The coefficient for $\bar{Y}_1$ is $(+1)(+1) = +1$, for $\bar{Y}_2$ is $(-1)(+1) = -1$, and so forth.

By utilizing the three orthogonal contrasts, sums of squares for the main and interaction effects can be computed by the methods of Chapter 2. The sum of squares for the A effect is

$$SS_A = \frac{n\left(\sum_{j=1}^{4} c_{1j}\overline{Y}_j\right)^2}{\sum_{j=1}^{4} c_{1j}^2}$$

where c_{1j} is the jth coefficient in the contrast for the A treatment effect. Similarly, letting c_{2j} be a coefficient for the B contrast,

$$SS_B = \frac{n\left(\sum_{j=1}^{4} c_{2j}\overline{Y}_j\right)^2}{\sum_{j=1}^{4} c_{2j}^2}$$

and letting c_{3j} be a coefficient for the AB interaction,

$$SS_{AB} = \frac{n\left(\sum_{j=1}^{4} c_{3j}\overline{Y}_j\right)^2}{\sum_{j=1}^{4} c_{3j}^2}$$

The SS_{error} for the one-dimensional layout corresponding to a 2×2 factorial design is the same as that from the 2×2 factorial itself since, in both cases, only within-cell information enters into the computation (also, the SS_{total} will be equivalent since the scores themselves are identical in the two arrangements).

To demonstrate the equivalence of the analysis based on orthogonal contrasts with the more usual 2×2 factorial analysis based on marginal sums, we utilize the fictitious data from Tables 3-5 and 3-6. Substituting, we obtain

$$SS_A = \frac{5(37.06 - 35.68 + 32.16 - 28.12)^2}{1^2 + 1^2 + 1^2 + 1^2}$$

$$= \frac{5(5.42)^2}{4} = \frac{5(29.3764)}{4} = 36.72$$

$$SS_B = \frac{5(37.06 + 35.68 - 32.16 - 28.12)^2}{1^2 + 1^2 + 1^2 + 1^2}$$

$$= \frac{5(12.46)^2}{4} = \frac{5(155.2516)}{4} = 194.06$$

$$SS_{AB} = \frac{5(37.06 - 35.68 - 32.16 + 28.12)^2}{1^2 + 1^2 + 1^2 + 1^2}$$

$$= \frac{5(-2.66)^2}{4} = \frac{5(7.0756)}{4} = 8.84$$

Each of these values is equivalent to that obtained by conventional methods and presented in Table 3-7. The within-cells-error term and the total sum of squares would be computed as shown earlier; F ratios are formed in the same manner also.

In this section we have demonstrated that the sums of squares for a 2×2 factorial design may be obtained by use of sets of orthogonal contrasts. These concepts will prove useful in analysis of more complex designs.

the 2^k series factorial designs

The 2×2 factorial design may be generalized to cases in which there are more than two treatment dimensions, but each dimension occurs at only two levels. Such designs are referred to as the 2^k *series;* the 2×2 factorial is, in this notation, a 2^2 factorial design. The 2^k series of factorial designs is relatively popular in research applications since the number of independent groups of subjects required in the experiment remains within reasonable limits even for fairly large numbers of experimental factors (e.g., for three dimensions, eight groups are required; for four factors, 16 groups are required; however, for five dimensions, a total of 32 independent groups is required, and this begins to constitute a rather large undertaking).

A 2^3 factorial design has three treatment dimensions, each applied at two levels. For representational purposes, such a design must be "collapsed" into two dimensions as in Table 3-9. Note that the first subscript on a score stands for the level of the C dimension, the second for the level of the B dimension, the third for the level of the A dimension, and the last for the position within the cell. The data representation assumes an equal number n of observations per cell. A later section of this chapter considers the problem of analysis when unequal numbers of cases are involved in the cells of a design.

In order to write a structural model for a score in a 2^3 factorial design, we must first identify the sources of variation which may affect an individual score. There are, of course, main effects corresponding to each of the three treatment dimensions. In addition, however, there are three "first-order" interactions obtained by taking the combinations of the three treatments, two at a time. Thus, there is an interaction of the A and B dimensions, of the A and C dimensions, and of the B and C dimensions. Since these are first-order, or two-factor, interactions, their

table 3-9 *Data representation for a 2^3 factorial design*

		DIMENSION A			
		1		2	
DIMENSION B:		1	2	1	2
DIMENSION C	1	Y_{1111}	Y_{1211}	Y_{1121}	Y_{1221}
		Y_{1112}	Y_{1212}	Y_{1122}	Y_{1222}
		....	....	....	....
		Y_{111n}	Y_{121n}	Y_{112n}	Y_{122n}
	2	Y_{2111}	Y_{2211}	Y_{2121}	Y_{2221}
		Y_{2112}	Y_{2212}	Y_{2122}	Y_{2222}
		....	....	....	....
		Y_{211n}	Y_{221n}	Y_{212n}	Y_{222n}

interpretations are the same as given for the single interaction from a 2×2 factorial design (however, in interpreting these interactions, the third factor, not entering into the interaction, is completely ignored; that is, cell means are computed for combinations of two factors by summing over the levels of the third factor). There is, also a "second-order," or three-factor, interaction which arises in a 2^3 factorial design. This interaction reflects the constancy of first-order interactions over the levels of the third factor; for example, the ABC interaction may be interpreted as the degree to which the AB interaction is constant for the two levels of the C dimension. This interpretation will be elaborated further when an illustrative analysis is carried out. Keeping in mind the types of interactions which may occur, we see that the complete model for a score in a 2^3 factorial design is

$$Y_{ijkm} = \mu + \alpha_k + \beta_j + \gamma_i + \alpha\beta_{kj} + \alpha\gamma_{ki} + \beta\gamma_{ji} + \alpha\beta\gamma_{kji} + \epsilon_{ijkm}$$

where μ = usual grand-mean effect
α_k = effect due to A dimension
β_j = effect due to B dimension
γ_i = effect due to C dimension
$\alpha\beta_{kj}$ = interaction of A and B
$\alpha\gamma_{ki}$ = interaction of A and C
$\beta\gamma_{ji}$ = interaction of B and C
$\alpha\beta\gamma_{kji}$ = second-order interaction of A, B, and C
ϵ_{ijkm} = random-error component, assumed to be normally and homogeneously distributed with expected value of 0

For each term in the model which represents an experimental effect, an appropriate null hypothesis can be formed. These are

$$H_{0_1}: \alpha_1 = \alpha_2 = 0 \qquad H_{0_2}: \beta_1 = \beta_2 = 0$$

$$H_{0_3}: \gamma_1 = \gamma_2 = 0 \qquad H_{0_4}: \alpha\beta_{11} = \alpha\beta_{12} = \alpha\beta_{21} = \alpha\beta_{22} = 0$$

$$H_{0_5}: \alpha\gamma_{11} = \alpha\gamma_{12} = \alpha\gamma_{21} = \alpha\gamma_{22} = 0 \qquad H_{0_6}: \beta\gamma_{11} = \beta\gamma_{12} = \beta\gamma_{21} = \beta\gamma_{22} = 0$$

$$H_{0_7}: \alpha\beta\gamma_{111} = \alpha\beta\gamma_{112} = \alpha\beta\gamma_{121} = \alpha\beta\gamma_{122} = \alpha\beta\gamma_{211} = \alpha\beta\gamma_{212} = \alpha\beta\gamma_{221} = \alpha\beta\gamma_{222} = 0$$

As the dimensionality of designs increases, it is apparent that the complexity of the hypothesis-testing routine increases exponentially. For complete factorial designs, there will be, in general, $2^m - 1$ main and interaction effects to test (e.g., for a 2×2 factorial there were $2^2 - 1 = 3$ effects; for a 2^3 factorial there are $2^3 - 1 = 7$ effects; for a 2^5 factorial design there would be a total of $2^5 - 1$, or 31 separate effects to test). By using combinatorial theory, the number of different types of effects can also be computed. For a design occurring at m levels, there will be m main effects, $m(m-1)/2$ first-order interactions, $m(m-1)(m-2)/6$ second-order interactions, and so forth.

As with the CRD and the 2×2 factorial design, the approach to analysis of a 2^3 design can take the line of partitioning the total sum of squares into seven components, one of which corresponds to each of the effect terms in the model for a score. However, in the case of all 2^k series designs, the simpler approach is to utilize orthogonal contrasts. With this procedure, the total sum of squares is computed directly from all the scores; the error sum of squares can be computed directly from the within-cells information or can be found by subtraction after computation of all other terms. In order to compute the sums of squares for the main and interaction effects, a tabular arrangement showing the contrast coefficients and cell means is highly convenient (Table 3-10). Columns are set up corresponding to each cell of the design (the method is general, but is illustrated for a 2^3 factorial design); for convenience, the cell means are also shown in the column headings. Rows of the table are set up in terms of the main and interaction effects which it is desired to estimate. If we always contrast level 1 of a factor with level 2 of a factor, in that order, the contrast coefficients can be entered in the table quite mechanically for the main effects. For each row effect, if the level code in a column is 1, enter $+1$ as the contrast coefficient; if the level code is 2, enter -1 as the contrast coefficient. First-order interaction terms are found by taking the product of the coefficients for the corresponding main effects (as illustrated previously for the 2×2 factorial design). The second-order (and higher-order interactions for higher-dimensional designs) interaction coefficients can be found from the product of coefficients for *all* main effects entering into the interaction (i.e., the ABC interaction coefficients are found by multiplying together the coefficients from the A, B, and C dimensions).

In addition to columns containing contrast coefficients, two columns are appended to the right of the table for computational purposes. The first of these computational columns (which is blank in Table 3-10) contains the linear combination of means which constitutes the contrast. Since these means have been written just under the cell codes in the heading of the table, the indicated combination of means is quite simple to compute. For simplicity, the single subscript h denotes the cell, and the summation is taken from the first cell (111) through the final cell (222). The number of cells for a 2^k design is also 2^k; thus, the h subscript runs from 1 through 2^k. In this notation, the operation $\Sigma_{h=1}^{2^3}$ replaces the more cumbersome triple summation $\Sigma_{i=1}^{2} \Sigma_{j=1}^{2} \Sigma_{k=1}^{2}$ for the 2^3 factorial design. Also, the subscript g refers to the specific contrast; C_1 is the contrast among means for the A effect, C_2 is the contrast for the B effect, and so forth through C_7, which is the contrast for the ABC effect. The second computational column completes the arithmetic for finding the sum of squares for the contrast. The steps embodied in these two computational columns are merely another format for carrying out the computation of the sum of squares for a contrast. The student should refer to the discussion of orthogonal contrasts in Chapter 2 and the example of the 2×2 factorial design earlier in this chapter to assure himself that these procedures are equivalent.

The computational format of Table 3-10 can easily be generalized to any 2^k series factorial design. It is necessary only to form a column corresponding to each cell of the design and a row corresponding to each main and interaction effect. The resulting matrix will have $2^k - 1$ rows

table 3-10 *Contrast-coefficient matrix for a 2^3 factorial design*

CELL:	(111)	(112)	(121)	(122)	(211)	(212)	(221)	(222)	$\sum_{h=1}^{2^3} c_{gh}\overline{Y}_h$	
MEAN:	$\overline{Y}_{111.}$	$\overline{Y}_{112.}$	$\overline{Y}_{121.}$	$\overline{Y}_{122.}$	$\overline{Y}_{211.}$	$\overline{Y}_{212.}$	$\overline{Y}_{221.}$	$\overline{Y}_{222.}$	$= C_g$	$nC_g^2/8$
effect										
A	+1	−1	+1	−1	+1	−1	+1	−1		SS_A
B	+1	+1	−1	−1	+1	+1	−1	−1		SS_B
C	+1	+1	+1	+1	−1	−1	−1	−1		SS_C
AB	+1	−1	−1	+1	+1	−1	−1	+1		SS_{AB}
AC	+1	−1	+1	−1	−1	+1	−1	+1		SS_{AC}
BC	+1	+1	−1	−1	−1	−1	+1	+1		SS_{BC}
ABC	+1	−1	−1	+1	−1	+1	+1	−1		SS_{ABC}

and 2^k columns. Coefficients for main effects can be entered by noting whether the cell subscript for that effect is 1 or 2; enter +1 as the coefficient whenever the subscript is 1, and enter −1 whenever the subscript is 2. Interaction-effect coefficients are most easily found by forming the product of the coefficients of the main effects entering into the interaction. Finally, the cell means are combined in accordance with the pattern of coefficients to form the contrast, and the corresponding sum of squares is easily computed. For any 2^k series factorial design, all main effects and interactions have only 1 degree of freedom. Thus, the sums of squares are equivalent to mean squares. The error mean square (which must be computed from within-cells information or by subtraction from the total sum of squares) will always have $2^k(n-1)$ degrees of freedom, and the degrees of freedom for the total sum of squares are $2^kn - 1$.

In the next section, a computation approach for factorial designs other than the 2^k series is presented. This procedure, which is based on two-dimensional summary tables, may also be utilized for 2^k series factorial designs, but is less efficient than the orthogonal-contrast technique.

In order to form appropriate F ratios to test the null hypotheses which are set up for a 2^k series factorial design, expectations for the mean squares must be known. As mentioned previously, random effects rarely enter into factorial designs since this would mean that the levels of one or more treatment dimensions were randomly selected from a population of available levels; such a procedure is highly unlikely with manipulated treatment variables. Table 3-11 presents expectations for the fixed model and random model of a 2^3 factorial design. For higher-order 2^k series designs, the pattern of expected values is similar. That is, for the fixed model, all mean squares estimate a term com-

table 3-11 *Expectations for mean squares in the 2^3 factorial design: fixed-effects and random-effects models*

mean square	FIXED-EFFECTS MODEL (all three dimensions fixed)	RANDOM-EFFECTS MODEL (all three dimensions random)
MS_A	$\sigma^2 + 4n\sigma_\alpha^2$	$\sigma^2 + 2n\sigma_{\alpha\beta}^2 + 2n\sigma_{\alpha\gamma}^2 + n\sigma_{\alpha\beta\gamma}^2 + 4n\sigma_\alpha^2$
MS_B	$\sigma^2 + 4n\sigma_\beta^2$	$\sigma^2 + 2n\sigma_{\alpha\beta}^2 + 2n\sigma_{\beta\gamma}^2 + n\sigma_{\alpha\beta\gamma}^2 + 4n\sigma_\beta^2$
MS_C	$\sigma^2 + 4n\sigma_\gamma^2$	$\sigma^2 + 2n\sigma_{\alpha\gamma}^2 + 2n\sigma_{\beta\gamma}^2 + n\sigma_{\alpha\beta\gamma}^2 + 4n\sigma_\gamma^2$
MS_{AB}	$\sigma^2 + 2n\sigma_{\alpha\beta}^2$	$\sigma^2 + n\sigma_{\alpha\beta\gamma}^2 + 2n\sigma_{\alpha\beta}^2$
MS_{AC}	$\sigma^2 + 2n\sigma_{\alpha\gamma}$	$\sigma^2 + n\sigma_{\alpha\beta\gamma}^2 + 2n\sigma_{\alpha\gamma}^2$
MS_{BC}	$\sigma^2 + 2n\sigma_{\beta\gamma}^2$	$\sigma^2 + n\sigma_{\alpha\beta\gamma}^2 + 2n\sigma_{\beta\gamma}^2$
MS_{ABC}	$\sigma^2 + n\sigma_{\alpha\beta\gamma}^2$	$\sigma^2 + n\sigma_{\alpha\beta\gamma}^2$
MS_{error}	σ^2	σ^2

posed of error variance and treatment effect (the limits of summation and the constant multiplier must, of course, be adjusted in terms of the specific design); for the random model, each main effect contains error variance, a specific treatment effect, and, also, interaction effects corresponding to each interaction involving the particular main effect (e.g., in a 2^5 factorial design, if the factors are labeled A, B, C, D, and E, the expectation for the mean square for C contains error variance, an effect due to the levels of C, and effects due to the $AC, BC, CD, CE, ABC, ACD, ACE, BCD, CDE, ABCD, ABCE, BCDE$, and $ABCDE$ interactions). Also, each interaction (first-order or higher-order) contains error variance, an effect specific to the interaction, and interaction effects corresponding to each higher-order interaction containing within it the same terms as the specific interaction (e.g., for the 2^5 factorial design, the expectation for the ABC interaction contains error variance, a specific ABC term, and interaction effects due to $ABCD$, $ABCE$, and $ABCDE$).

The mixed model is more difficult to detail since for the 2^3 factorial there are six different ways in which a mixture of fixed and random treatment dimensions can occur (that is, A fixed, B and C random; B fixed, A and C random; C fixed, A and B random; A and B fixed, C random; A and C fixed, B random; and B and C fixed, A random). Table 3-12 presents expectations for the cases with A and B fixed and C random, and A fixed and B and C random. Although a general procedure for finding expected values for mean squares in factorial designs will be presented in the next section of this chapter, the pattern of terms entering into the expectations in the mixed models is not difficult to discern. Any mean square based on an effect which contains a random element (e.g., in the first example of Table 3-12, any mean square containing C, such as MS_C and MS_{AC}) has as its expectation only error variance, a term specific to the effect, and interactions with

table 3-12 *Expectations for mean squares in the 2^3 factorial design: examples of two mixed models*

mean square	MIXED MODEL (A and B fixed, C random)	MIXED MODEL (A fixed, B and C random)
MS_A	$\sigma^2 + 2n\sigma_{\alpha\gamma}^2 + 4n\sigma_\alpha^2$	$\sigma^2 + 2n\sigma_{\alpha\beta}^2 + 2n\sigma_{\alpha\gamma}^2 + n\sigma_{\alpha\beta\gamma}^2 + 4n\sigma_\alpha^2$
MS_B	$\sigma^2 + 2n\sigma_{\beta\gamma}^2 + 4n\sigma_\beta^2$	$\sigma^2 + 2n\sigma_{\beta\gamma}^2 + 4n\sigma_\beta^2$
MS_C	$\sigma^2 + 4n\sigma_\gamma^2$	$\sigma^2 + 2n\sigma_{\beta\gamma}^2 + 4n\sigma_\gamma^2$
MS_{AB}	$\sigma^2 + n\sigma_{\alpha\beta\gamma}^2 + 2n\sigma_{\alpha\beta}^2$	$\sigma^2 + n\sigma_{\alpha\beta\gamma}^2 + 2n\sigma_{\alpha\beta}^2$
MS_{AC}	$\sigma^2 + 2n\sigma_{\alpha\gamma}^2$	$\sigma^2 + n\sigma_{\alpha\beta\gamma}^2 + 2n\sigma_{\alpha\gamma}^2$
MS_{BC}	$\sigma^2 + 2n\sigma_{\beta\gamma}^2$	$\sigma^2 + 2n\sigma_{\beta\gamma}^2$
MS_{ABC}	$\sigma^2 + n\sigma_{\alpha\beta\gamma}^2$	$\sigma^2 + n\sigma_{\alpha\beta\gamma}^2$
MS_{error}	σ^2	σ^2

other random effects. A mean square which is based on an effect containing fixed elements has as its expectation error variance, a term specific to the effect, and terms corresponding to each interaction involving the particular combination of fixed effects and a random effect. Thus, in the first example in Table 3-12, the mean square for A, in addition to error variance and a term based on A, contains an AC interaction effect. Also, the AB interaction mean square has the second-order interaction ABC in its expectation.

In order to set up F ratios to test the various null hypotheses, it is apparent that, in the case of a fixed-effects model, the within-cells-error mean square is the appropriate denominator in all instances. The random-effects model, however, presents an anomalous situation; there is no appropriate denominator for any of the main effects! This startling outcome should warn the reader of the necessity to determine expectations for mean squares *prior* to carrying out an experiment. In the random-effects model, the error mean square can be used to test the second-order interaction, and the second-order interaction can be used to test each of the first-order interactions; however, each main effect contains five terms, and no other source of variation contains more than three terms. Thus, any ratio which is set will not necessarily have an expectation of 1 when the relevant null hypothesis is correct. For example, if MS_A were tested against MS_{AB}, a large value for the F ratio could be due to (1) an interaction of A and C, but a negligible effect due to A; (2) an effect due to A, but a negligible interaction of A and C; or (3) both an interaction of A and C, and an effect due to A. Thus, this approach to testing the main effects would lead to ambiguous results whenever the null hypothesis appeared to be rejected by the F ratio.

To test the main effects in the random-effects model, the experimenter must either set up *quasi* F ratios or use a sequential strategy, neither of which gives a mathematically correct test of these main effects. The use of quasi F ratios involves combining two or more sources of variation into one *quasi mean square* which contains the correct terms in its expectation. For example, an appropriate quasi mean square for testing the mean square for the A treatment can be generated from $\text{MS}_{AB} + \text{MS}_{AC} - \text{MS}_{ABC}$. Note that the expectations, when combined, yield

$$\begin{aligned}\sigma^2 + n\sigma^2_{\alpha\beta\gamma} + 2n\sigma_{\alpha\beta}{}^2 + \sigma^2 + n\sigma^2_{\alpha\beta\gamma} + 2n\sigma_{\alpha\gamma}{}^2 - \sigma^2 - n\sigma^2_{\alpha\beta\gamma}& \\ = \sigma^2 + 2n\sigma_{\alpha\beta}{}^2 + 2n\sigma_{\alpha\gamma}{}^2 + n\sigma^2_{\alpha\beta\gamma}&\end{aligned}$$

This quasi mean square differs in expectation from MS_A only in its lack of a term specific to the A treatment effect. Thus, when the null hypothesis $H_{0_1}: \alpha_1 = \alpha_2 = 0$ is correct, the ratio $\text{MS}_A/(\text{MS}_{AB} + \text{MS}_{AC} - \text{MS}_{ABC})$ has an expected value of 1. Unfortunately, the distribution of this ratio is not exactly the usual tabled F distribution. The tabled F distributions may, however, be used as approximations for testing the null hypothesis. A reasonably good approximation can be obtained if the degrees of freedom for the denominator are computed as

$$\frac{(\text{MS}_{AB} + \text{MS}_{AC} - \text{MS}_{ABC})^2}{\dfrac{\text{MS}_{AB}{}^2}{df_{ab}} + \dfrac{\text{MS}_{AC}{}^2}{df_{ac}} + \dfrac{\text{MS}^2_{ABC}}{df_{abc}}}$$

where df_{ab}, df_{ac}, and df_{abc} are the ordinary degrees of freedom associated with the AB, AC, and ABC interaction terms (for the 2^3 factorial design, each of these is 1). An inconvenient problem which may arise is that, if MS_{ABC} is large, the quasi mean square may be a negative quantity, and the resulting F is impossible to evaluate since a ratio of variances can only be a positive quantity. In this case, it is usually possible to combine additional terms in the numerator and denominator of the ratio to guarantee that the ratio is positive.

An alternative to the construction of quasi F ratios for the random-effects model is to establish that certain terms in the expectation of the mean square for a main effect are 0; if these terms are chosen judiciously, the remaining terms may be such that an appropriate denominator exists among the other mean squares. For example, consider the mean square for the C effect. If we could assume that $\sigma_{\alpha\gamma}{}^2$ is 0, the expectation for MS_C would reduce to $\sigma^2 + 2n\sigma_{\beta\gamma}{}^2 + n\sigma^2_{\alpha\beta\gamma} + 4n\sigma_\gamma{}^2$, and MS_{BC} would provide a mean square for testing MS_C. We can show that $\sigma_{\alpha\gamma}{}^2$ is or is not equal to 0 by testing MS_{AC} against MS_{ABC}. If the AC interaction term is nonsignificant, MS_{AC} is used to test MS_C. If, however, the AC interaction is significant, a similar strategy can still be tried with MS_{BC} since $\sigma_{\beta\gamma}{}^2$ is contained in the expectation of MS_C. Should both the AC and the BC interaction prove to be significant, there is little left to do except construct a quasi mean square. In the final analysis, random-effects models are best avoided in factorial designs.

Turning to the expectations for mixed models reported in Table 3-12, we see that there is no problem in establishing appropriate F ratios for the first case (that is, A and B fixed, C random). However, in the second case (that is, A fixed, B and C random), there is no appropriate denominator for an F ratio to test the effect due to A. Again, either we can attempt to establish that the AB (or AC) interaction is negligible, or we can turn to the use of quasi mean squares. Note, also, in this design that the B effect is tested against the BC interaction, the C effect is tested against the BC interaction, the AB and AC interactions are tested against the ABC interaction, the BC interaction is tested against within-cells error, and the ABC interaction is tested against within-cells error.

Since randomly selected levels are rarely utilized in factorial designs, the researcher will seldom be faced with the difficulties discussed above. However, the principles developed here have great generality and will be highly relevant to randomized block designs and nested designs.

As was the case with the 2×2 factorial design, a significant main effect in a 2^3 or higher-order design can be interpreted easily in terms of mean differences since each factor is applied at only two levels. Also, first-order interactions can be evaluated by means of graphs based on two-way summary tables. This involves computing means for the cells of two-way tables in which the scores have been summed over the levels of all other factors. For example, in a 2^3 design, a significant interaction for the AB effect can be graphed by forming a two-way summary table showing only levels of the A and B treatment dimensions. These means would be computed by summing over the levels of the C dimension. These procedures will be illustrated when a worked example of a 2^3 factorial design is presented. The interpretation of a second- or higher-order interaction presents some problem since a three-dimensional (or higher-dimensional) graph cannot be easily constructed. For these cases, the interaction can be plotted separately for each level (or combination of levels) of the remaining factors not involved in the significant interaction. For a significant ABC interaction in a 2^3 design, the combination of means for the A and B dimensions can be plotted separately for the two levels of the C dimension. The degree of interaction is reflected in the lack of similarity of the profiles for the means at the two levels of C (equally well, the A and C means could be plotted for the levels of B, or the B and C means for the levels of A). This approach to interpreting higher-order interactions will be demonstrated for the example which follows.

One issue remains to be discussed prior to turning to a practical example of the analysis and interpretation of a 2^k series factorial design. Because of the relatively large number of subjects required for a high-dimensionality design, experiments are sometimes conducted with only one experimental subject per cell. That is, $n = 1$ for all cells of the factorial design. Under these circumstances, all main effects and interactions can still be computed by the orthogonal-contrast procedure summarized in Table 3-10. Also, a total sum of squares can be computed from all the scores in the data table. There is, however, no within-cell variance since only one observation appears in each cell of the design. Thus, no error mean square can be computed, and there is no direct estimate of experimental error. If we refer to Table 3-11 for the expectations of mean squares in a 2^3 factorial design, it is apparent that the table of expectations for the $n = 1$ case is identical *except* that the final row of the table, MS_{error}, does not exist. There is, then, no appropriate term against which to test main and interaction effects in the fixed-effects model (although with the random-effects model, the only new difficulty added to an already difficult analysis is the lack of a term for testing MS_{ABC}). The ordinary resolution in the case of the fixed-effects model is to use either MS_{ABC} or one of the first-order interactions to test the main effects and to use MS_{ABC} to test the first-order interactions. Such an approach always results in a *conservative* test. That is, the expectation for the resulting F ratios will be *less than* 1, and fewer Type I errors will be committed than indicated by the nominal level of significance if the standard F distribution tables are used for checking significance. Of course, the resulting test has less power than if an appropriate error term were available; however, if a null hypothesis for a main or interaction effect is rejected, we can be sure that it would be rejected at, at least, as high a level of significance if the suitable error term were used. It is apparent (see Table 3-12) that a similar strategy would be required for certain effects in mixed models.

We turn now to an example of the analysis and interpretation of 2^3 factorial design. Once again, the experimental situation and the data reported are fictitious. This experiment is similar to the one described and analyzed in connection with the 2×2 factorial design. However, an additional treatment dimension has been added.

> This experiment, like the 2×2 factorial design described earlier, contrasted open-book and closed-book examination procedures; also, some subjects had review prior to the examination, and others had no review. In addition, two forms of the 50-item multiple-choice test were developed; one form had the items graded in difficulty from easy to difficult, and the second form contained items in random order of difficulty. Thus the design was a 2^3 factorial:

		TREATMENT A		
	open-book		closed-book	
TREATMENT B:	review	no review	review	no review
TREATMENT C Ordered form	(111)	(121)	(112)	(122)
TREATMENT C Random form	(211)	(221)	(212)	(222)

> The examination was fairly highly speeded (40-minute time limit). The instructional unit was comprised of material on inequalities presented to intact classroom groups of ninth-grade algebra students. The unit of analysis was classroom means.

table 3-13 *Data from a fictitious* 2^3 *factorial design*

	TREATMENT A			
	open-book		closed-book	
TREATMENT B:	review	no review	review	no review
	38.1	32.1	37.4	25.1
Ordered form	40.7	33.8	31.4	27.7
	36.2	29.9	33.5	30.3
TREATMENT C				
	36.2	30.2	34.7	26.2
Random form	35.7	31.6	36.1	24.7
	37.1	34.6	34.3	28.5

Note that the appropriate unit of analysis is the classroom mean since classrooms, but not individual students, were randomly assigned to the treatment combinations. Because of the requirement for a total of eight treatment groups, only three classrooms were assigned to each cell of the design. The data resulting from application of the fictitious experiment are presented in Table 3-13.

In order to provide summary data for the ensuing analysis, cell information is presented in Table 3-14. For each cell, we have summarized the sum of scores, sum of squared scores, mean score, and variance of the scores.

Although there is an equal number of observations per cell of the design, it is desirable to test the homogeneity-of-variance assumption. Utilizing the simple $F_{\max}$ test, we have $S^2_{\max} = 9.27$, $S^2_{\min} = .50$, and $F_{\max} = 9.27/.50 = 18.54$. For a total of eight variances and 2 degrees of freedom per variance, this value does not approach significance; thus, we may conclude that the treatments do not differentially affect the variability of the response measures.

table 3-14 *Cell summary for a fictitious* 2^3 *factorial design*

	CELL							
	(111)	(121)	(112)	(122)	(211)	(221)	(212)	(222)
Sum of scores	115.0	95.8	102.3	83.1	109.0	96.4	105.1	79.4
Sum of squared scores	4,418.54	3,066.86	3,506.97	2,315.39	3,961.34	3,107.76	3,683.79	2,108.78
Mean	38.33	31.93	34.10	27.70	36.33	32.13	35.03	26.47
Variance	5.10	3.82	9.27	6.76	.50	5.05	.89	3 66

Turning to the analysis of variance for the 2^3 design, we see that there is a total of seven hypotheses to test: three main effects, three first-order interactions, and one second-order interaction. The total sum of squares can be found from the total sum of all scores and the total sum of squared scores. These can be computed from the cell summary in Table 3-14:

$$\sum_{i=1}^{2}\sum_{j=1}^{2}\sum_{k=1}^{2}\sum_{m=1}^{n} Y_{ijkm} = 115.0 + 95.8 + \cdots + 79.4 = 786.1$$

$$\sum_{i=1}^{2}\sum_{j=1}^{2}\sum_{k=1}^{2}\sum_{m=1}^{n} Y^2_{ijkm} = 4{,}418.54 + \cdots + 2{,}108.78 = 26{,}169.43$$

Then,

$$\text{SS}_{\text{total}} = 26{,}169.43 - \frac{786.1^2}{24} = 26{,}169.43 - 25{,}748.05 = 421.38$$

The error sum of squares can also be found from the cell summary. The sum of squares for a given cell is

$$\sum_{m=1}^{n} Y^2_{ijkm} - \frac{\left(\sum_{m=1}^{n} Y_{ijkm}\right)^2}{n}$$

The sum of these quantities over the eight cells of the design comprises the sum of squares for error. This sum is

$$10.21 + 7.65 + 18.54 + 13.52 + 1.01 + 10.11 + 1.79 + 7.33 = 70.16 = \text{SS}_{\text{error}}$$

table 3-15 *Contrast matrix for a fictitious 2^3 factorial design*

CELL:	(111)	(112)	(121)	(122)	(211)	(212)	(221)	(222)	$\sum_{h=1}^{8} c_{gh}\overline{Y}_h = C_g$	$nC_g^2/8$
MEAN:	38.33	34.10	31.93	27.70	36.33	35.03	32.13	26.47		
effect										
A	+1	−1	+1	−1	+1	−1	+1	−1	15.43	89.32
B	+1	+1	−1	−1	+1	+1	−1	−1	25.57	245.13
C	+1	+1	+1	+1	−1	−1	−1	−1	2.10	1.65
AB	+1	−1	−1	+1	+1	−1	−1	+1	−4.37	7.15
AC	+1	−1	+1	−1	−1	+1	−1	+1	1.50	.84
BC	+1	+1	−1	−1	−1	−1	+1	+1	.03	.00
ABC	+1	−1	−1	+1	−1	+1	+1	−1	4.37	7.15

table 3-16 *Analysis-of-variance summary table for a fictitious 2^3 factorial design*

source	df	sum of squares	mean square	F†	p
A	1	89.32	89.32	20.35	<.005
B	1	245.13	245.13	55.84	<.005
C	1	1.65	1.65	.38	ns
AB	1	7.15	7.15	1.63	ns
AC	1	.84	.84	.19	ns
BC	1	.00	.00	.00	ns
ABC	1	7.15	7.15	1.63	ns
Error	16	70.16	4.39		
Total	23	421.38			

†F ratios computed assuming a fixed-effects model.

In order to find sums of squares based on the main and interaction effects, the tabular arrangement shown in Table 3-10 is utilized. For the specific design, this matrix of coefficients and the required computations are shown in Table 3-15.

The terms appearing in the final column of Table 3-15 are, in fact, the required sums of squares for the main and interaction effects. To check the accuracy of the work, the total of these sums of squares should be $SS_{total} - SS_{error}$, or $421.38 - 70.16 = 351.22$. The actual total is 351.24, which is within an accumulated rounding error of the result obtained by subtraction. The results are summarized in an analysis-of-variance table (Table 3-16).

The main effects for both open-book versus closed-book and review versus no review are significant at well beyond conventional levels. The main effect for the form of the examination and all interaction effects are nonsignificant. The lack of interaction implies that the significant effects for open-book versus closed-book and review versus no review

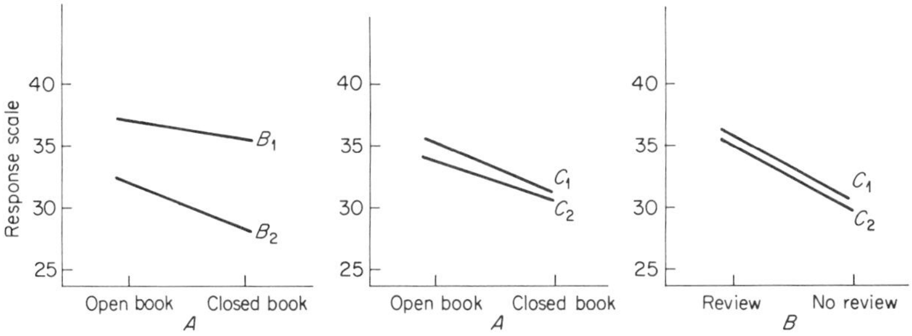

figure 3-3 *Graphs of first-order interactions for fictitious 2^3 factorial design.*

table 3-17 *Two-way summary tables for a fictitious 2^3 factorial design*

		TREATMENT A open-book	closed-book
TREATMENT B	Review	37.33	34.57
	No review	32.03	27.08
		TREATMENT A open-book	closed-book
TREATMENT C	Ordered form	35.13	30.90
	Random form	34.23	30.75
		TREATMENT B review	no review
TREATMENT C	Ordered form	36.22	29.82
	Random form	35.68	29.30

are additive. It is instructive to plot each of the first-order interactions even though additivity holds in this example; Figure 3-3 shows these first-order interactions. The interaction graphs are based on two-way summary tables of means. Each of these tables represents the data collapsed over one of the treatment dimensions (Table 3-17).

The lack of interaction between pairs of the main dimensions of the 2^3 factorial design is readily apparent from the graphs. Although the lines deviate slightly from parallelism, each mean is based on only six scores, and the observed result is well within the limits of sampling error from a population in which the graphs do, in fact, show parallelism. The problem of representing the second-order interaction (which is also nonsignificant) can be attacked by using the original table of cell means (Table 3-14). In Figure 3-4, the interaction of the A and B dimensions has been graphed separately for each level of the C dimension. The profiles of the means in the two graphs are quite similar and reflect the relatively small F value obtained for the ABC interaction. In interpreting Figure 3-4, it should be remembered that the left graph represents the interaction of the A and B dimensions only at the first level of the C dimension; the right graph represents this interaction only at the second level of the C dimension.

In summary, the analysis of the fictitious 2^3 factorial design revealed highly significant main effects for the A and B dimensions. For the A dimension it is apparent, from inspection of the means, that the open-

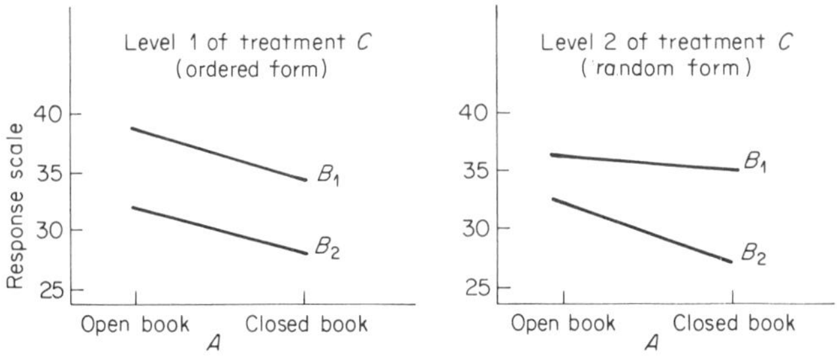

figure 3-4 *Second-order interaction graph for fictitious 2^3 factorial design.*

book test-taking condition yielded the higher average performance. For the B dimension, the use of a review session prior to the examination was superior to the lack of such a review session. No effect could be attributed to the use of an ordered form rather than a random form with respect to the organization of test items. Also, the lack of significant interactions leads to the conclusion that all effects are additive; that is, there are no effects uniquely attributable to the cells of the experiment.

designs at more than two levels

Principles underlying the construction and analysis of designs in which each dimension exists at only two levels (that is, 2^k series factorial designs) can be generalized to the entire class of factorial designs with no restrictions on the number of levels per experimental factor. As noted earlier, such designs are described in terms of the numbers of levels of the dimensions entering into the design; e.g., a $3 \times 4 \times 2$ design has one dimension at three levels, one at four levels, and one at two levels, resulting in a total of 24 cells or combinations of treatments.

The number of different main and interaction effects of various orders can be found by use of combinatorial formulas and does not depend upon the numbers of levels for the dimensions of the design. Thus, any m-dimensional design has m main effects, $m(m-1)/2$ first-order interactions, $m(m-1)(m-2)/6$ second-order interactions, $m(m-1)(m-2)(m-3)/24$ third-order interactions, and so forth. For each main effect and interaction an appropriate null hypothesis must be set up and an appropriate term placed in the structural model for a score in the design. These hypotheses differ from those appropriate to 2^k series factorial designs only in the number of levels which must be shown in the subscripts of the parameters, and the structural model for a score needs no modification at all. When large numbers of levels are involved, it is easier to utilize some form of shortened notation. For a $3 \times 4 \times 2$ design, for example, the hypotheses might be written

$$H_{0_1}: \alpha_k = 0 \quad k = 1, 2 \qquad H_{0_2}: \beta_j = 0 \quad j = 1, 4$$

$$H_{0_3}: \gamma_i = 0 \quad i = 1, 3 \qquad H_{0_4}: \alpha\beta_{kj} = 0 \quad \begin{matrix} k = 1, 2 \\ j = 1, 4 \end{matrix}$$

$$H_{0_5}: \alpha\gamma_{ki} = 0 \quad \begin{matrix} k = 1, 2 \\ i = 1, 3 \end{matrix} \qquad H_{0_6}: \beta\gamma_{ji} = 0 \quad \begin{matrix} j = 1, 4 \\ i = 1, 3 \end{matrix}$$

$$H_{0_7}: \alpha\beta\gamma_{kji} = 0 \quad \begin{matrix} k = 1, 2 \\ j = 1, 4 \\ i = 1, 3 \end{matrix}$$

The use of the subscripts implies that all combinations of subscripts, within the indicated ranges, are included in the hypothesis.

The major practical problem in the use of higher-dimensional factorial designs is the rapid increase in the number of independent treatment groups required. It is often difficult, in large experiments, to maintain close control over environmental conditions which may influence the outcome of the research. A $3 \times 4 \times 2$ factorial design requires 24 independent groups of experimental subjects. It may be impossible to apply all the treatment combinations to all groups at the same time or during the same time period. Thus, possible differential effects may be introduced by time of day, day of the week, etc. Also, a variety of research assistants may be required to conduct the experiment, and minor variations in their handling of experimental subjects, application of treatments, etc., may also introduce unique effects. Furthermore, if an experiment covers a protracted period of time, there is always the possibility of communication among experimental subjects, which may bias the responses of groups tested later than others. In some circumstances, factors related to conducting the experiment can themselves be made part of the experimental design. Thus, if an experiment must be run on two different days of the week, half the subjects may be used to run the entire design on the first day and the other half on the second day. There is a complete crossing of day of the week with the other treatment dimensions, and day of the week becomes an additional dimension of the design (it would not be proper to call day of the week a manipulated treatment variable; thus, the implied design is more properly a randomized blocks design, which will be described in Chapter 5; the analysis, however, is identical with that for factorial designs).

Two methods for computing sums of squares corresponding to the main and interaction effects of the general class of factorial designs are considered in this section. The first method is a summary table approach; the second method utilizes sets of patterned orthogonal contrasts and is highly adaptable to computerization.

SUMMARY TABLE ANALYSIS OF FACTORIAL DESIGNS

The summary table approach depends upon constructing two-, three-, and higher-dimensional summary tables as required by the dimensionality of the design. From each table, an among-cells sum of squares can be computed (i.e., a sum of squares where the cells are considered to be equivalent to a one-way design, or CRD). This sum of squares can then be partitioned to find some main effects and interactions. By judicious choices of summary tables, all effects can be computed in an efficient manner. The procedure is most readily understood in terms of an example. The experimental situation, which is fictitious, is in the area of counseling research.

The school dropout is considered a major problem in public education today since the future opportunities for these individuals are severely limited by lack of a high school diploma. Furthermore, there are positive indicators of the tendency to drop out of school (e.g., parental background, achievement below indicated ability level, number of absences from school, and lack of participation in extracurricular activities at school). The purpose of this study was to investigate the effects upon standardized achievement of programs designed to reinforce the potential dropout's desire to remain in school. Potential dropouts, all of whom were male ninth-graders, were identified in terms of the indicators listed above. Three types of programs were concurrently operated in all junior high schools in one large county:

A. Special vocational counseling (SVC). One group received no special counseling, one group received 1 hour of counseling per week, and one group received 2 hours of counseling per week.

B. Parental involvement (PI). One group had no parental involvement; one group had monthly meetings of student, parent(s), and school counselor; and one group had monthly meetings of student, parent(s), school counselor, and vocational counseling expert.

C. Outside employment program (OEP). One group had no outside employment program through the school, and one group participated in a special related-time work-study program under school supervision.

All combinations of treatments were applied to separate, randomly formed groups of students. The design is a $2 \times 3 \times 3$ factorial:

	TREATMENT A								
	no SVC			SVC—1 hour			SVC—2 hours		
TREATMENT B:	no PI	PI	PI+ expert	no PI	PI	PI+ expert	no PI	PI	PI+ expert
No OEP	(111)	(121)	(131)	(112)	(122)	(132)	(113)	(123)	(133)
TREATMENT C									
OEP	(211)	(221)	(231)	(212)	(222)	(232)	(213)	(223)	(233)

There are 18 treatment groups, and across the county a total of 100 students were placed in each treatment combination. At the end of the ninth grade, after a full academic year of program participation, a standardized achievement battery was administered to all students. The composite score was used in the analysis.

The data and analysis presented here will assume that the full sample of 100 students yielded usable data in each cell of the design. In reality, were such a program to be tested in schools, there would be attrition from the program for a variety of reasons (e.g., students dropping out of school, illness, transfers to other schools, or incomplete or unusable test results). In the final section of this chapter, we consider analysis of the same design but with a more realistic, unequal number of observations per cell.

Because of the large amount of data comprising the results from this design (that is, 100 observations in each of 18 cells, or 1,800 scores), only certain summary data are shown in Table 3-18. We have included the sum of scores, sum of squared scores, mean, and variance for each cell. These data are sufficient to carry out the analytical steps; in practice, of course, the researcher must compute these quantities directly from the original data.

Prior to carrying out the analysis of variance on the data, we must test the homogeneity-of-variance assumption. Despite the fact that there is an equal number of observations per cell, the researcher would want to know whether or not the treatments produced unusual differences in the variability of response measures. Hartley's $F_{\max}$ statistic can be appropriately computed on the cell variances. Thus, $S^2_{\max} = 148.84$, $S^2_{\min} = 67.24$, and $F_{\max} = 148.84/67.24 = 2.21$; for 18 variances

table 3-18 *Cell summary for a fictitious 2 × 3 × 3 factorial design*

cell	sum of scores	sum of squared scores	mean	variance
(111)	5,238	283,114	52.4	88.36
(121)	5,561	323,982	55.6	148.84
(131)	5,614	329,665	56.1	146.41
(112)	5,818	345,813	58.2	73.96
(122)	6,237	397,381	62.4	84.64
(132)	6,066	381,516	60.7	136.89
(113)	5,794	347,683	57.9	121.00
(123)	6,008	375,455	60.1	146.41
(133)	5,922	357,358	59.2	67.24
(211)	6,040	374,915	60.4	102.01
(221)	6,511	437,024	65.1	132.25
(231)	6,424	421,993	64.2	94.09
(212)	5,838	348,664	58.4	79.21
(222)	6,615	449,780	66.2	123.21
(232)	6,101	386,478	61.0	144.00
(213)	5 633	330,173	56.3	129.96
(223)	6,076	382,271	60.8	132.25
(233)	5,944	363,611	59.4	104.04

each based on 99 degrees of freedom, this value is nonsignificant (although it approaches significance at the .05 level).

The total sum of squares and the error sum of squares can be computed directly from the information in Table 3-18. The sum of all the scores is

$$Y_{....} = 107{,}440$$

(found by summing the "Sum of Scores" column in Table 3-18). Also, the total sum of squared scores is

$$\sum_{i=1}^{r} \sum_{j=1}^{q} \sum_{k=1}^{p} \sum_{m=1}^{n} Y_{ijkm}^2 = 6{,}636{,}876$$

(found by summing the "Sum of Squared Scores" column in Table 3-18). From these quantities,

$$\text{SS}_{\text{total}} = 6{,}636{,}876 - \frac{107{,}440^2}{1{,}800} = 6{,}636{,}876 - 6{,}412{,}974.22 = 223{,}901.78$$

Also, the error sum of squares can be found by computing the sum of squares per cell, $\Sigma_{m=1}^{n} Y_{ijkm}^2 - (Y_{ijk.})^2/n$, for the cell (ijk) and then pooling over all cells. It is computationally more convenient to use the formula

$$\sum_{i=1}^{r}\sum_{j=1}^{q}\sum_{k=1}^{p}\sum_{m=1}^{n} Y_{ijkm}^2 - \sum_{i=1}^{r}\sum_{j=1}^{q}\sum_{k=1}^{p}\frac{(Y_{ijk.})^2}{n}$$

Thus

$$\begin{aligned} SS_{\text{error}} &= 6{,}636{,}876 - \left[\frac{5{,}238^2}{100} + \frac{5{,}561^2}{100} + \cdots + \frac{5{,}944^2}{100}\right] \\ &= 6{,}636{,}876 - 6{,}433{,}453.62 = 203{,}422.38 \end{aligned}$$

The summary table method will now be utilized to find the sums of squares for the main and interaction effects. The sum of squares for each first-order interaction can be found from a summary table representing the combination of levels of the dimensions entering into that interaction. For SS_{AB}, a summary table of dimensions A and B is set up, and so forth. At the same time, main effects can be computed from the marginal totals of these summary tables. Starting with the AB interaction, Table 3-19 presents cell *sums* for the A and B dimensions [each cell entry is found by summing across the two levels of the C dimension; e.g., the first cell sum, 11,278, was found by adding the sums of scores for cells (111) and (211) from Table 3-18]. The marginal sums (i.e., row and column totals) are also shown since these are used in computing main effects. The AB summary table contains information concerning only the A and B dimensions; it can be analyzed as a two-dimensional factorial design to find SS_A, SS_B, and SS_{AB}. The main effects can be found from the marginal totals; thus

$$SS_A = \frac{\sum_{k=1}^{p}\left(\sum_{i=1}^{r}\sum_{j=1}^{q}\sum_{m=1}^{n} Y_{ijkm}\right)^2}{qrn} - \frac{\left(\sum_{i=1}^{r}\sum_{j=1}^{q}\sum_{k=1}^{p}\sum_{m=1}^{n} Y_{ijkm}\right)^2}{pqrn}$$

table 3-19 *AB summary table for a fictitious* 2 × 3 × 3 *factorial design*

		TREATMENT A			
		no SVC	SVC—1 hr	SVC—2 hr	total
TREATMENT B	No PI	11,278	11,656	11,427	34,361
	PI	12,072	12,852	12,084	37,008
	PI+expert	12,038	12,167	11,866	36,071
	Total	35,388	36,675	35,377	107,440

or

$$\text{SS}_A = \frac{\sum_{k=1}^{p} (Y_{..k.})^2}{qrn} - \frac{(Y_{....})^2}{pqrn}$$

In the present instance, $p = 3$, $q = 3$, $r = 2$, and $n = 100$. The terms in parentheses in the left-hand expression are simply the column sums from the AB summary table (that is, $Y_{..k.}$). The right-hand term is the same as that used in computing the total sum of squares. Similarly, the main effect for the B treatment can be computed from

$$\text{SS}_B = \frac{\sum_{j=1}^{q} (Y_{.j..})^2}{prn} - \frac{(Y_{....})^2}{pqrn}$$

The terms in parentheses on the left-hand side of the expression are simply the row sums from the AB summary table. Finally, the interaction effect can be found by

$$\text{SS}_{AB} = \frac{\sum_{j=1}^{q} \sum_{k=1}^{p} (Y_{.jk.})^2}{rn} - \frac{(Y_{....})^2}{pqrn} - \text{SS}_A - \text{SS}_B$$

Note that the term $(Y_{....})^2/pqrn$ appears in all three computational formulas and has already been computed in connection with the SS_{total}. Thus, it can be "saved" from that computation and reused. To illustrate the use of these formulas:

$$\begin{aligned}\text{SS}_A &= \frac{35{,}388^2}{600} + \frac{36{,}675^2}{600} + \frac{35{,}377^2}{600} - 6{,}412{,}974.22 \\ &= 6{,}414{,}830.50 - 6{,}412{,}974.22 = 1{,}856.28\end{aligned}$$

$$\begin{aligned}\text{SS}_B &= \frac{34{,}361^2}{600} + \frac{37{,}008^2}{600} + \frac{36{,}071^2}{600} - 6{,}412{,}974.22 \\ &= 6{,}418{,}979.04 - 6{,}412{,}974.22 = 6{,}004.82\end{aligned}$$

$$\begin{aligned}\text{SS}_{AB} &= \frac{11{,}278^2}{200} + \frac{11{,}656^2}{200} + \cdots + \frac{11{,}866^2}{200} - 6{,}412{,}974.22 - 1{,}856.28 \\ &\qquad - 6{,}004.82 \\ &= 6{,}421{,}566.91 - 6{,}412{,}974.22 - 1{,}856.28 - 6{,}004.82 = 731.59\end{aligned}$$

In order to find the AC interaction term, and also the sum of squares for the C dimension, an AC summary table must be set up

table 3-20 *AC summary table for a fictitious 2 × 3 × 3 factorial design*

	TREATMENT A no SVC	SVC—1 hour	SVC—2 hours	total
TREATMENT C No OEP	16,413	18,121	17,724	52,258
OEP	18,975	18,554	17,653	55,182
Total	35,388	36,675	35,377	107,440

(Table 3-20). Although the sum of squares for the A dimension could also be found from this table, it is unnecessary since the AB summary table has already been utilized for that purpose. The sum of squares for the C treatment can be found from the formula

$$\mathrm{SS}_C = \frac{\sum_{i=1}^{r} (Y_{i...})^2}{pqn} - \frac{(Y_{....})^2}{pqrn}$$

The sum of squares for the interaction of the A and C treatments is

$$\mathrm{SS}_{AC} = \frac{\sum_{i=1}^{r}\sum_{k=1}^{p} (Y_{i.k.})^2}{qn} - \frac{(Y_{....})^2}{pqrn} - \mathrm{SS}_A - \mathrm{SS}_C$$

Substituting from Table 3-20, we obtain

$$\mathrm{SS}_C = \frac{52{,}258^2}{900} + \frac{55{,}182^2}{900} - 6{,}412{,}974.22 = 6{,}417{,}724.10 - 6{,}412{,}974.22$$
$$= 4{,}749.88$$

$$\mathrm{SS}_{AC} = \frac{16{,}413^2}{300} + \frac{18{,}121^2}{300} + \cdots + \frac{17{,}653^2}{300} - 6{,}412{,}974.22 - 1{,}856.28$$
$$- 4{,}749.88$$
$$= 6{,}426{,}091.12 - 6{,}412{,}974.22 - 1{,}856.28 - 4{,}749.88 = 6{,}510.74$$

Finally, the BC interaction can be found from a BC summary table (Table 3-21). The formula for this term is

$$\mathrm{SS}_{BC} = \frac{\sum_{i=1}^{r}\sum_{j=1}^{q} (Y_{ij..})^2}{rn} - \frac{(Y_{....})^2}{pqrn} - \mathrm{SS}_B - \mathrm{SS}_C$$

table 3-21 *BC summary table for a fictitious 2 × 3 × 3 factorial design*

		TREATMENT B		
	no PI	PI	PI+expert	total
No OEP	16,850	17,806	17,602	52,258
TREATMENT C				
OEP	17,511	19,202	18,469	55,182
Total	34,361	37,008	36,071	107,440

Substituting, we obtain

$$\text{SS}_{BC} = \frac{16{,}850^2}{300} + \frac{17{,}806^2}{300} + \cdots + \frac{18{,}469^2}{300} - 6{,}412{,}974.22 - 6{,}004.82 - 4{,}749.88$$

$$= 6{,}424{,}208.09 - 6{,}412{,}974.22 - 6{,}004.82 - 4{,}749.88 = 479.17$$

At this point, all main effects and all first-order interactions have been computed. The only remaining term is the second-order interaction SS_{ABC}. This term can be found from the cells of the original data table. If all main effects and first-order interactions are subtracted from the among-cells sum of squares, the remainder will be second-order interaction. Following this strategy, the formula is

$$\text{SS}_{ABC} = \frac{\sum_{i=1}^{r}\sum_{j=1}^{q}\sum_{k=1}^{p}(Y_{ijk.})^2}{n} - \frac{(Y_{....})^2}{pqrn} - \text{SS}_A - \text{SS}_B - \text{SS}_C - \text{SS}_{AB} - \text{SS}_{AC} - \text{SS}_{BC}$$

For the present example, this becomes

$$\text{SS}_{ABC} = \frac{5{,}238^2}{100} + \frac{5{,}561^2}{100} + \cdots + \frac{5{,}944^2}{100} - 6{,}412{,}974.22 - 1{,}856.28 - 6{,}004.82 - 4{,}749.88 - 731.59 - 6{,}510.74 - 479.17$$

$$= 6{,}433{,}453.62 - 6{,}412{,}974.22 - 1{,}856.28 - 6{,}004.82 - 4{,}749.88 - 731.59 - 6{,}510.74 - 479.17$$

$$= 146.92$$

At last, we are ready to summarize the complete analysis in an analysis-of-variance summary table (Table 3-22). All three main effects are highly significant, as is the interaction of the A and C dimensions. In line with the descriptions of the treatment variables, we have assumed that all three treatment variables represent fixed effects. The

table 3-22 *Analysis-of-variance summary table for a fictitious 2 × 3 × 3 factorial design*

source	df	sum of squares	mean square	F†	p
A	2	1,856.28	928.14	8.13	<.005
B	2	6,004.82	3,002.41	26.30	<.005
C	1	4,749.88	4,749.88	41.61	<.005
AB	4	731.59	182.90	1.60	ns
AC	2	6,510.74	3,255.37	28.52	<.005
BC	2	479.17	239.58	2.10	ns
ABC	4	146.92	36.73	.32	ns
Error	1,782	203,422.38	114.15		
Total	1,799	223,901.78			

†F values computed by assuming fixed-effects model.

interpretation of the significant effect for C is straightforward since this factor occurs at only two levels. Referring to Table 3-20, we can convert the marginal sums for the C dimension to means by dividing by 900. Thus, the mean achievement for the group with no employment program is $52{,}258/900 = 58.06$ and for the group with the special employment program it is $55{,}182/900 = 61.31$. Thus, the higher achievement was clearly on the part of those potential dropouts who were involved in the released-time work-study program. The interpretation of the significant effects for the A and B treatment dimensions is less direct since these factors each exist at three levels. The means for both of these factors can be found from the marginal sums of Table 3-19.

For A we find $\bar{Y}_{..1.} = 35{,}388/600 = 58.98$, $\bar{Y}_{..2.} = 36{,}675/600 = 61.12$, and $\bar{Y}_{..3.} = 35{,}377/600 = 58.96$. From inspection of these values it appears likely that level 2 (special vocational counseling for 1 hour per week) is superior to either no special vocational counseling or special counseling for 2 hours per week (which do not differ). However, this impression should be substantiated by a test of significance. Since we did not indicate any specific contrasts of interest prior to completing the analysis, we must utilize procedures appropriate for postmortem testing of hypotheses. For this example, we should want to test all possible pairwise differences among the means for the A treatment dimension. To illustrate its application, we utilize the Newman-Keuls technique for these data. For each pairwise contrast, we compute $(\bar{Y}_i - \bar{Y}_j)/\sqrt{\text{MS}_{\text{error}}/n}$ for the ith and jth means. In the present application, n is the number of observations entering into the computation of *one* of the means. Since each mean is based on 600 observations, $n = 600$. From Table 3-22, $\text{MS}_{\text{error}} = 114.15$. Thus, the denominator (which is constant for all three pairwise comparisons) is $\sqrt{\text{MS}_{\text{error}}/n} = \sqrt{114.15/600} = \sqrt{.1902}$

= .4361. In applying the Newman-Keuls test, it should be remembered that the choice of critical values from Appendix Table B-7 depends upon the degree of rank-order separation between the pair of means being tested in the contrast. For the A treatment, the means, in rank order, are $\bar{Y}_{..3.} = 58.96$, $\bar{Y}_{..1.} = 58.98$, $\bar{Y}_{..2.} = 61.12$. There are two contrasts involving differences of one rank position, or covering two ranks:

$$\frac{\bar{Y}_{..3.} - \bar{Y}_{..1.}}{.4361} = \frac{58.96 - 58.98}{.4361} = \frac{-.02}{.4361} = -.46$$

$$\frac{\bar{Y}_{..1.} - \bar{Y}_{..2.}}{.4361} = \frac{58.98 - 61.12}{.4361} = \frac{-2.14}{.4361} = -4.91$$

Referring to Appendix Table B-7, we have 599 degrees of freedom for each group (that is, 600 − 1); although the table does not show values between 120 and ∞, it is clear that for adjacent pairs of means (i.e., for two steps between ordered means), the difference between levels 1 and 3 is nonsignificant, and the difference between levels 1 and 2 is significant at the .01 level, with level 2 yielding the higher mean value. Turning to the contrast of levels 2 and 3, we have the test statistic

$$\frac{\bar{Y}_{..2.} - \bar{Y}_{..3.}}{.4361} = \frac{61.12 - 58.96}{.4361} = \frac{2.16}{.4361} = 4.95$$

This value is significant at the .01 level for ordered means differing by three steps (i.e., by two rank positions). Thus, these tests confirm our impressions from inspection of the results. Levels 1 and 3 of treatment A produce no differential effects, but both are significantly below the degree of achievement posted by level 2.

Turning to the B dimension, we can carry out similar tests. Since each of these means is also based on 600 observations, the standard-error term remains the same (that is, .4361). The means (computed from the marginals of Table 3-19) in rank order are $\bar{Y}_{.1..} = 57.27$, $\bar{Y}_{.3..} = 60.12$, $\bar{Y}_{.2..} = 61.68$. These means are spaced out more than were those for the A dimension; however, at least levels 1 and 2, the extremes, must differ significantly. For adjacent pairs, the statistics are

$$\frac{\bar{Y}_{.1..} - \bar{Y}_{.3..}}{.4361} = \frac{57.27 - 60.12}{.4361} = \frac{-2.85}{.4361} = -6.54$$

which is significant at the .01 level, and

$$\frac{\bar{Y}_{.3..} - \bar{Y}_{.2..}}{.4361} = \frac{60.12 - 61.68}{.4361} = \frac{-1.56}{.4361} = -3.58$$

which is significant at the .05 level but not at the .01 level of significance. The final contrast, involving the extremes, is

$$\frac{\bar{Y}_{.1..} - \bar{Y}_{.2..}}{.4361} = \frac{57.27 - 61.68}{.4361} = \frac{-4.41}{.4361} = -10.11$$

which is significant at the .01 level. In summary, all pairwise contrasts for the B dimension are significant at conventional levels. However, in terms of the descriptions of the levels of this dimension, the best achievement was gained by students whose parents were involved in conferences with the counselor and the student; the second-best performance occurred when a vocational expert was also involved in the conference. The worst performance did occur in the absence of any parental involvement.

The only remaining significant effect from the analysis of the $2 \times 3 \times 3$ design is the interaction of the A and C dimensions. An indispensable aid in the interpretation of this effect is a graph of the profile of the means from the AC summary table (Figure 3-5); each cell sum has been converted to a cell mean by dividing by 300, the number of observations entering into the computation of the sums. The graph plainly reveals that the employment program has a substantial effect on student achievement in the absence of special vocational counseling. When there is either 1 hour or 2 hours per week of special vocational counseling, there is very little differential between students in and students not in the employment program. The interpretation of this

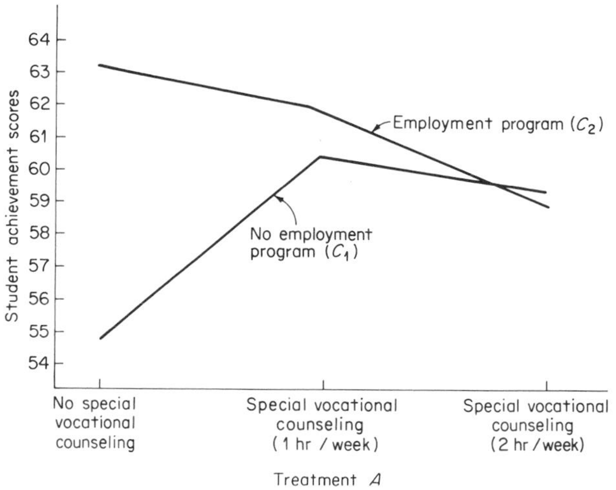

figure 3-5 *AC interaction graph for fictitious $2 \times 3 \times 3$ factorial design.*

interaction makes a major difference in the type of recommendation which the experimenter might be willing to make concerning the variables studied in this design.

The analysis of the interaction effect can be carried beyond plotting of the profile of means. The Newman-Keuls, or other multiple-comparison, technique can be applied to the cell means of the AC summary table. Each mean is based on 300 observations, and the standard error would be $\sqrt{\text{MS}_{\text{error}}/300}$. It would be of interest to test the significance of the difference between the levels of C for each level of the A treatment dimension. These tests are

$$\frac{\bar{Y}_{1.1.} - \bar{Y}_{2.1.}}{\sqrt{\text{MS}_{\text{error}}/300}} = \frac{54.71 - 62.25}{\sqrt{114.15/300}} = \frac{-8.54}{\sqrt{.3805}} = \frac{-8.54}{.6168} = -13.84$$

In the set of six means, this pair covers six rank positions; reference to Appendix Table B-7 for 299 degrees of freedom per cell makes it apparent that this value is significant at the .01 level of significance.

$$\frac{\bar{Y}_{1.2.} - \bar{Y}_{2.2.}}{.6168} = \frac{60.40 - 61.85}{.6168} = \frac{-1.45}{.6168} = -2.35$$

This pair of means covers two rank positions, and the difference is not significant at the .05 level of significance.

$$\frac{\bar{Y}_{1.3.} - \bar{Y}_{2.3.}}{.6168} = \frac{59.08 - 58.84}{.6168} = \frac{.24}{.6168} = .39$$

This pair of means covers two rank positions, and the difference is obviously nonsignificant.

The analysis of differences between the levels of C for the three levels of A confirms the impression gained from inspection of the interaction graph. That is, the employment program makes a significant difference only in the absence of special vocational counseling.

In the analysis of the fictitious $2 \times 3 \times 3$ factorial design, we assumed that all treatment levels were fixed values, arbitrarily set by the experimenter. In the fixed-effects model, the within-cells-error term is the proper mean square against which all other effects are tested. If one or more of the treatment dimensions represents a random effect, this must be taken into account when F ratios are set up. Any three-dimensional factorial design has expectations for mean squares which parallel those for the 2^3 design presented in Table 3-11. Except for differences in constant multipliers, the expectations would hold for the $2 \times 3 \times 3$ design or any other design with three treatment dimensions (regardless of the numbers of levels of the different treatments). Although the student may refer to tables such as 3-11 or 3-12, he is in a better position to approach the analysis of any factorial design if he knows an algorithm for generating the expected values themselves rather than relying on the availability of tables which provide these expectations. The procedure presented here is a modification of a technique due to Cornfield and Tukey (1956). Their procedure is general and applies to designs other than factorials. Our modified procedure is aimed specifically at finding expected values of mean squares for factorial designs. In later sections, the procedure will be augmented to cover additional designs.

The Cornfield and Tukey algorithm allows us to find general expressions for the expectations of mean squares; these expressions are relevant whether the dimensions are fixed, random, or some mixture of fixed and random. The general expressions can be particularized to suit actual experimental needs by suitable operations. These operations depend upon a set of "operators" which take on the value 0 for a fixed-effects treatment and the value 1 for random-effects treatment. For each treatment factor, we define an operator $D_g = 1 - g/G$, where g is the number of levels of the treatment actually used in the experimental design, and G is the universe of possible levels from which g was actually chosen. If the treatment levels are fixed, $g = G$ since the actual levels are the only ones of interest to the researcher and there is no selection of levels from a wider universe of levels. Thus, for a fixed treatment, $D_g = 1 - g/G = 1 - 1 = 0$. For a random treatment, G is ordinarily very large relative to g, and g/G is approximately 0. Thus, for a random treatment dimension, $D_g = 1 - g/G = 1 - 0 = 1$. Therefore, D_g becomes 0 for a fixed-treatment dimension and 1 for a random-treatment dimension. In the case of a particular design, the operators are chosen to conform to the treatments. For example, in a three-dimensional design, assume that treatment A exists at p levels, treatment B exists at q levels, and treatment C exists at r levels Our operators would be $D_p = 1 - p/P$, $D_q = 1 - q/Q$, and $D_r = 1 - r/R$. That is, the potential numbers of levels are P, Q, and R, respectively. Of course, for a fixed-effects treatment dimension, the potential and actual numbers of levels are the same. The algorithm can be presented in a number of steps:

1. Write the structural model for a score in order to determine all main and interaction effects.

2. Construct a table using the main and interaction effects as row headings and the different *subscripts* appearing on terms in the model as column headings. These subscripts are, typically, k, j, i, etc., as needed for treatments A, B, C, etc.

3. The cell entries in column k are found from:

a. Enter D_p for each effect which contains k in its subscripts.

b. Enter p for each effect which does not contain k in its subscripts.

4. The entries in column j are found by the same process, except that the operator is D_q and the constant is q; this process is continued for all columns of the table. The final column, which contains the subscript corresponding to subjects, or individuals within cells of the design, will always have the entry n since there are n "levels" of subjects per cell.

5. When all columns are filled, append the error term as a final row to the table and place 1s for it in each column.

6. The expected value of MS_A is the weighted sum of variances due to all effects which contain k among their subscripts; the weight is the product of all entries in the row, except the entry in column k.

7. The mean squares for all other main effects are found by the same procedure, utilizing the column with subscript corresponding to the treatment.

8. The expected value of MS_{AB} is the weighted sum of variances due to effects which contain *both* k and j among their subscripts; the weights are computed by omitting both column k and column j.

9. All remaining interactions follow the same pattern; also, three-factor and higher-order interactions have variances in their expectations which contain all relevant subscripts, and the weights are computed by omitting all columns headed by these subscripts.

These rules will be applied to the $2 \times 3 \times 3$ factorial design based on the fictitious research situation. The model, in accordance with step 1, is

$$Y_{ijkm} = \mu + \alpha_k + \beta_j + \gamma_i + \alpha\beta_{kj} + \alpha\gamma_{ki} + \beta\gamma_{ji} + \alpha\beta\gamma_{kji} + \epsilon_{ijkm}$$

Table 3-23 has been set up with the seven main and interaction effects as rows and the different subscripts i, j, k, and m as column headings. In accordance with step 3, the entries in column k are either D_p or p; D_p has been placed wherever there is a k in the subscripts of the row effects; otherwise, p has been placed in the cell. Similarly, the entries in columns j and i were found. The entries for the final column, labeled m, are all n since m appears in none of the main or interaction effects and since the number of "levels" of subjects is n (that is, there are n subjects per cell of the design). The final row of the table, for ϵ_{ijkm}, was appended after finding all other entries. The expected values were found by utilizing steps 6 through 9. For example, $E(MS_A)$ is the weighted sum of variances from all effects which have a k among their subscripts; the

table 3-23 *Expected values for mean squares for a* $2 \times 3 \times 3$ *factorial design*

effect	k	j	i	m	E(MS)
α_k	D_p	q	r	n	$\sigma^2 + nD_qD_r\sigma_{\alpha\beta\gamma}{}^2 + nqD_r\sigma_{\alpha\gamma}{}^2 + nrD_q\sigma_{\alpha\beta}{}^2 + nqr\sigma_{\alpha}{}^2$
β_j	p	D_q	r	n	$\sigma^2 + nD_pD_r\sigma_{\alpha\beta\gamma}{}^2 + npD_r\sigma_{\beta\gamma}{}^2 + nrD_p\sigma_{\alpha\beta}{}^2 + npr\sigma_{\beta}{}^2$
γ_i	p	q	D_r	n	$\sigma^2 + nD_pD_q\sigma_{\alpha\beta\gamma}{}^2 + npD_q\sigma_{\beta\gamma}{}^2 + nqD_p\sigma_{\alpha\gamma}{}^2 + npq\sigma_{\gamma}{}^2$
$\alpha\beta_{kj}$	D_p	D_q	r	n	$\sigma^2 + nD_r\sigma_{\alpha\beta\gamma}{}^2 \quad + nr\sigma_{\alpha\beta}{}^2$
$\alpha\gamma_{ki}$	D_p	q	D_r	n	$\sigma^2 + nD_q\sigma_{\alpha\beta\gamma}{}^2 \quad + nq\sigma_{\alpha\gamma}{}^2$
$\beta\gamma_{ji}$	p	D_q	D_r	n	$\sigma^2 + nD_p\sigma_{\alpha\beta\gamma}{}^2 \quad + np\sigma_{\beta\gamma}{}^2$
$\alpha\beta\gamma_{kji}$	D_p	D_q	D_r	n	$\sigma^2 + n\sigma_{\alpha\beta\gamma}{}^2$
ϵ_{ijkm}	1	1	1	1	σ^2

weights are computed by omitting the entry in column k. Since ϵ_{ijkm} contains all four subscripts, it appears in every expectation. The remaining terms containing a k are the ABC interaction, the AC interaction, the AB interaction, and the A effect itself. Also, for the BC interaction, the terms entering the expectation must contain both the subscripts j and i. These are error, the ABC interaction, and the BC interaction itself. The weights are computed by omitting the entries in both column j and column i.

Once the generalized table of expectations has been found, the D_g operators can be given specific numerical values (that is, 0 or 1) in order to arrive at the expectations for a particular model. For the fixed-effects model, it is apparent that all terms with a D_p, D_q, or D_r disappear from the expectations since all three operators are equal to 0. Thus, each mean square has in its expectation only random error and a variance unique to the effect itself. Also, for a random-effects model, $D_p = D_q = D_r = 1$, and all terms are retained in the model. The student should confirm that, except for constant multipliers, these expectations are identical with those presented in Table 3-11 for the 2^3 factorial design. Similarly, results equivalent to those in Table 3-12 can be obtained by appropriate choices for the operators. The first example has A and B fixed and C random. Thus, $D_p = 0$, $D_q = 0$, and $D_r = 1$. In the second example, A is fixed, while both B and C are random; thus, $D_p = 0$, $D_q = 1$, and $D_r = 1$. Substituting these values, and deleting terms containing multipliers of 0, will generate the models in Table 3-12.

ORTHOGONAL-CONTRAST ANALYSIS OF FACTORIAL DESIGNS

We now turn to the problem of generating the sums of squares for any factorial design from an appropriate set of orthogonal contrasts built upon the cells of the factorial design. The procedure presented here depends upon the use of sets of patterned contrasts and will be illustrated on the $2 \times 3 \times 3$ factorial design previously analyzed by the summary table method. In the case of 2^k series factorial designs, there was no problem in writing one contrast to account for each main and interaction effect, since each of these had 1 degree of freedom and a single orthogonal contrast accounts for exactly 1 degree of freedom. When the treatments exist at three or more levels, we must still absorb the available degrees of freedom by means of orthogonal contrasts; however, it will take two or more contrasts to account for some of the main and interaction effects. In connection with the completely randomized design, it was mentioned that a set of $p - 1$ orthogonal contrasts will account for the treatment sum of squares based on p levels of a single treatment dimension. In the present context we are utilizing this same fact, but reversing the procedure. Initially, we do not know the value of the sum of squares for a treatment effect; we can, however, build a set of orthogonal contrasts, find the sum of squares corresponding to each contrast, and then add these to find the sum of squares for the treatment effect. In practice, this method is about as efficient as the summary table procedure; it is easily adaptable to calculating-machine computation and presents an especially powerful tool for computer analysis of experimental designs.

The procedure is best understood by illustration, and so the $2 \times 3 \times 3$ factorial design will be used for this purpose. In Table 3-24, the cells have been laid out as columns of a contrast-coefficient matrix. For ease of reference, the actual cell means are also included in the column headings. The rows of the table are contrasts which have been chosen to provide the various sums of squares of interest. For the treatment which exists at only two levels (i.e., dimension C), the contrast is simply the difference between the two levels, and $+1$ and -1 have been appropriately inserted as coefficients for this contrast. Since treatment A exists at three levels, it requires two orthogonal contrasts to absorb its degrees of freedom and to construct the sum of squares for A. These two contrasts follow a definite pattern. The first is a comparison of level 1 of A with level 2 of A, ignoring level 3 (note that 0s are used as coefficients for all cells at level 3 of A). The second contrast is the average of levels 1 and 2 compared with level 3 of A (the coefficients $+1$ have been used for all cells at levels 1 and 2, and the coefficient -2 has been used for all cells at level 3; thus, each cell at level 1 or 2 is weighted one-half what a cell at level 3 is weighted). It is apparent from studying the pattern of $+1$s and -1s for the two contrasts that they are orthogonal. The -2s in the second contrast are always paired with 0s in the second contrast; therefore they do not add to the sum of cross products. The two contrasts for treatment B have been constructed in the same manner; that is, the first contrast compares levels 1 and 2 of B, and the second contrast is levels 1 and 2 versus level 3. The interaction contrasts

table 3-24 *Patterned orthogonal-contrast matrix for a fictitious* 2 × 3 × 3 *factorial design*

CELL: MEAN†		(111) 52.38	(121) 55.61	(131) 56.14	(112) 58.18	(122) 62.37	(132) 60.66	(113) 57.94	(123) 60.08	(133) 59.22	(211) 60.40	(221) 65.11	(231) 64.24	(212) 58.38	(222) 66.15	(232) 61.01	(213) 56.33	(223) 60.76	(233) 59.44
effect																			
A	1	+1	+1	+1	−1	−1	−1	0	0	0	+1	+1	+1	−1	−1	−1	0	0	0
	2	+1	+1	+1	+1	+1	+1	−2	−2	−2	+1	+1	+1	+1	+1	+1	−2	−2	−2
B	1	+1	−1	0	+1	−1	0	+1	−1	0	+1	−1	0	+1	−1	0	+1	−1	0
	2	+1	+1	−2	+1	+1	−2	+1	+1	−2	+1	+1	−2	+1	+1	−2	+1	+1	−2
C	1	+1	+1	+1	+1	+1	+1	+1	+1	+1	−1	−1	−1	−1	−1	−1	−1	−1	−1
AB	1	+1	−1	0	−1	+1	0	0	0	0	+1	−1	0	−1	+1	0	0	0	0
	2	+1	+1	−2	−1	−1	+2	0	0	0	+1	+1	−2	−1	−1	+2	0	0	0
	3	+1	−1	0	+1	−1	0	−2	+2	0	+1	−1	0	+1	−1	0	−2	+2	0
	4	+1	+1	−2	+1	+1	−2	−2	−2	+4	+1	+1	−2	+1	+1	−2	−2	−2	+4
AC	1	+1	+1	+1	−1	−1	−1	0	0	0	−1	−1	−1	+1	+1	+1	0	0	0
	2	+1	+1	+1	+1	+1	+1	−2	−2	−2	−1	−1	−1	−1	−1	−1	+2	+2	+2
BC	1	+1	−1	0	+1	−1	0	+1	−1	0	−1	+1	0	−1	+1	0	−1	+1	0
	2	+1	+1	−2	+1	+1	−2	+1	+1	−2	−1	−1	+2	−1	−1	+2	−1	−1	+2
ABC	1	+1	−1	0	−1	+1	0	0	0	0	−1	+1	0	+1	−1	0	0	0	0
	2	+1	+1	−2	−1	−1	+2	0	0	0	−1	−1	+2	+1	+1	−2	0	0	0
	3	+1	−1	0	+1	−1	0	−2	+2	0	−1	+1	0	−1	+1	0	+2	−2	0
	4	+1	+1	−2	+1	+1	−2	−2	−2	+4	−1	−1	+2	−1	−1	+2	+2	+2	−4

†To ensure additional accuracy, means were computed from sums in Table 3-18 to two decimal places.

have been constructed by the mechanical procedure of multiplying together corresponding coefficients from a contrast in each of the entering treatment dimensions. There are as many contrasts in an interaction as there are possible combinations of contrasts from the two or more treatments entering into the interaction. Thus, there are 4 degrees of freedom and four contrasts for the AB interaction. The coefficients for the first AB contrast were generated by multiplying together corresponding coefficients from the first contrast for A and the first contrast for B. The remaining AB contrasts result from combining the first contrast in A with the second contrast in B, the second contrast in A with the first contrast in B, and the second contrast in A with the second contrast in B. A similar procedure was used to find the coefficients for the AC and BC interactions. The triple interaction ABC can be found in a variety of ways. It was generated here by multiplying together corresponding coefficients from the AB interaction and the C main effect.

Once the matrix of contrast coefficients has been found for all main and interaction effects, sums of squares can be computed by the usual procedures. In order to identify each contrast, the effect letter name and the contrast number within the effect will be used. Thus, A_1 is the first contrast for the A treatment, AB_2 is the second contrast for the AB interaction, and so forth. The sum of squares for the contrast whose identification is M is

$$\text{SS}_M = \frac{n\left(\sum_{h=1}^{d} c_{Mh}\overline{Y}_h\right)^2}{\sum_{h=1}^{d} c_{Mh}{}^2}$$

assuming that there is a total of d cells in the design (i.e., for a three-dimensional design, $pqr = d$). The sum of squares for a particular effect may be the sum of two or more such contrast sums of squares. Thus, $\text{SS}_A = \text{SS}_{A_1} + \text{SS}_{A_2}$, and $\text{SS}_{AB} = \text{SS}_{AB_1} + \text{SS}_{AB_2} + \text{SS}_{AB_3} + \text{SS}_{AB_4}$. However, there is only one contrast for the C treatment, and $\text{SS}_C = \text{SS}_{C_1}$.

Table 3-25 presents the contrast sums (that is, the terms $\Sigma_{h=1}^{d} c_{Mh}\overline{Y}_h$) and the sums of squares for each of the contrasts. At the right in Table 3-25 the sum of squares for each of the main and interaction effects is shown as sums of appropriate contrast sums of squares. None of these values deviates by more than .02 from the values recorded in Table 3-22, which were computed by the summary table method. To complete the analysis, the total sum of squares would be computed directly from the individual scores, and the error sum of squares would be computed from within-cells information. Since this has already been done in connection with the summary table analysis, it will not be repeated here.

table 3-25 *Sum of squares for contrasts from a 2 × 3 × 3 factorial design*

contrast	contrast sum	contrast sum of squares	pooled sum of squares
A_1	−12.87	1,380.31	SS_A = 1,380.31 + 475.97
A_2	13.09	475.97	= 1,856.28
B_1	−26.47	5,838.84	SS_B = 5,838.84 + 165.98
B_2	−7.73	165.98	= 6,004.82
C_1	−29.24	4,749.88	SS_C = 4,749.88
AB_1	4.02	202.00	SS_{AB} = 202.00 + 337.50 + 190.41 + 1.68
AB_2	−9.00	337.50	= 731.59
AB_3	−6.76	190.41	
AB_4	−1.10	1.68	
AC_1	−21.29	3,777.21	SS_{AC} = 3,777.21 + 2,733.55
AC_2	−31.37	2,733.55	= 6,510.76
BC_1	7.35	450.19	SS_{BC} = 450.19 + 28.98
BC_2	−3.23	28.98	= 479.17
ABC_1	−2.10	55.12	SS_{ABC} = 55.12 + 16.01 + .96 + 74.83
ABC_2	1.96	16.01	= 146.92
ABC_3	.48	.96	
ABC_4	−7.34	74.83	

The analysis of factorial designs by means of sets of orthogonal contrasts is most efficient when the researcher has planned a set of a priori contrasts; the direct computation of main effects can be initially omitted, and the sums of squares associated with the preplanned orthogonal contrasts can be carried out. These contrast sums of squares can then be combined as in Table 3-25 to find main effects.

In experimental designs having treatment dimensions with more than three levels, the construction of patterned orthogonal contrasts follows the model for the case illustrated. We begin by comparing levels 1 and 2 of the dimension; then, the average of levels 1 and 2 is compared with level 3; next, the average of levels 1 to 3 is compared with level 4 (in this instance, the coefficient for a cell at level 1, 2, or 3 is +1, and the coefficient for a cell at level 4 is −3). This process is repeated until all levels have been included in the system of contrasts. This procedure guarantees that the contrasts will be orthogonal and generates just the right number of contrasts to consume the degrees of freedom for the treatment dimension.

analysis with unequal cell sizes

All the computational procedures for factorial designs considered up to this point have been based on the assumption that each cell of the experimental design contained the same number n of subjects. Unfortunately, in practice, unequal cell sizes are the rule rather than the exception. Even in experiments originally designed for equal cell sizes, there is often loss of data from such causes as subject mortality (including subjects' failing to show up for experimental sessions, subjects' becoming ill or bored and leaving an experimental session, etc.), improperly completed data-collection forms (such as recording errors by research assistants or failures to follow instructions by subjects), and misapplications of treatments. Also, some experimental designs may dictate the sampling of subjects from preexisting strata; if proportional sampling is utilized, the samples may vary in size. Thus, it is highly likely that, in practical research situations, the user of statistical methods will be faced with analyzing data from a factorial design based on groups of different sizes. This section presents some tools for dealing with this problem. We begin with a general discussion of some more or less undesirable, but sometimes expediently used, procedures. Next, an approximate technique known as the method of *unweighted means* is described and illustrated. Finally, an elegant, but computationally involved and tedious, procedure based on the least-squares principle is introduced. However, full utilization of this last-mentioned approach depends upon skill in calculus and matrix algebra; since it is assumed that this degree of mathematical training does not necessarily characterize the reader, outside references are given in lieu of a full treatment of the technique.

There is one condition under which departures from equal cell sizes have no real effect on the analysis of factorial designs. If the cell sizes are proportional by rows (or, equivalently, by columns), the computational formulas based on the summary table procedure can be slightly modified to yield a straightforward solution. The condition of proportionality means that constant ratios can be set up for all pairs of rows (or columns). For example, Table 3-26 presents the number of cases in each cell of a 4×3 factorial design. Note that a continued proportion formed by ratios of any pair of rows (or columns) is a true proportion. Thus, for rows 1 and 3, $5{:}20 = 10{:}40 = 15{:}60$; and rows 2 and 3, $15{:}20$

table 3-26 *Factorial design with proportional cell sizes*

		TREATMENT A		
		1	2	3
TREATMENT B	1	$n_{11} = 5$	$n_{12} = 10$	$n_{13} = 15$
	2	$n_{21} = 15$	$n_{22} = 30$	$n_{23} = 45$
	3	$n_{31} = 20$	$n_{32} = 40$	$n_{33} = 60$
	4	$n_{41} = 10$	$n_{42} = 20$	$n_{43} = 30$

$= 30{:}40 = 45{:}60$. Similarly, any pair of columns represents a true proportion; e.g., columns 1 and 2 are $5{:}10 = 15{:}30 = 20{:}40 = 10{:}20$. When proportionality holds, certain of the constants must be adjusted to give a correct solution by the summary table procedure (the summary table procedure can also be applied to 2^k series factorial designs under conditions of unequal but proportional sample sizes in the cells). These adjustments will be presented in terms of a three-dimensional factorial design, but can be easily generalized to other designs. The total sum of squares can be found from

$$SS_{\text{total}} = \sum_{i=1}^{r} \sum_{j=1}^{q} \sum_{k=1}^{p} \sum_{m=1}^{n_{ijk}} Y_{ijkm}^2 - \frac{(Y_{....})^2}{n_t}$$

where

$$n_t = \sum_{i=1}^{r} \sum_{j=1}^{q} \sum_{k=1}^{p} n_{ijk}$$

is the total number of scores in all cells of the factorial design, and n_{ijk} is the number of observations in cell (ijk). Note that this formula differs from that for equal n only in the substitution of n_t for $pqrn$ in the denominator of the right-hand term and the appearance of the cell size n_{ijk} as a limit of summation for the subscript m. The error sum of squares is given by

$$SS_{\text{error}} = \sum_{i=1}^{r} \sum_{j=1}^{q} \sum_{k=1}^{p} \sum_{m=1}^{n_{ijk}} Y_{ijkm}^2 - \sum_{i=1}^{r} \sum_{j=1}^{q} \sum_{k=1}^{p} \frac{(Y_{ijk.})^2}{n_{ijk}}$$

Again, the only change is replacing n by n_{ijk} to take explicit account of the differing cell sizes. In these two formulas, the substitution of n_{ijk} for n limits summation to the actual number of observations in a cell; for the equal cell size case, this number is always n. From the AB summary table, the sums of squares for A, B, and AB can be computed. The unequal (but proportional) cell size formulas are

$$SS_A = \sum_{k=1}^{p} \left[\frac{(Y_{..k.})^2}{n_{..k}} \right] - \frac{(Y_{....})^2}{n_t}$$

where $n_{..k}$ is the number of scores at level k of the A treatment dimension (that is, $n_{..1}$ is the number of scores at level 1 of A, $n_{..2}$ the number at level 2 of A, and so forth).

$$SS_B = \sum_{j=1}^{q} \left[\frac{(Y_{.j..})^2}{n_{.j.}} \right] - \frac{(Y_{....})^2}{n_t}$$

where $n_{.j.}$ is the number of scores at level j of the B treatment dimension.

$$\text{SS}_{AB} = \sum_{j=1}^{q} \sum_{k=1}^{p} \left[\frac{(Y_{.jk.})^2}{n_{.jk}} \right] - \frac{(Y_{....})^2}{n_t} - \text{SS}_A - \text{SS}_B$$

where $n_{.jk}$ is the number of scores entering into cell (jk) of the AB summary table. For the AC and BC summary tables, similar changes are necessary, and, by following the pattern for the AB summary table, the student can easily determine these.

In effect, if the cell sizes are proportional, the usual formulas based on the summary table analysis may be used if changes are introduced to take into account the unequal cell sizes. The degrees of freedom for main and interaction effects are unchanged. However, the degrees of freedom for the total sum of squares become $n_t - 1$, and the degrees of freedom for error are $n_t - pqr$.

A factorial design with unequal *and* disproportionate cell sizes poses more than computational difficulties. Indeed, the partitioning of the total sum of squares into independent, additive components is impossible under these circumstances (if a researcher erroneously utilized the formulas presented above to compute sums of squares for a design with unequal and disproportionate cell sizes, each sum of squares thus computed would actually contain a mixture of sources of variation; this mixture would include two or more *main* effects, and meaningful interpretation would be impossible). In order to sidestep this problem, some researchers eliminate data from cells of the design in order to achieve balance in the sense of equal or proportional cell sizes. This practice does, of course, arrive at a tractable computational situation and allows a straightforward analysis. In relatively large experiments with only modest departures from the equal or proportional n condition, the practice of discarding data to arrive at a balanced design is difficult to fault. Naturally, the cases chosen for deletion from the experiment should be arrived at by a random process. In smaller experiments, the researcher is reluctant to throw away data since this may appreciably reduce the power of the experiment by reducing its size substantially.

A not uncommon educational practice is to utilize classroom groups to compose the cells of a factorial design. Since classrooms typically differ in size, the design is unbalanced. However, this practice is fundamentally unsound from the point of view of the establishment of a randomization unit. That is, the individual students in an intact classroom cannot be considered independent experimental subjects since they did not arrive in the classroom by a random process. Thus, the mean of the classroom scores should be used; to achieve replication of cases within cells, a number of different classes must be randomly assigned to the cells of the experimental design. Under these conditions, the design can be made to conform to the equal or proportional cell size requirements.

Let us assume that the design is correctly set up and experimental subjects (or classes) have been randomly assigned to cells of the design. Further, assume that although equal cell sizes were initially planned, some data were lost and the analysis must be carried out on unequal and disproportionate cell sizes. A computationally feasible method is presented by the technique of unweighted means. This method is relevant when equal cell sizes were planned since it computes sums of squares based on an average cell size. The first step in an unweighted-means analysis is to compute cell means based on the available observations. Analysis of variance is then performed on these cell means; that is, we consider the design as having only one observation per cell, the cell mean. From this analysis, a sum of squares for A, B, and AB (assuming a two-dimensional design) can be found by the use of ordinary formulas. Each of these terms is then "extended" by multiplication by the harmonic mean of the sample sizes in the cells of the design. If, for a two-dimensional design, n_{ij} is the number of observations in cell (ij), then the harmonic mean of the cell sizes is

$$n_h = \frac{pq}{(1/n_{11}) + (1/n_{12}) + \cdots + (1/n_{pq})}$$

That is, n_h is the reciprocal of the mean of the reciprocals of the cell sizes. The harmonic mean, rather than the arithmetic mean, is utilized in the unweighted-means analysis since the variance of a sampling distribution is proportional to $1/n_{ij}$ rather than to n_{ij}. Then, if we use the notation SS'_A, SS'_B, and SS'_{AB} for the sums of squares from the table of means, the adjusted sums of squares for main and interaction effects are $\text{SS}_A = n_h\text{SS}'_A$, $\text{SS}_B = n_h\text{SS}'_B$, and $\text{SS}_{AB} = n_h\text{SS}'_{AB}$. The sum of squares for error can be found in the usual manner from within-cells information (from the original data table, not from the table of cell means). Thus,

$$\text{SS}_{\text{error}} = \sum_{i=1}^{q} \sum_{j=1}^{p} \sum_{k=1}^{n_{ij}} Y_{ijk}^2 - \sum_{i=1}^{q} \sum_{j=1}^{p} \frac{(Y_{ij.})^2}{n_{ij}}$$

for a two-dimensional design. The principle underlying the unweighted-means analysis is to adjust all sums of squares to a constant cell size n_h.

To illustrate the unweighted-means analysis, we return to the example of a $2 \times 3 \times 3$ factorial design presented earlier. This design had 100 observations per cell; assume that some of the data were lost be-

table 3-27 *Cell summary for a fictitious* 2 × 3 × 3 *factorial design assuming unequal cell sizes*

cell	number of cases	sum of scores	sum of squared scores	mean	variance
(111)	98	5,106	274,424	52.1	86.51
(121)	100	5,561	323,982	55.6	148.84
(131)	92	5,134	299,323	55.8	139.39
(112)	99	5,782	345,043	58.4	75.01
(122)	100	6,237	397,381	62.4	84.64
(132)	95	5,748	360,521	60.5	135.50
(113)	89	5,144	307,924	57.8	120.60
(123)	91	5,487	343,478	60.3	140.33
(133)	98	5,792	349,226	59.1	71.21
(211)	99	5,920	363,800	59.8	99.96
(221)	92	5,980	401,220	65.0	137.58
(231)	100	6,424	421,993	64.2	94.09
(212)	97	5,665	338,474	58.4	79.44
(222)	97	6,441	439,046	66.4	118.23
(232)	96	5,837	369,326	60.8	151.83
(213)	99	5,564	325,086	56.2	126.31
(223)	100	6,076	382,271	60.8	132.25
(233)	94	5,574	340,605	59.3	108.37

cause of teachers' withdrawing their classes from the experiment, records' being incorrectly kept, administration errors in giving examinations, etc. At the end of the experiment, the numbers of cases per cell, sums of scores, etc., are as shown in Table 3-27.

For the unweighted-means analysis, we consider the design to have only one observation per cell, the cell mean value. Table 3-28 presents the design in this format. Marginal sums are shown for each of the three treatment dimensions. In order to compute main effects and inter-

table 3-28 *Cell means for a fictitious* 2 × 3 × 3 *factorial design*

		TREATMENT A: 1			2			3			
TREATMENT B:		1	2	3	1	2	3	1	2	3	sum (C)
TREATMENT C	1	52.1	55.6	55.8	58.4	62.4	60.5	57.8	60.3	59.1	522.0
	2	59.8	65.0	64.2	58.4	66.4	60.8	56.2	60.8	59.3	550.9
Sum (A)			352.5			366.9			353.5		1,072.9
Sum (B)		342.7	370.5	359.7							

table 3-29 *Two-way summary tables for a fictitious* $2 \times 3 \times 3$ *factorial design*

		AB Summary Table				*AC* Summary Table				*BC* Summary Table		
		TREATMENT *A*				TREATMENT *A*				TREATMENT *B*		
		1	2	3		1	2	3		1	2	3
	1	111.9	116.8	114.0	*C* 1	163.5	181.3	177.2	*C* 1	168.3	178.3	175.4
TREATMENT *B*	2	120.6	128.8	121.1	*C* 2	189.0	185.6	176.3	*C* 2	174.4	192.2	184.3
	3	120.0	121.3	118.4								
Sum		352.5	366.9	353.5		352.5	366.9	353.5		342.7	370.5	359.7

actions (unadjusted), we set up two-way summary tables of the same sort used in an equal-cell-size analysis. Table 3-29 contains all three two-way summary tables. In computing sums of squares from these summary tables, we use $n = 1$ for the cell size. Thus,

$$\text{SS}'_A = \frac{(352.5)^2}{6} + \frac{(366.9)^2}{6} + \frac{(353.5)^2}{6} - \frac{(1{,}072.9)^2}{18}$$
$$= 63{,}972.35 - 63{,}950.80 = 21.55$$

$$\text{SS}'_B = \frac{(342.7)^2}{6} + \frac{(370.5)^2}{6} + \frac{(359.7)^2}{6} - \frac{(1{,}072.9)^2}{18}$$
$$= 64{,}016.27 - 63{,}950.80 = 65.47$$

$$\text{SS}'_{AB} = \frac{(111.9)^2}{2} + \frac{(116.8)^2}{2} + \cdots + \frac{(118.4)^2}{2} - \frac{(1{,}072.9)^2}{18} - \text{SS}'_A - \text{SS}'_B$$
$$= 64{,}045.56 - 63{,}950.80 - 21.55 - 65.47 = 7.74$$

$$\text{SS}'_C = \frac{522.0^2}{9} + \frac{550.9^2}{9} - \frac{1{,}072.9^2}{18} = 63{,}997.20 - 63{,}950.80 = 46.40$$

$$\text{SS}'_{AC} = \frac{163.5^2}{3} + \frac{181.3^2}{3} + \cdots + \frac{176.3^2}{3} - \frac{1{,}072.9^2}{18} - \text{SS}'_A - \text{SS}'_C$$
$$= 64{,}083.94 - 63{,}950.80 - 21.55 - 46.40 = 65.19$$

$$\text{SS}'_{BC} = \frac{168.3^2}{3} + \frac{178.3^2}{3} + \cdots + \frac{184.3^2}{3} - \frac{1{,}072.9^2}{18} - \text{SS}'_B - \text{SS}'_C$$
$$= 64{,}067.88 - 63{,}950.80 - 65.47 - 46.40 = 5.21$$

$$\text{SS}'_{ABC} = \frac{52.1^2}{1} + \frac{55.6^2}{1} + \cdots + \frac{59.3^2}{1} - \frac{1{,}072.9^2}{18} - \text{SS}'_A - \text{SS}'_B - \text{SS}'_C$$
$$- \text{SS}'_{AB} - \text{SS}'_{AC} - \text{SS}'_{BC}$$
$$= 64{,}164.13 - 63{,}950.80 - 21.55 - 65.47 - 46.40 - 7.74 - 65.19$$
$$- 5.21 = 1.77$$

The unadjusted sums of squares based on the cell means must now be "extended" to represent a constant cell size n_h:

$$n_h = \frac{18}{\frac{1}{98} + \frac{1}{100} + \cdots + \frac{1}{94}} = \frac{18}{.186875} = 96.32$$

Thus,

$$SS_A = (96.32)SS'_A = (96.32)(21.55) = 2{,}075.70$$

$$SS_B = (96.32)(65.47) = 6{,}306.07$$

$$SS_C = (96.32)(46.40) = 4{,}469.25$$

$$SS_{AB} = (96.32)(7.74) = 745.52$$

$$SS_{AC} = (96.32)(65.19) = 6{,}279.10$$

$$SS_{BC} = (96.32)(5.21) = 501.83$$

$$SS_{ABC} = (96.32)(1.77) = 170.49$$

Finally, the sum of squares for error can be obtained directly from within-cells information from the original data table. Thus,

$$SS_{\text{error}} = 6{,}383{,}123 - \left(\frac{5{,}106^2}{98} + \frac{5{,}561^2}{100} + \cdots + \frac{5{,}574^2}{94}\right)$$
$$= 6{,}383{,}123 - 6{,}187{,}975.58 = 195{,}147.42$$

It should be noted that the sums of squares computed by the unweighted-means procedure compare fairly well with those from the analysis based on complete data and summarized in Table 3-22. The degrees of freedom for the main and interaction effects are the same in the unweighted-means analysis and the analysis based on complete data. However, the degrees of freedom for error are $n_t - pqr = 1{,}736 - 18 = 1{,}718$. The complete unweighted-means analysis for the fictitious $2 \times 3 \times 3$ factorial design is summarized in Table 3-30.

Once the analysis of variance is completed and if a significant F occurred, it may be desirable to complete multiple comparisons in order to locate significantly different pairs of means for treatment dimensions existing at three or more levels. The average cell size n_h (that is, the harmonic mean of the cell sizes) can be used in place of n in the test procedures. Similarly, if preplanned orthogonal contrasts have been set up, n_h may be used in lieu of n in the formulas for the sum of squares for a linear contrast.

table 3-30 *Analysis-of-variance summary table based on unweighted-means analysis of a fictitious* 2 × 3 × 3 *factorial design*

source	df	sum of squares	mean square	F	p
A	2	2,075.70	1,037.85	9.14	<.005
B	2	6,306.07	3,153.04	27.76	<.005
C	1	4,469.25	4,469.25	39.35	<.005
AB	4	745.52	186.38	1.64	ns
AC	2	6,279.10	3,139.55	27.64	<.005
BC	2	501.83	250.92	2.21	ns
ABC	4	170.49	42.62	.38	ns
Error	1,718	195,147.42	113.59		

The unweighted-means procedure presents an attractive computational tool for the researcher since its application involves very little additional computation when compared with an analysis based on equal cell sizes. It is, however, an approximate solution based on the device of adjusting all cell sizes to a constant quantity. An exact solution based on the least-squares principle is possible but does present some computational hardships. For this reason, it is best carried out with the aid of an electronic computer and suitable programs. The general principles of the least-squares solution will be sketched in this section; however, details of the computational routines and illustration will be left to outside references.

For a two-dimensional analysis-of-variance design, for example, with unequal and disproportionate cell sizes, the least-squares solution works directly from the structural model for a score; the model is, however, written without an explicit interaction term. That is, initially, a completely additive model is assumed. For the two-dimensional design, the model for a score is $Y_{ijk} = \mu + \alpha_j + \beta_i + \epsilon_{ijk}$. The least-squares principle involves selecting the parameters μ, α_j, and β_i so as to minimize the sum of squared errors as estimated from the data. The quantity to be minimized is

$$\sum_{i=1}^{q} \sum_{j=1}^{p} \sum_{k=1}^{n_{ij}} (Y_{ijk} - \mu - \alpha_j - \beta_i)^2$$

The usual conditions are imposed on the parameters corresponding to effects; that is, $\sum_{j=1}^{p} \alpha_j = 0$ and $\sum_{i=1}^{q} \beta_i = 0$. Least-squares estimates for the α_j, for β_i, and for μ are found by expanding the square, taking partial derivatives with respect to each parameter, and setting each of the resulting equations (which are called *normal equations*) equal to zero. This system of normal equations must be simultaneously solved for the estimates of the parameters. Efficient solutions depend upon the utilization of the procedures of matrix algebra. Once estimates of μ, the α_j, and β_i are found by the least-squares method, sums of squares for the treatment effects can be computed from expressions of the form $SS_A = \Sigma_{j=1}^{p} \alpha_j a_j$, where a_j is a term based upon a function of the cell sizes n_{ij}. The calculus procedures required to find the partial derivatives and the matrix algebra procedures required to solve the system of normal equations are quite tedious for hand or calculating-machine computation: in this age, they are best left to computers, which have a much higher threshold for boredom than does man. The student with the necessary mathematical sophistication who wishes to study the least-squares solution can consult Winer (1962, pp. 224–227) or Peng (1967, Chapter 9).

Under some circumstances the researcher can avoid the problems attendant upon computation with unequal sample sizes; if the cell sizes are unequal because of one or a small number of "missing" observations, there are formulas for estimating the missing values, replacing them in the data table, and then carrying out the usual computations for sums of squares. The values which are computed and inserted in the data table are selected so as to have no influence on interaction terms in the analysis. That is, if an observation is missing from cell (ij) of a two-dimensional design, it is replaced by a value which does not contribute to the row-by-column interaction sum of squares. These formulas are most useful and most commonly used when there is only one observation per cell and the unweighted-means analysis cannot be applied (i.e., with $n = 1$ per cell, there is no cell mean for a cell with missing data). Also, designs with one observation per cell are more likely to arise in connection with randomized blocks (Chapter 5) than with pure factorial designs. To illustrate this approach, consider the following data table for a 3×4 design with $n = 1$ per cell:

		TREATMENT A				
		1	2	3	4	sum
	1	8	11	9	6	34
TREATMENT B	2	10	14	9	10	43
	3	14	17	X	15	46
Sum		32	42	18	31	123

Note that the observation in cell (33) is missing. The task is to fill in a value for the missing observation which will not affect the AB interaction; then, computation of the usual sums of squares can be accomplished by including this value as if it were an actual observation. However, the total number of observations is 11, and the total degrees of freedom are 10; thus, the AB interaction will be based on 5, not 6, degrees of freedom.

The general formula for a single missing value can be developed from the structural model: $Y_{ij} = \mu + \alpha_j + \beta_i + \alpha\beta_{ji} + \epsilon_{ij}$. Since interaction and experimental error are completely confounded when $n = 1$ per cell, the terms $\alpha\beta_{ji}$ and ϵ_{ij} are each set equal to 0. Then $Y_{ij} = \mu + \alpha_j + \beta_i$. In order to find estimates for μ, α_j, and β_i, we introduce the notation [assuming the missing observation is in cell (km)]

B = total of all scores for row k
T = total of all scores for column m
G = total of all scores in the table

Then our estimate of μ is $(G + X)/pq$, where X is the missing value to be filled in, and p and q are, respectively, the number of columns and rows of the table. Similarly, the estimate of β_i is $(B + X)/p - (G + X)/pq$ and of α_j is $(T + X)/q - (G + X)/pq$. Thus our estimate of $\mu + \alpha_j + \beta_i$ is

$$\frac{T + X}{q} + \frac{B + X}{p} - \frac{G + X}{pq}$$

Rearranging terms and simplifying,

$$X = \frac{pT + qB - G}{(p - 1)(q - 1)}$$

A proof that this procedure does, in fact, minimize the AB interaction is given by Jaech (1966).

Returning to the example, note that the missing value is in row 3 and column 3. The sums are $T = 18$, $B = 46$, and $G = 123$. Since $p = 4$ and $q = 3$, our estimate is

$$X = \frac{4(18) + 3(46) - 123}{(4 - 1)(3 - 1)} = \frac{87}{6} = 14.5$$

To complete the analysis, this value is inserted in cell (33), and the usual partitioning of the total sum of squares is carried out. The breakdown of degrees of freedom will be:

source	df
A	3
B	2
AB	5
Total	10

Note that, in effect, replacing the missing value by the above procedure is simply a convenient device to allow use of the standard computing formulas. It should never be assumed that the estimated value is, in fact, equal to the value which would actually be obtained in the cell containing the missing observation.

If additional analysis, such as multiple range tests, are carried out on the data, standard-error formulas must be adjusted to account for the missing observation. For example, in the Newman-Keuls test, MS_{AB} would enter into the formula for computing the standard-error term. However, in all comparisons involving a column containing the missing observation, MS_{AB} must be replaced by $MS_{AB}\{(2/q) + p/[q(q-1)(p-1)]\}$; similarly, for a comparison involving the row containing the missing observation, we must use $MS_{AB}\{(2/p) + q/[p(p-1)(q-1)]\}$.

In the case of more than one missing observation, an iterative procedure must be utilized. First, arbitrary values are entered for all missing observations except one; the above formula is used to estimate the value for this one missing score. Then this value is entered in the data table; the remaining missing values, except one, are arbitrarily entered, and the one is estimated from the formula. This process is continued until an *initial* estimate for all missing values is determined. Then the *entire* procedure is repeated as often as necessary to arrive at estimated values which show convergence. That is, if the values from two succeeding iterations are very close (say within .01), the process can be stopped. A numerical example of a case with two missing observations is presented in Cochran and Cox (1957, pp. 111–112).

exemplary applications of factorial designs

example 1 Lang and Hochman (1966) report an investigation of three methods of instruction as applied in a beginning educational psychology course. The study was motivated by the practical goal of improving the instruction in this course, which is typically taken by prospective teachers. The three instructional variations were (1) two hours of lecture per week for 15 weeks, plus two hours per week for 14 weeks serving as teacher aides, plus one hour per week for 14 weeks participating in seminars conducted by experienced teachers; (2) two hours of lecture per week for

15 weeks, plus one hour per week for 14 weeks taking part in small group discussions led by their instructor; and (3) three hours of lecture per week for 15 weeks. The experimental subjects consisted of a group of 124 enrollees in an educational psychology course at an eastern university. Students were randomly divided among six sections of the course; two instructors each taught three sections of the course, one section per treatment level. If we consider the instructor factor as one dimension of a design, the experimental plan is a 2 × 3 factorial design:[2]

		TREATMENT 1	2	3
INSTRUCTOR	1	G_1	G_2	G_3
	2	G_4	G_5	G_6

A battery of pretests, comprising an achievement test in educational psychology, a test on case studies in teaching, and a student questionnaire, was administered to all students. The same battery was utilized at the end of the semester. In this summary, only results from the first two instruments, the achievement test and the test on case studies, are presented. The authors chose to base their analysis upon gain scores from pretest to posttest. Thus, for each student, a gain score in achievement and a gain score from the case-studies test were computed; then analysis of variance was applied to these gain scores. The results for the achievement test in educational psychology were:

source	*df*	sum of squares	mean square	*F*	*p*
Treatment	2	126.99	63.49	1.44	>.05
Instructor	1	21.51	21.51	.49	>.05
Interaction	2	16.66	8.33	.19	>.05
Error	118	5,206.94	44.13		

Both main effects and their interaction were nonsignificant at conventional levels. Thus, in terms of gains in knowledge of educational psy-

[2]Unfortunately, the authors do not report the number of students per cell of the design; they do, however, indicate that the totals for the three treatment levels were 53, 39, and 32, respectively.

chology concepts, etc., the three arrangements of instruction produced no differential effect. Also, the two instructors were, apparently, equally effective in their roles, and this effectiveness was similar for the three instructional methods. For the test on case studies in teaching, the summary was:

source	*df*	sum of squares	mean square	*F*	*p*
Treatment	2	.79678	.39839	3.019	$>.05$
Instructor	1	.00061	.00061	.004	$>.05$
Interaction	2	.32131	.16065	1.220	$>.05$
Error	118	15.56621	.13192		

Once again, none of the effects were significant, and the authors conclude that neither the treatments nor the instructors have differential effects with respect to increasing students' abilities to apply psychological principles to the analysis of case studies. Since the results of both analyses were null, no additional tests can be applied to the treatment or interaction effects.

example 2 Jester and Travers (1967) investigated the effects of presentation patterns upon comprehension of speeded speech. Speeded speech involves, essentially, increasing the rate of presentation of tape-recorded messages by removing silent segments from the tape. The amount of material removed determines the rate at which a message can be presented. The design utilized was a 3×4 factorial with presentation patterns as one dimension and number of repetitions of the message as the second dimension. Three presentation patterns were used:

1. Increasing-rate condition. One presentation at the base rate of 200 words per minute, followed by one presentation at each of the speeded rates 250, 300, and 350 words per minute.
2. Decreasing-rate condition. One presentation at each of the speeded rates 350, 300, and 250 words per minute, followed by one presentation at the base rate of 200 words per minute.
3. Constant-rate condition. All four presentations at 263 words per minute (to keep total time the same as in conditions 1 and 2).

The number-of-repetitions dimension involved varying the actual number of presentations given a subject; thus, some subjects in the increasing-rate condition received only the 200 word per minute presentation; others received only the 200 and 250 word per minute presentation; and so forth. Introductory educational psychology students at a state university were used as subjects; 10 students were assigned per treatment combination, and the total experiment comprised 120 subjects. Comprehension of the speeded speech material was measured by a 40-item test administered at completion of the experimental listening sessions.

The analysis-of-variance summary table for the effects from the 3×4 factorial design was:

source	*df*	sum of squares	mean squares	*F*	*p*
Repetition patterns	2	323.76	161.88	4.31	$<.05$
Number of repetitions	3	1,992.49	664.16	17.68	$<.01$
Interaction	6	629.09	104.85	2.79	$<.05$
Error	108	4,057.02	37.56		
Total	119	7,002.36			

The significant effect due to number of repetitions of the speeded speech material is, of course, expected since students varied from one to four exposures to the material prior to taking the comprehension examination. The patterns of repetitions also showed significant differences, and the mean scores for the three conditions (i.e., ascending, descending, and constant) were 16.82, 12.88, and 14.11. Although no contrasts were reported, it is apparent that the extreme outcomes, the ascending and descending repetition pattern conditions, indicate significantly different outcomes. The interaction, which was also significant, is plotted in Figure 3-6. Although differences appear during the first and second repetitions, the final levels of comprehension after four exposures to the material seem highly comparable for the three repetition patterns.

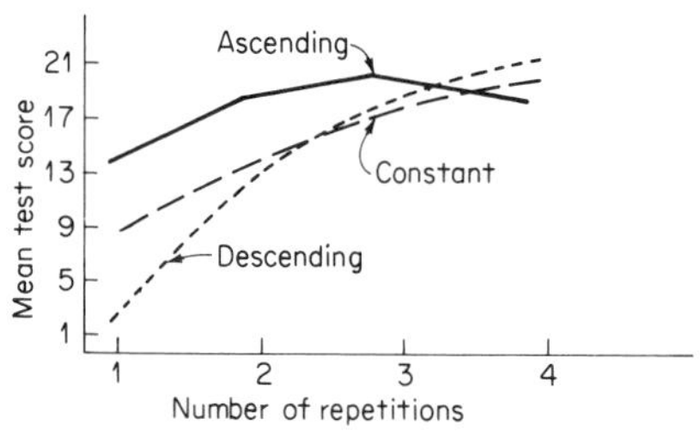

figure 3-6 *Interaction graph for repetition patterns and number of repetitions.*

chapter four

incomplete factorial designs

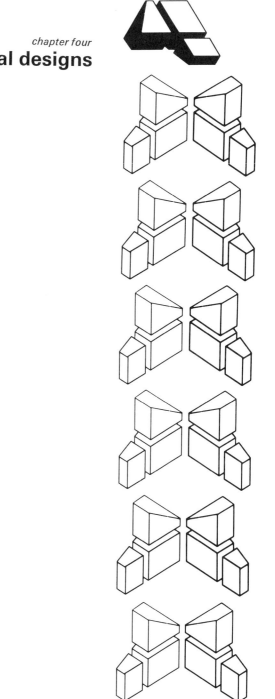

general principles

Factorial designs with only two or three treatment dimensions are usually experimentally feasible in the sense that the number of treatment combinations and the number of experimental subjects stay within reasonable limits. However, when a large number of dimensions are to be studied simultaneously, the size of the total experiment can easily exceed available resources. Incomplete factorial designs, also known as *fractional factorial designs*, are available which reduce the required size of the experiment. This reduction in size is gained at a price; in incomplete factorial designs some interactions cannot be estimated at all, and there is *confounding* of effects. Confounding means that the same computed sum of squares has two or more different interpretations in terms of two or more different experimental effects. This is best understood by illustration. Consider a 2^3 factorial design; there is a total of eight cells, or treatment combinations. By studying the matrix of contrast coefficients for this design (Table 4-1), it is apparent that just

table 4-1 *Matrix of contrast coefficients for a* 2^3 *factorial design*

	CELL							
effect	(111)	(121)	(112)	(122)	(211)	(221)	(212)	(222)
A	+1	+1	−1	−1	+1	+1	−1	−1
B	+1	−1	+1	−1	+1	−1	+1	−1
C	+1	+1	+1	+1	−1	−1	−1	−1
AB	+1	−1	−1	+1	+1	−1	−1	+1
AC	+1	+1	−1	−1	−1	−1	+1	+1
BC	+1	−1	+1	−1	−1	+1	−1	+1
ABC	+1	−1	−1	+1	−1	+1	+1	−1

half the eight cells, or four cells, are sufficient to estimate all effects except the second-order interation ABC. For example, by choosing just the cells (111), (122), (221), and (212), it is still possible to build a set of orthogonal contrasts that conforms to the main effects and the first-order interactions. For the main effects for A, B, and C, the pattern of contrast coefficients is:

	(111)	(221)	(212)	(122)
A	+1	+1	−1	−1
B	+1	−1	+1	−1
C	+1	−1	−1	+1

The required coefficients for contrasts corresponding to the first-order interactions can be found by multiplying together coefficients from the entering main effects. Thus, for AB we have:

	(111)	(221)	(212)	(122)
A	$+1$	$+1$	-1	-1
B	$+1$	-1	$+1$	-1
AB	$+1$	-1	-1	$+1$

Note, however, that the pattern of coefficients for the AB interaction term is identical with the pattern for the main effect for C. Thus, the AB interaction and the C main effect are confounded. If a sum of squares were computed for the third contrast from the means of these four cells, that sum of squares would reflect both the main effect due to the C treatment and the interaction of the A and B treatments. There is, furthermore, no way of separating the two sources of variation from the information contained in just these four cells. Similarly, the AC interaction contrast is identical with the main effect for B, and the BC interaction contrast is the same as that for A. Thus, there is complete confounding of main effects and first-order interactions.

The four cells chosen from the 2^3 factorial design [i.e., cells (111), (221), (212), and (122)] represent the four cells which have $+1$ coefficients for the ABC interaction effect in the matrix of contrast coefficients for the complete design. The four-cell design is described as a $\frac{1}{2}$ *replicate* of the complete 2^3 factorial design. A different $\frac{1}{2}$ replicate could be formed by taking the cells with -1 coefficients for the ABC interaction effect. If this were done, the contrasts for the four-cell design would be:

	(211)	(121)	(112)	(222)
A	$+1$	$+1$	-1	-1
B	$+1$	-1	$+1$	-1
C	-1	$+1$	$+1$	-1

To locate coefficients for the AB interaction, we proceed as before to multiply together corresponding coefficients from the entering main effects:

	(211)	(121)	(112)	(222)
A	+1	+1	−1	−1
B	+1	−1	+1	−1
AB	+1	−1	−1	+1

In this case, the coefficients of the AB interaction have the same pattern, but with reversed sign, as the main effect for C. Since the reversal of signs for the contrast coefficients has no influence on the interpretation of the contrast or the resulting sum of squares computed from the contrast, AB and C are still confounded. Similarly, AC and B, and BC and A are completely confounded. If we choose the positive half of the ABC interaction to build the $\frac{1}{2}$ replicate, the confounding is expressed as $C = AB, B = AC$, and $A = BC$. If we choose the negative half, these become $C = -AB$, $B = -AC$, and $A = -BC$. In the first case, if the AB interaction is a positive term [i.e., the cell sum for (111) + (122) is greater than that for (221) + (212)], it will inflate the sum of squares for A; if it is a negative term, it decreases the sum of squares for A. In the second case, the reverse conditions hold. When effects are confounded in fractional factorial designs, they are described as *aliases*. Thus, in the first $\frac{1}{2}$ replicate of the 2^3 factorial design, A and BC are aliases, B and AC are aliases, and C and AB are aliases. Also, the original ABC contrast from the complete factorial design is called the *defining contrast* since the $\frac{1}{2}$ replicates are formed by choosing either the positive or negative half of the cells entering into the ABC contrast. By utilizing the defining contrast, the alias for any effect can be found by a mechanical rule: Algebraically, multiply together the effect and the defining contrast. Then, drop from the resulting expression any term with power 2. The remainder is the alias of the effect in question. For example, for the 2^3 factorial design, the defining contrast is ABC. To find the alias of A, multiply A by ABC: $A(ABC) = A^2BC = BC$; the A^2 term was dropped from the expression since its power is 2. Similarly, $B(ABC) = AB^2C = AC$, and $C(ABC) = ABC^2 = AB$.

It should be noted that the four cells constituting a $\frac{1}{2}$ replicate of a 2^3 factorial design can be interpreted as a *complete* 2^2 factorial design. Only three separate sums of squares can be computed (i.e., only one sum of squares for each alias pair); these may be represented as A, B, and AB. This suggests that, from a computational point of view, the analysis of a $\frac{1}{2}$ replicate of a 2^3 factorial design is identical with the analysis of a 2^2 factorial design. In fact, by utilizing the orthogonal-contrast computing procedure, contrasts based on the four cells may be used directly. The problem, of course, in using and interpreting fractional or incomplete factorial designs is the confounding of effects. These designs are useful only when there is considerable prior information concerning the magnitudes of interactions. In the $\frac{1}{2}$ replicate of the 2^3 factorial design, interpretation of the main effects hinges upon knowledge that the first-order interactions are negligible. The category of fractional factorial designs known as Latin squares is often utilized in the behavioral sciences, and these are treated in some detail in this section. In order to build the necessary foundation for a consideration of Latin-square designs, some additional material on fractional replication in general is covered.

Although the 2^3 factorial design was chosen to illustrate the principles involved in fractional replication of factorial designs, there would, in practice, be little need to employ fractional designs when the complete design contains only eight cells. However, even the 2^k series factorial designs begin to require inconvenient numbers of cells for five or six experimental treatment factors. Let us consider a 2^6 factorial design, which, in its complete form, would require 64 treatment combinations, or cells. A $\frac{1}{2}$ replicate of the 2^6 factorial design requires 32 cells and could be built by using the $ABCDEF$ interaction as the defining contrast. If this were done, each main effect would be confounded with a fourth-order interaction (for example, A would be confounded with $BCDEF$, the alias for C would be $ABDEF$, and so forth). Also, each first-order interaction would be confounded with a third-order interaction (e.g., the alias for AB is $CDEF$), and second-order interactions would be confounded with other second-order interactions (for example, ABC is the alias for DEF). If third- and higher-order interactions were known to be negligible, all main effects and first-order interactions could be calculated and would be interpretable. The treatment combinations required to actually set up such a design could be either the positive-valued or the negative-valued cells in the contrast for the $ABCDEF$ interaction. Since this requires writing down the contrasts for a design with 64 cells, it is rather inconvenient. A shortcut procedure is available which avoids this tedious detailing of the contrast matrix; the positive-valued half of the cells will contain either 6, 4, 2, or no 1s. For example, cell (111111) contains six 1s and is included in the positive-valued half of the interaction; similarly, cells (121211), (112222), (211222), (222222), etc., each contain an even number of 1s and are included. Cells such as (121111) and (111112) contain an odd number of 1s and fall in the negative-valued half of the interaction. It is easy to see that the 32 cell codes with an even number of 1s could be written systematically to define the $\frac{1}{2}$ replicate of the 2^6 factorial design. Of course, either the positive-valued or the negative-valued $\frac{1}{2}$ replicate could actually be used to plan an experiment. The analysis of such a design presents no problem since the procedures for a 2^5 factorial design would yield the desired effects (or their aliases).

The $\frac{1}{2}$ replicate of a 2^6 factorial design still requires 32 treatment combinations, and this may also be too large an experiment for practical consideration. If this were the case, the researcher might consider a $\frac{1}{4}$ replicate rather than a $\frac{1}{2}$ replicate. A $\frac{1}{4}$ replicate of the 2^6 factorial design would contain 16 treatment combinations or cells. In constructing such a design, we might wish to sacrifice the fourth-order interactions. However, if this strategy is pursued, some undesirable consequences occur. For example, after using the $ABCDEF$ interaction to form a $\frac{1}{2}$ replicate, it seems natural to choose a fourth-order interaction, such as $ABCDE$, to further divide the 32 cells in half and form a $\frac{1}{4}$ replicate. However, when a defining contrast is selected, the alias for that defining contrast also becomes a defining contrast and cannot be estimated from the data. The alias of $ABCDE$ is $ABCDE(ABCDEF) = A^2B^2C^2D^2E^2F$. Eliminating all terms with power of 2 gives the alias of $ABCDE$ as the main effect F. Thus, F is a defining contrast and is lost in the same way that $ABCDEF$ is lost in the $\frac{1}{2}$ replicate. We must, therefore, seek some other method to locate 16 cells to form the $\frac{1}{4}$ replicate of the 2^6 factorial design.

A workable approach is to avoid altogether the first step of forming a $\frac{1}{2}$ replicate; rather, use two third-order interactions as the two defining contrasts required to reduce the total number of cells from 64 to 16. If these two third-order interactions are chosen so that they share two letters, the alias of the first interaction with respect to the second interaction (or vice versa) will also be a third-order interaction. For example, we might choose $ABDF$ and $ABCE$ as defining contrasts. Then, $CDEF = ABDF(ABCE)$ is also a defining contrast, but is still a third-order interaction. (The student may wish to confirm, by trial and error, that there is no combination of the $ABCDEF$ interaction with a fourth-order interaction as a defining contrast which does not result in a main effect's also becoming a defining contrast.) With $ABDF$ and $ABCE$ as defining contrasts, and $CDEF$ also becoming a defining contrast, each effect now has three, not one, aliases. Thus, the main effect for A has aliases $A(ABDF) = BDF$, $A(ABCE) = BCE$, and $A(CDEF) = ACDEF$. Note that two of A's aliases are second-order interactions, and the third is a fourth-order interaction. Each first-order interaction will also have three aliases; for example, the AB interaction has aliases $AB(ABDF) = DF$, $AB(ABCE) = CE$, and $AB(CDEF) = ABCDEF$. Note that two of these aliases are themselves first-order interactions.

Thus, the $\frac{1}{4}$ replicate of a 2^6 factorial design provides meaningful estimates of main effects only if second- and higher-order interactions can safely be assumed to be negligible. However, first-order interactions are confounded with other first-order interactions, and there is no hope of finding meaningful estimates for these. In order to find the cells (i.e., treatment combinations) which enter into a $\frac{1}{4}$ replicate, we can use a generalization of the rule previously presented for finding a $\frac{1}{2}$ replicate. That is, starting with the $ABDF$ defining contrast, we choose all cells which have either 4, 2, or 0 (that is, an even number) level codes *for* A, B, D, *or* F which are 1s. This will produce 32 cells, or a $\frac{1}{2}$ replicate of the 2^6 factorial design. Next, using the $ABCE$ defining contrast, we select, from among the 32 cells, those which have an even number (or 0) of 1s as level codes *for* A, B, C, *or* E. The resulting 16 cells constitute the $\frac{1}{4}$ replicate. To illustrate, the 32 cells resulting from the $ABDF$ defining contrast are:

(111111)	(111212)	(221111)	(211112)	(121211)	(211211)	(121112)	(221212)
(112111)	(112212)	(222111)	(212112)	(122211)	(212211)	(122112)	(222212)
(112121)	(112222)	(222121)	(212122)	(122221)	(212221)	(122122)	(222222)
(111121)	(111222)	(221121)	(211122)	(121221)	(211221)	(121122)	(221222)

We eliminate from these 32 cells the 16 which do not have an even number of 1s for the treatments A, B, C, or E in applying the defining contrast $ABCE$. The remainder is:

(111111)	(221111)	(122211)	(122112)
(112121)	(222121)	(121221)	(121122)
(111212)	(212112)	(212211)	(221212)
(112222)	(211122)	(211221)	(222222)

The 16 treatment combinations implied by these 16 cells would constitute the entire experiment utilizing the $\frac{1}{4}$ replicate. Thus, an experiment which, in its complete form, would require 64 independent groups of experimental subjects, can be run as a $\frac{1}{4}$ replicate with only 16 groups of experimental subjects. Of course, the researcher must bear in mind the considerable confounding of effects which occurs; the potential saving can be reaped only if there is a good deal of knowledge concerning the interactions among the treatment dimensions.

For 2^k series factorial designs other than those mentioned in this section, fractional replicates of various sizes can be constructed by following the principles which have been developed. Since this may be tedious, the researcher considering the use of fractional replication can consult relatively exhaustive plans for the 2^k series designs in Chapter 6A of the experimental design book by Cochran and Cox (1957).

In designs other than the 2^k series, the same principles apply, but their application is complicated by the fact that higher-order interactions are based on more than 1 degree of freedom. Consider the 3^k series of factorial designs; here, each of the k treatment variables exists at exactly three levels. Thus, each main effect has 2 degrees of freedom, each first-order interaction has 4 degrees of freedom, each second-

table 4-2 *Cell codes for a 3^2 factorial design*

		TREATMENT A		
		1	2	3
	1	(11)	(12)	(13)
TREATMENT B	2	(21)	(22)	(23)
	3	(31)	(32)	(33)

order interaction has 8 degrees of freedom, and so forth. Since 3^k is an odd integer, any design in the 3^k will have an odd number of cells (for example, $3^3 = 27$, $3^5 = 243$, etc.); for this reason $\frac{1}{2}$ replicates of 3^k series factorial designs cannot be constructed. However, 3^k is always divisible by 3, and $\frac{1}{3}$ replicates can be constructed. Unfortunately, the approach of utilizing the contrast for a high-order interaction to divide the cells into fractional replicates is no longer applicable since each high-order interaction is based on several contrasts, not one. For simple designs, however, a straightforward, rational approach will locate the required cells. For a 3^2 factorial design, there is a total of nine cells, or treatment combinations, in the experiment. A $\frac{1}{3}$ replicate of this design would contain three cells (of course, such a design would have no practical appeal). Since the levels of the main effects must be evenly represented in the fractional replicate, the possible $\frac{1}{3}$ replicates involve taking one and only one cell from each row and each column of the design. From Table 4-2, some possibilities are (11), (22), and (33); (13), (21), and (32); and (12), (23), and (31). The pattern of these cells is best seen by omitting the remaining cells in the table (Table 4-3). Any one of these three sets of cells constitutes a $\frac{1}{3}$ replicate of the 3^2 factorial design. There is, of course, considerable confounding of effects in such a design. In fact, each of the main effects is confounded with some portion of the AB interaction.

Fractional replicates for higher-order 3^k series designs may be similarly constructed. Because of the large number of possibilities, the use of prepared plans is preferred; the previous reference (Cochran

table 4-3 *Cell codes for $\frac{1}{3}$ replicates of a 3^2 factorial design*

		TREATMENT A				TREATMENT A				TREATMENT A		
		1	2	3		1	2	3		1	2	3
	1	(11)			1			(13)	1		(12)	
TREATMENT B	2		(22)		B 2	(21)			B 2			(23)
	3			(33)	3		(32)		3	(31)		

and Cox, 1957) indexes a variety of these and illustrates the analytical procedures necessary to obtain sums of squares for the effects which can be estimated. If one leaves simple series such as the 2^k or the 3^k series, the problem of constructing fractional replicates multiplies; the previous reference cites selected plans for designs which do not have all treatments at the same number of levels.

Latin squares

Latin-square designs represent a special variety of fractional replicates. A Latin square is always a $1/k$ replicate of a k^3 series factorial design. Thus, a Latin square always involves three treatment dimensions, and each dimension exists at k levels. Further, the combinations of treatments are chosen so that there is a complete balance of main effects. Traditionally, the Latin-square designs are represented by $k \times k$ tables with letters in the cells for levels of the third treatment. In order to achieve balance, the letters must be arranged so that each letter is represented once and only once in each row and each column of the table. For example, a 3×3 Latin-square design might take the form of Table 4-4. The columns stand for one of the treatments, the rows for the second, and the letters in the cells for the third. Note that each letter a, b, or c occurs exactly once in each row and column of the table. Of course, in this usage, the letter a stands for the first level of the third dimension, b stands for the second level, and c for the third level. Since the 3×3 Latin-square design is a $\frac{1}{3}$ replicate of a 3^3 factorial design, it is instructive to study the complete design in order to determine which cells are included in the Latin square. Table 4-5 presents the complete design with the Latin-square cells underlined. Note that in the complete design, the pattern of Latin-square cells is such that one cell is drawn from each row, column, and tier of the design. There is more than one arrangement of the letters in the cells of the 3×3 table that conforms to the definition of a Latin square. A Latin square is said to be in a

table 4-4 *Illustrative* 3×3 *Latin-square design*†

		TREATMENT A		
		1	2	3
	1	a	b	c
TREATMENT B	2	b	c	a
	3	c	a	b

†Treatment C is represented by the letters in the cells of the design.

table 4-5 *Complete 3^3 factorial design with Latin-square cells underlined*

		TREATMENT A								
		1			2			3		
TREATMENT B:		1	2	3	1	2	3	1	2	3
	1	$\underline{(111)}$	(121)	(131)	(112)	(122)	$\underline{(132)}$	(113)	$\underline{(123)}$	(133)
TREATMENT C	2	(211)	$\underline{(221)}$	(231)	$\underline{(212)}$	(222)	(232)	(213)	(223)	$\underline{(233)}$
	3	(311)	(321)	$\underline{(331)}$	(312)	$\underline{(322)}$	(332)	$\underline{(313)}$	(323)	(333)

standard form if the letters occur in natural order in the first row and first column. Thus, the Latin square represented in Table 4-4 is in standard form. For a 3 × 3 Latin square, it is apparent that there is only one standard form since the letters in the four cells not in the first row or first column cannot be changed without violating the definition of a Latin square; however, for higher-order Latin squares, there will be more than one standard form. From the one standard Latin square of size 3 × 3, a total of 11 more Latin squares can be constructed by simply interchanging rows and/or columns of the table. For simplicity, we shall represent the Latin square by the cell letters only, allowing the row and column labeling to be understood. If the first and third columns of the Latin square in standard form are interchanged, we obtain:

c	*a*	*b*
b	*c*	*a*
a	*b*	*c*

By a similar process, the remaining 3 × 3 Latin squares can be generated.

Let us consider, now, 4 × 4 Latin squares. With a simplified notational scheme, a square in standard form is:

a	*b*	*c*	*d*
b	*c*	*d*	*a*
c	*d*	*a*	*b*
d	*a*	*b*	*c*

By rearranging the letters not in the first row or first column, three additional Latin squares in standard form can be located. These are:

a	*b*	*c*	*d*		*a*	*b*	*c*	*d*		*a*	*b*	*c*	*d*
b	*a*	*d*	*c*		*b*	*d*	*a*	*c*		*b*	*a*	*d*	*c*
c	*d*	*b*	*a*		*c*	*a*	*d*	*b*		*c*	*d*	*a*	*b*
d	*c*	*a*	*b*		*d*	*c*	*b*	*a*		*d*	*c*	*b*	*a*

The first Latin square presented above is of particular interest because its cell letters were generated by a process of cyclic permutation. A standard Latin square of any given order can be developed in this manner. Begin with the first row of letters, which are always in natural order. To find the entries for the second row, shift each letter one position to the left and bring the letter which "falls off the edge" over to the far right. Continue the process until all rows have been generated. For example, to generate a standard 6 × 6 Latin square, we start with the first six letters in natural order:

a b c d e f

Then, each letter is shifted one position to the left; the *a* is moved to the far right. Thus, the first two rows are:

a b c d e f
b c d e f a

By continuing this process, an entire standard 6 × 6 Latin square is formed:

a b c d e f
b c d e f a
c d e f a b
d e f a b c
e f a b c d
f a b c d e

To illustrate the range of selection for Latin squares, Table 4-6 shows the numbers of standard forms and total numbers of different Latin squares from order 2 through order 7. For relatively large squares, the number of possibilities is enormous.

table 4-6 *Summary of available Latin squares*

order	standard forms	total number
2 × 2	1	2
3 × 3	1	12
4 × 4	4	576
5 × 5	56	161,280
6 × 6	9,408	812,851,200
7 × 7	16,942,080	61,479,419,904,000

Latin-square designs are often used in behavioral science settings when some factors involved in the design of an experiment prevent the use of complete factorial designs, but balance in the administration of the treatments is desired. In research in public school settings, it is often impossible to replicate an entire factorial design within one school. A Latin square may provide sufficient balance to warrant the experiment (however, as will be described below, there is confounding of effects in Latin-square designs). Consider an experiment in secondary school curriculum. The researcher is interested in comparing two versions of a new tenth-grade geometry course of study with the traditional course of study. Since high schools in his region track students by general ability, he also wants to take this factor into consideration. If we assume that there are three tracks of interest (say high, middle, and low), a complete design calls for a 3^2 factorial with nine treatment combinations. Further, assume that he is unable to set up all nine cells within individual high schools because of administrative reasons (e.g., the assigning of classes to teachers trained in the new curricula may pose hardships unless the experiment is kept small within the school). The researcher may consider, as an alternative, the use of a 3×3 Latin square utilizing a total of three high schools. The design appears as in Table 4-7; in school 1, the high track receives the a curriculum (the usual course of study), the middle track receives the b curriculum (version 1 of the new, experimental curriculum), and the low track receives the c curriculum (version 2 of the new, experimental curriculum). In school 2, the order of curricula for the three tracks is b, c, and a; in school three, it is c, a, and b. The data from the Latin-square design allow estimation of all three main effects *if* the first-order interactions of the main effects are negligible. Each of the main effects is confounded with the interaction of the remaining two main effects. Thus, the main effect for the curricula is a valid estimate of the impact of the courses of study *only if* the interaction of schools and ability tracks is negligible, Similarly, the interpretations of effects for schools and ability tracks are meaningful only if the other two first-order interactions are nonexistent. There is, of course, no way of estimating the magnitudes of these interactions with-

table 4-7 3×3 *Latin square for an experiment in curriculum*

		TREATMENT A (ability tracks)		
		1	2	3
TREATMENT B (schools)	1	a	b	c
	2	b	c	a
	3	c	a	b

in the framework of the single experiment since they are confounded with main effects. In an experiment such as the one proposed, there may be reasonable outside evidence for a lack of interaction between schools and ability tracks. For example, an examination of records of previous achievement by students in the three ability tracks may show similar profiles for the three schools. Since the major interest in carrying out the experiment is to estimate the effects of the new curricula, tests of significance among the curricula may safely be assumed to reflect only treatment effect. It would be much more difficult to rule out interactions between the A and C factors or the B and C factors; this may be, however, of little concern since these effects per se are of minor interest. The use of the Latin-square design in this example does achieve a balance in the sense that each course of study is applied to one and only one ability-track group within each school.

The nature of the confounding between main effects and first-order interactions in Latin squares can be revealed from studying a contrast matrix for the effects. Table 4-8 displays such a matrix for the 3×3 Latin square based upon the design in Table 4-7. The column headings are in terms of the cells from the complete 3^3 factorial design which are actually involved in the 3×3 Latin-square design. Since each main

table 4-8 *Contrast matrix for a* 3×3 *Latin-square design*

	CELL								
effect	(111)	(132)	(123)	(221)	(212)	(233)	(331)	(322)	(313)
A_1	+1	−1	0	+1	−1	0	+1	−1	0
A_2	+1	+1	−2	+1	+1	−2	+1	+1	−2
B_1	+1	0	−1	−1	+1	0	0	−1	+1
B_2	+1	−2	+1	+1	+1	−2	−2	+1	+1
C_1	+1	+1	+1	0	0	0	−1	−1	−1
C_2	+1	+1	+1	−2	−2	−2	+1	+1	+1
AB_1	+1	0	0	−1	−1	0	0	+1	0
AB_2	+1	+2	0	+1	−1	0	−2	−1	0
AB_3	+1	0	+2	−1	+1	0	0	−1	−2
AB_4	+1	−2	−2	+1	+1	+4	−2	+1	−2
AC_1	+1	−1	0	0	0	0	−1	+1	0
AC_2	+1	−1	0	−2	+2	0	+1	−1	0
AC_3	+1	+1	−2	0	0	0	−1	−1	+2
AC_4	+1	+1	−2	−2	−2	+4	+1	+1	−2
BC_1	+1	0	−1	0	0	0	0	+1	−1
BC_2	+1	0	−1	+2	−2	0	0	−1	+1
BC_3	+1	−2	+1	0	0	0	+2	−1	−1
BC_4	+1	−2	+1	−2	−2	+4	−2	+1	+1

effect has 2 degrees of freedom, two contrasts have been shown in order to absorb these degrees of freedom; as earlier, these have been designated by the subscripts 1 and 2. For all main effects, patterned contrasts have been used. For treatments A and B, we compare levels 1 and 2 for the first contrast and the average of 1 and 2 versus 3 for the second contrast. For the letters, treatment C, we compare levels 1 and 3 for the first contrast and their average versus level 2 for the second contrast. Each first-order interaction has 4 degrees of freedom, and these have been labeled 1 through 4. From the pattern of coefficients for AB_1, we see that AB_1 is part of the C_2 contrast Although the weights are different, both AB_1 and C_2 contain a comparison of cells (111) and (322) versus cells (221) and (212). Also, AB_2 is part of the C_1 contrast since they both involve a comparison of cells (111) and (132) versus cells (331) and (322). AB_3 is likewise part of the C_1 contrast since both contain a comparison of cells (111) and (123) versus (322) and (313). The fourth component of AB is not represented in either of the C contrasts. It is interesting to note that, for this arrangement of the matrix coefficients, the sum of AB_2 and AB_3 is equivalent to C_1. The coefficients for the sum of AB_2 and AB_3 are $+2$, $+2$, $+2$, 0, 0, 0, -2, -2, -2. Dividing each of these coefficients by 2 yields exactly the C_1 contrast coefficients. Similar relationships hold for the other two interactions. AC_1 is part of B_2, $AC_2 + AC_3$ is equivalent to B_1, and AC_4 is unaccounted for in the B contrasts. BC_1 is part of A_2, $BC_2 + BC_3$ is equivalent to A_1, and BC_4 is unaccounted for in the A contrasts.

The analysis of Latin-square designs is relatively straightforward since there is a limited number of effects to test. The model for a score in a Latin square requires some special attention; in a 3×3 Latin square, for example, there is a total of nine cells. Thus, there are 8 degrees of freedom which can be attributed to experimental effects. However, each of the treatments has three levels and 2 degrees of freedom; thus, the main effects consume 6 of these 8 degrees of freedom. The remaining 2 degrees of freedom are a mixture of inextricably confounded interactions. For explicitness of notation, these 2 degrees of freedom are described as *residual* variability. In writing the model for a score, we add a term to account for the residual variance; we do not, however, show any first-order interactions since these are partially confounded with main effects and partially included in the residual term. The model is

$$Y_{ijkm} = \mu + \alpha_k + \beta_j + \gamma_i + \text{res} + \epsilon_{ijkm}$$

The term "res" is included to indicate the residual source of variability. The model is based on the restriction that all first-order interactions are 0. By assuming that each cell has the same number n of individuals, computation of sums of squares can be accomplished rather directly. The sum of squares for the columns (i.e., for the A treatment) can be most easily computed from the column totals:

$$\text{SS}_A = \sum_{k=1}^{p} \frac{(Y_{..k.})^2}{pn} - \frac{(Y_{....})^2}{p^2n}$$

The summation usage in this formula is somewhat misleading since the cell codes are included within the rows and columns of the table. Thus, the sum of the first column only involves summing over the n individuals in each of p rows of the table. Similarly:

$$\text{SS}_B = \sum_{j=1}^{p} \frac{(Y_{.j..})^2}{pn} - \frac{(Y_{....})^2}{p^2n}$$

and

$$\text{SS}_C = \sum_{i=1}^{p} \frac{(Y_{i...})^2}{pn} - \frac{(Y_{....})^2}{p^2n}$$

The sum of squares for the residual term can be found by subtraction

$$\text{SS}_{\text{res}} = \sum_{i=1}^{p} \sum_{j=1}^{p} \sum_{k=1}^{p} \frac{(Y_{ijk.})^2}{n} - \frac{(Y_{....})^2}{p^2n} - \text{SS}_A - \text{SS}_B - \text{SS}_C$$

Finally, the sum of squares for within-cells error is computed directly from the within-cell information (or can be found by subtraction after computation of a total sum of squares):

$$\text{SS}_{\text{error}} = \sum_{i=1}^{p} \sum_{j=1}^{p} \sum_{k=1}^{p} \sum_{m=1}^{n} Y_{ijkm}^2 - \sum_{i=1}^{p} \sum_{j=1}^{p} \sum_{k=1}^{p} \frac{(Y_{ijk.})^2}{n}$$

To complete the analysis-of-variance summary table and to act as a check on the sum of other effects, the total sum of squares is

$$\text{SS}_{\text{total}} = \sum_{i=1}^{p} \sum_{j=1}^{p} \sum_{k=1}^{p} \sum_{m=1}^{n} Y_{ijkm}^2 - \frac{(Y_{....})^2}{p^2n}$$

Since all factors exist at p levels, the degrees of freedom for each main effect are $p - 1$. The degrees of freedom for the residual sum of squares are $p^2 - 3p + 2 = (p - 1)(p - 2)$ [that is, among cells there are $p^2 - 1$ degrees of freedom; $p - 1$ of these are lost for each of the three main effects; the remainder is $p^2 - 1 - 3(p - 1) = p^2 - 3p + 2$].

The error term has $n - 1$ degrees of freedom per cell; thus, the degrees of freedom for the error sum of squares are $p^2(n-1)$. The total sum of squares has $np^2 - 1$ degrees of freedom. Because of the restrictive assumptions concerning interactions, the expected values for mean squares based on a Latin-square design contain only an effect due to the treatment and a random-error component. For example, assuming negligible first-order interactions,

$$E(\text{MS}_A) = \sigma^2 + np\sigma_\alpha{}^2 \qquad E(\text{MS}_B) = \sigma^2 + np\sigma_\beta{}^2$$

$$E(\text{MS}_C) = \sigma^2 + np\sigma_\gamma{}^2 \qquad E(\text{MS}_{\text{res}}) = \sigma^2 + \sigma^2_{\text{res}}$$

Of course, $E(\text{MS}_{\text{error}}) = \sigma^2$. Thus, the mean square for error is the appropriate choice for testing all main effects as well as the residual component. If negligible interactions are assumed, the interpretations of the main effects are straightforward. The test of the residual mean square presents a partial check on the magnitudes of interaction effects. This test is not sufficient to warrant the conclusion that all first-order interactions are negligible; some components of these interactions, as shown earlier, are inextricably confounded with main effects. From a practical point of view, the test of MS_{res} serves no useful experimental purpose. The analysis can be summarized as in Table 4-9.

To illustrate the computations and interpretation of Latin-square designs, fictitious data are presented for the curriculum experiment previously proposed; the design was summarized in Table 4-7. In actual practice, the standard form should be chosen randomly from the available standard forms of the relevant size. Of course, when the number of standard forms is enormous, this is impractical; in this circumstance, a random selection can be made from tables of standard Latin squares. Once a standard form has been selected, the rows and columns should be randomly permuted to arrive at the actual design. This randomization process can be easily carried out; using ordinary random-number tables, select one-digit numbers until the first p integers (for a $p \times p$

table 4-9 *Analysis-of-variance summary table for a Latin-square design*

source	df	sum of squares	mean square	F
A (columns)	$p-1$	SS_A	$\text{SS}_A/(p-1)$	$\text{MS}_A/\text{MS}_{\text{error}}$
B (rows)	$p-1$	SS_B	$\text{SS}_B/(p-1)$	$\text{MS}_B/\text{MS}_{\text{error}}$
C (letters)	$p-1$	SS_C	$\text{SS}_C/(p-1)$	$\text{MS}_C/\text{MS}_{\text{error}}$
Residual	$(p-1)(p-2)$	SS_{res}	$\text{SS}_{\text{res}}/(p-1)(p-2)$	
Error	$p^2(n-1)$	SS_{error}	$\text{SS}_{\text{error}}/p^2(n-1)$	
Total	np^2-1	SS_{total}		

Latin square) have occurred in some order; use these digits to rearrange the columns of the standard Latin square; repeat this process for both the rows and the letters to generate a random permutation of the original square. This process of randomization assures that there is no bias in the assignment of groups to treatments and introduces an additional element of randomization into the experimental design.

Starting with the standard Latin square from Table 4-7, we enter a table of random numbers and select one-digit numbers until, for example, we arrive at 3, 1, 2 as the random order for columns. By interchanging the indicated columns, the permuted Latin square is now:

$$\begin{matrix} c & a & b \\ a & b & c \\ b & c & a \end{matrix}$$

Focusing on the rows, we reenter the table of random numbers and generate the series 3, 2, 1. The square becomes:

$$\begin{matrix} b & c & a \\ a & b & c \\ c & a & b \end{matrix}$$

The final step involves permuting the letters. Assuming that the random series is 1, 3, 2, we interchange the letters b and c, but leave the letter a intact. The final random permutation of the Latin square is:

$$\begin{matrix} c & b & a \\ a & c & b \\ b & a & c \end{matrix}$$

Referring to Table 4-7, we see that this means that in school 1, the high-track students (level 1) receive the c curriculum (i.e., the second version of the new, experimental curriculum), the middle-track students receive the b curriculum (i.e., version 1 of the new, experimental curriculum), and the low-track students receive the usual course of study. Similarly, in school 2, the high track receives the usual course of study, the middle track receives version 2, and the low track version 1 of the new, experimental curriculum. The third school has version 1 of the new, experimental curriculum for the high track and the usual course of study for the middle track; the low-track students receive version 2 of the new, experimental curriculum.

To provide a randomization base within the schools, assume that three classes of students in each track were randomly selected from the available classes in each school. Thus, $n = 3$ per cell of the design. The

table 4-10 *Fictitious experiment based on a 3 × 3 Latin-square design*

		TREATMENT A (ability tracks) 1	2	3
	1	28.2 25.4 (c) 30.7	30.0 31.1 (b) 26.9	22.1 21.9 (a) 27.1
TREATMENT B (schools)	2	22.6 25.5 (a) 23.7	25.9 27.1 (c) 28.2	34.4 29.1 (b) 29.5
	3	27.7 30.7 (b) 32.4	20.9 26.5 (a) 25.3	25.5 28.4 (c) 24.4

fictitious data are displayed in Table 4-10. Sums of squares for the main effects can be found from row, column, and letter sums. The sums for each of the three levels of A are 246.9, 241.9, and 242.4. The sum of squares based on treatment A is

$$\begin{aligned} SS_A &= \frac{246.9^2}{9} + \frac{241.9^2}{9} + \frac{242.4^2}{9} - \frac{731.2^2}{27} \\ &= 19{,}803.66 - 19{,}801.98 = 1.68 \end{aligned}$$

The row sums for B are 243.4, 246.0, and 241.8. Thus, the sum of squares for B is

$$\begin{aligned} SS_B &= \frac{243.4^2}{9} + \frac{246.0^2}{9} + \frac{241.8^2}{9} - \frac{731.2^2}{27} \\ &= 19{,}802.98 - 19{,}801.98 = 1.00 \end{aligned}$$

The sums for the letters are 215.6, 271.8, and 243.8. Then the sum of squares for letters is

$$\begin{aligned} SS_C &= \frac{215.6^2}{9} + \frac{271.8^2}{9} + \frac{243.8^2}{9} - \frac{731.2^2}{27} \\ &= 19{,}977.45 - 19{,}801.98 = 175.47 \end{aligned}$$

The residual sum of squares is found by subtracting main effects from the sum of squares for the cells:

$$\begin{aligned} SS_{\text{res}} &= \frac{84.3^2}{3} + \frac{88.0^2}{3} + \cdots + \frac{78.3^2}{3} - \frac{731.2^2}{27} - SS_A - SS_B - SS_C \\ &= 19{,}988.07 - 19{,}801.98 - 1.68 - 1.00 - 175.47 \\ &= 7.94 \end{aligned}$$

The total sum of squares is found in the usual way by use of the individual scores:

$$SS_{total} = 28.2^2 + 25.4^2 + \cdots + 24.4^2 - \frac{(28.2 + 25.4 + \cdots + 24.4)^2}{27}$$
$$= 20{,}090.76 - 19{,}801.98$$
$$= 288.78$$

From this, the error sum of squares can be found by subtraction:

$$SS_{error} = SS_{total} - SS_A - SS_B - SS_C - SS_{res} = 102.69$$

These results can be summarized as in Table 4-11. Since the treatments all represent fixed dimensions, the F ratios are each set up using MS_{error}. From the F ratios, only the curricula show significant differences; neither the schools nor the ability tracks significantly affect the level of student achievement. Since there are three curricula, we must continue the analysis to determine where the significant difference for the curricula arises. The Newman-Keuls test can be used to test all pairwise contrasts on a *post hoc* basis. The means for the three curricula are $\overline{Y}_a = 23.96$, $\overline{Y}_b = 30.20$, and $\overline{Y}_c = 27.09$. Since each of these means is based on nine observations, the standard error for a mean is $\sqrt{MS_{error}/9} = \sqrt{5.70/9} = \sqrt{.6333} = .7958$. In rank order, the means are $\overline{Y}_b = 30.20$, $\overline{Y}_c = 27.09$, and $\overline{Y}_a = 23.96$. Testing adjacent pairs first, we have

$$\frac{\overline{Y}_b - \overline{Y}_c}{\sqrt{MS_{error}/9}} = \frac{30.20 - 27.09}{.7958} = \frac{3.11}{.7958} = 3.91$$

From Appendix Table B-7 for adjacent means based on 8 degrees of freedom, this value is significant at the .05 level but not at the .01 level. The second adjacent pair is

$$\frac{\overline{Y}_c - \overline{Y}_a}{.7958} = \frac{27.09 - 23.96}{.7958} = \frac{3.13}{.7958} = 3.93$$

table 4-11 *Analysis-of-variance summary table for a* 3 × 3 *Latin-square design*

source	df	sum of squares	mean square	F	p
A (ability track)	2	1.68	.84	.15	ns
B (schools)	2	1.00	.50	.09	ns
C (curricula)	2	175.47	87.74	15.41	<.01
Residual	2	7.94	3.97	.70	ns
Error	18	102.69	5.70		
Total	26	288.78			

From the table, this value is also significant between the .05 and .01 levels. The final test involves means spanning three rank positions:

$$\frac{\bar{Y}_b - \bar{Y}_a}{.7958} = \frac{30.20 - 23.96}{.7958} = \frac{6.24}{.7958} = 7.84$$

This value is significant beyond the .01 level. Thus, adopting the conventional .05 level of significance, we find that all three pairwise contrasts indicate reliable population differences between the curricula. The rank-ordering implies that the most effective procedure is version 1 of the new, experimental curriculum; the next most effective is version 2 of the new, experimental curriculum; and the standard curriculum is the least effective among those evaluated. This conclusion is tenable only to the extent that we can be assured that there is no interaction between schools and ability tracks since this interaction is confounded with the differences between curricula; if it is assumed that outside evidence points to a negligible interaction, the conclusions concerning the curricula can be given creditability.

Given the circumstances of this fictitious research study, the tests for ability tracks and schools have very limited interpretability, whether they are accepted or rejected. There is no way of assessing a possible interaction between either curriculum and ability track or curriculum and schools; if these interactions exist, they may either increase or decrease the value of the sum of squares for their alias among the main effects.

In behavioral science research, Latin-square designs are sometimes utilized in experiments involving repeated measures on the same group of subjects. This application falls within the limits of the material covered in Chapter 7, and discussion will be deferred until that point. Also, Latin squares are sometimes applied to provide some counterbalancing of effects which are incidental to the purpose of the experiment. For example, it may be necessary to conduct an experiment over several different days and at several different time periods. If the number of days and number of time periods can be made equal to the number of levels of the manipulated treatment variable, these two diurnal factors may constitute rows and columns of a Latin-square design. To make the example more concrete, suppose that there are four levels of a treatment variable that are of primary concern. The administration of these treatments to independent groups of experimental subjects might be planned along the lines of Table 4-12 to counterbalance the days and hours of the day on which treatments are applied. The counterbalancing is, however, only partial; although each treatment level is administered at each time period on only one day, the *sequence* of administrations is not counterbalanced. Thus, level b of the

table 4-12 4 × 4 *Latin-square design used to control diurnal factors*

		DAYS OF THE WEEK			
		Monday	Tuesday	Wednesday	Friday
TIME OF DAY	9 A.M.	*a*	*b*	*c*	*d*
	10 A.M.	*b*	*c*	*d*	*a*
	1 P.M.	*c*	*d*	*a*	*b*
	2 P.M.	*d*	*a*	*b*	*c*

treatment is applied later in the day than level *a* on 3 of the 4 days. Similar lacks of balance in order of treatment application apply to other pairs of levels and to higher-order sequences of treatment levels. This type of lack of balance may not be of concern in many experimental situations.

Greco-Latin squares and hypersquares

The logic of the Latin square can be easily generalized by adding dimensions to the design. A Latin square is always a $1/k$ replicate of a k^3 factorial design; if one dimension is added, the design will become a $1/k^2$ replicate of a k^4 design. When the same kind of balance that typifies the Latin square is present, the design is known as a Greco-Latin square. To represent such a design, Greek letters are imposed on the cells with the restriction that a Greek letter occur once and only once in each row and column of the design *and once and only once in combination with each Latin letter*. The construction of Greco-Latin squares is rather tedious, as the student can confirm by attempting to build a 5 × 5 square. Fortunately, some examples are presented in standard reference works (e.g., Peng, 1967, p. 104). Greco-Latin square designs are seldom found in practical applications since the requirement that all treatments exist at the same number of levels is rarely met.

To illustrate this class of designs, consider the Greco-Latin square in Table 4-13. Note that if the Greek and Latin letters are separated and displayed as at the left of Table 4-13, we obtain two Latin squares. Latin squares which, when superimposed, fulfill the requirements of a Greco-Latin square are described as *orthogonal*. In certain cases, there exist complete sets of $k - 1$ mutually orthogonal Latin squares of order $k \times k$.

The analysis of a Greco-Latin square design follows the same lines as that for a Latin-square design, except that a sum of squares for Greek letters must be computed. This sum of squares is found from the sums of scores for each of the Greek letters. Otherwise, the formulas presented for Latin-square designs may be utilized without alteration.

table 4-13 *Example of a 4 × 4 Greco-Latin square design*

entering Latin squares								Greco-Latin square			
A	B	C	D	α	γ	δ	β	$A\alpha$	$B\gamma$	$C\delta$	$D\beta$
B	A	D	C	β	δ	γ	α	$B\beta$	$A\delta$	$D\gamma$	$C\alpha$
C	D	A	B	γ	α	β	δ	$C\gamma$	$D\alpha$	$A\beta$	$B\delta$
D	C	B	A	δ	β	α	γ	$D\delta$	$C\beta$	$B\alpha$	$A\gamma$

If each treatment exists at k levels, the rows, columns, Latin letters, and Greek letters each have $k - 1$ degrees of freedom. The residual term has $(k - 1)(k - 3)$ degrees of freedom, and there are $k^2(n - 1)$ degrees of freedom for the within-cells-error sum of squares. There is, of course, confounding of main effects and interactions; however, it is even more severe than in Latin squares. Each main effect is confounded with first-order interactions and second-order interactions; the residual term is a mixture of components also due to these interactions.

The Latin-square principle can be further generalized to produce *hypersquares*. For example, one dimension can be added to a Greco-Latin square to form a $1/k^3$ replicate of a k^5 factorial design. An example of such a design is presented in Table 4-14; numbers have been added to represent the fifth treatment dimension. Needless to say, such a design has very limited practical applicability. Note that a hypersquare is a series of superimposed Latin squares; the entering Latin squares are three from a set of mutually orthogonal Latin squares of the given size. The analysis, once again, closely parallels that for a Latin square and will be left as a problem for the interested student. There is, of course, confounding: each main effect is confounded with first-, second-, and third-order interactions.

table 4-14 *Example of a 4 × 4 hypersquare*

$A\alpha 1$	$B\beta 2$	$C\gamma 3$	$D\delta 4$
$B\gamma 4$	$A\delta 3$	$D\alpha 2$	$C\beta 1$
$C\delta 2$	$D\gamma 1$	$A\beta 4$	$B\alpha 3$
$D\beta 3$	$C\alpha 4$	$B\delta 1$	$A\gamma 2$

chapter five

randomized block designs

introduction

A fundamental principle for increasing the precision of an experiment is to employ homogeneous material; that is, the experimental subjects should be highly similar to one another so that only a small part of their responses is a product of individual characteristics, the bulk being due to the effects of the treatments. In the physical sciences, and other more exact scientific areas, the use of homogeneous material in experiments is so refined that statistical methods of analysis are rarely required since the amount of experimental error is negligible. In the behavioral sciences, the situation is completely opposite. The behavioral scientist must work with relatively heterogeneous material, and experimental error plays a fundamental role in data analysis. The category of experimental designs referred to as *randomized blocks* provides some control over the diversity of experimental subjects. In these designs, one or more characteristics of the subjects which are known to be related to performance in the experimental situation are explicitly taken into account in the structure of the design. In effect, these variables operate in the same capacity as treatment variables, although they are not directly manipulated by the experimenter.

Structurally, randomized blocks designs are identical with factorial designs. That is, the experiment consists of two or more dimensions, and all possible combinations of these dimensions are applied to independent, random groups of experimental subjects. However, in the randomized blocks designs one or more of the dimensions represent classification variables used to group subjects into relatively homogeneous subclasses. In some experiments, the classification dimensions, or *blocking variables*, are introduced solely for the purpose of increasing the precision of the experiment relative to the manipulated treatment dimensions. In other cases, these blocking variables may be of intrinsic interest. For example, in an experiment concerned with classroom learning, the researcher may block students on the basis of previous academic performance since he is aware that new learning will be related to previous performance. He may not, however, be at all interested in studying this relationship per se, since it is well established; his aim is merely to introduce homogeneous subgroups in order to increase the precision of comparisons involving manipulated treatment variables. On the other hand, in the same experiment, the researcher may block students by sex; in this instance, his concern may be with whether or not performance is differential for the two sexes and whether or not sex interacts with other treatment dimensions.

In this chapter, any design which is structurally a factorial design but in which one or more of the dimensions is a classification variable is called a randomized blocks design. In other sources, the student may encounter different terminology. For example, an experiment involving one manipulated treatment dimension and one classification dimension is sometimes called a *levels-by-treatment* design; if there are two manipulated dimensions and one classification dimension, the design is sometimes called a *levels-by-treatment-by-treatment* design; and so forth.

A randomized blocks design may be conceptualized as a basic factorial design to which one or more classification dimensions have been appended. The researcher may, for example, have planned a 2×2 factorial experiment in which both the treatment dimensions represent manipulated variables. However, it may also be relevant to study the sex of experimental subjects since this may influence their performance. Thus, sex may be added as a third dimension of the design; the resulting design is structurally identical with a 2^3 factorial design since both male and female subjects will be administered each of the four treatment combinations from the basic 2×2 factorial design.

It is important to clarify the manner in which the introduction of one or more blocking dimensions may increase the precision of an experiment. Consider the simplest experimental design, the completely randomized design. In its basic form, the variability of scores is due either to effects from the treatment levels or to experimental error. Within any one treatment group, the only source of variability among scores is experimental error. Of course, the smaller the amount of experimental error, the more sensitive will be the test for treatment effects, since MS_{error} will be estimating a smaller variance. If a blocking dimension is added to the CRD, there will be formed, within each treatment group and across all treatment groups, relatively homogeneous subclasses of experimental subjects. The sum of squares associated with the blocking variable is removed directly from experimental error, and the treatment effects are unchanged. Thus, the MS_{error} is smaller, and the reduction in the amount of experimental error is related to the degree to which the blocking classifications do, in fact, result in homogeneous subclasses. The formulation of this relationship in more precise terms is discussed later in this section.

For a basic CRD, the reduction in experimental error which may occur from the introduction of a blocking variable can be inferred from the two designs represented in Table 5-1. The first design is the CRD; the error mean square is computed by pooling the variances from the separate treatment groups. These variances are S_1^2, S_2^2, and S_3^2. Assume that the subjects in the treatment groups differ in terms of some outside variable which is related to performance in the experiment (e.g., intelligence, anxiety, or sociability). Since subjects are randomly assigned to the treatment groups, this outside variable is randomly distributed over the groups. Thus, some of the variability *within* groups

table 5-1 *Accounting of variability in a completely randomized design and a randomized blocks design*

COMPLETELY RANDOMIZED DESIGN			RANDOMIZED BLOCKS DESIGN				
	treatment					treatment	
1	2	3			1	2	3
S_1^2	S_2^2	S_3^2	BLOCKS	1	S_{11}^2	S_{12}^2	S_{13}^2
				2	S_{21}^2	S_{22}^2	S_{23}^2

is due to differences in this outside variable; in the CRD, however, it is unaccounted for in any explicit sense and merely becomes part of the within-cell-error variance. Consider, now, the second design in Table 5-1. For simplicity, two levels of the outside variable are shown as comprising two blocks of experimental subjects (e.g., students in a classroom learning experiment have been blocked into two levels of intellectual ability, say high and low). Within treatment level 1 there are two cells corresponding to the two levels of the blocking variable. If the blocking is successful, then S_{11}^2 and S_{21}^2 will each be smaller than the variance in the CRD, S_1^2. A similar relationship will hold for levels 2 and 3 of the treatment dimension. In the second design, the error mean square is a result of pooling the within-cell variances S_{11}^2, S_{12}^2, S_{13}^2, and so forth. Thus, the introduction of the blocking variable will reduce the size of experimental error in the sense that MS_{error} will be estimating a smaller quantity. As will be demonstrated later, the sum of squares for error is reduced by an amount equal to the *among-blocks* variability. The total variability of scores in the design is the same as for the CRD; it is, however, allocated differently. That is, the total variability is split among treatment, blocks, and error, whereas in the CRD it is split between only treatment and error. Of course, more than one blocking dimension can be added to an experimental design. From a design point of view, the researcher must search for blocking variables which will, in fact, result in more homogeneous subgroups than would occur through the use of completely randomized groups. This consideration is mitigated somewhat by the fact that in some applications the researcher may introduce a blocking variable not so much to reduce error variability but because he is basically interested in the effects of this blocking variable. In this event, the design is essentially no different from a factorial design except that the blocking dimension is a selection rather than a manipulated variable.

When the intent is to reduce error variability, the choice of blocking variables and the decision concerning how to establish the levels of such variables become critical. Often a variable which a researcher wishes to utilize as a blocking variable is continuous and not naturally occurring in discrete classes. Thus, arbitrary cutting points must be established to form the levels for a randomized blocks design.[1] These two problems, choosing outside variables for blocking purposes and then establishing cutting points, are discussed in turn.

The major considerations in the choice of blocking variables are their relationship to the criterion variable and their availability prior to the formation of the experimental treatment groups. Randomization in a randomized blocks design must involve random assignment of subjects within each block to the various treatment levels, or combinations of treatment levels. For example, the design depicted in Table 5-1 requires that the subjects in block 1 be randomly assigned to the levels of the treatment and, similarly, that the subjects in block 2 be randomly assigned to the levels of the treatment. This can be accomplished only if the blocks are formed prior to actually setting up the experimental conditions required for application of the treatment variable. Thus, the measurements which form the basis for blocking subjects must be taken before the experiment is executed. In education, this may mean setting up special testing sessions, or it may only require going to available school records (e.g., prior achievement scores, intelligence scores, etc., which are frequently used for blocking, are generally available from student cumulative records). The choice of a blocking variable should be made on the basis of its predictive value with respect to the criterion variable. In general, the greater the degree of correlation between the blocking variable and the criterion variable, the greater will be the increase in power of the randomized blocks design relative to the completely randomized design. The direction of relationship is irrelevant; that is, a positive or a negative correlation can typify a satisfactory blocking variable. Thus, the researcher should select a blocking variable which is most likely to bear a close relationship to the criterion variable. An obvious choice is the use of a premeasure which is either the measure of the criterion variable itself or some closely parallel form of that measure. For example, in a classroom learning study, a pretest may be given to all subjects; after blocks are formed and the experimental treatments applied, the same test may be used as the measure of the criterion. In most practical situations, the use of a blocking variable which is highly similar to or identical with the measure of the criterion is a simple way of assuring a high degree of correlation between the blocking and criterion variables (indeed, the treatment effects would be very unusual if this procedure did not produce the desired homogeneity of subjects within subgroups; only if the criterion measure and pretest were inherently very unreliable could problems arise). Ordinarily, the

[1]An alternative procedure is to use the outside variable in a regression model to reduce within-cell-error variability. This technique is known as the analysis of covariance and is treated in Chapter 8.

researcher's knowledge concerning the criterion measure will suggest one or more relevant blocking variables. If the experiment is focused on some aspect of academic performance involving achievement, the researcher can select or build a relevant aptitude or preachievement instrument; measures of general intelligence are commonly used, although a measure of a specific and relevant aptitude is generally a better predictor of academic criteria. Studies in which the criterion involves personality measures can pose special problems. Pretesting with a personality test may, in some experimental conditions, result in a sensitization of the experimental subjects with respect to the purposes of the experiment. For example, if the criterion test were a measure of anxiety, pretesting with an anxiety test might be a cue to the subjects that the purpose of the experiment was to have an effect on their responses to the anxiety scale. This is an important consideration since personality tests are generally susceptible to faking and facade effects which do not pose problems when achievement is being measured. It is, of course, possible to assess experimentally the operation of sensitization in experiments of this type (the Solomon four-group design, described in Campbell and Stanley, 1963, is an experimental design especially constructed to provide control for sensitization effects in personality-change experiments).

If the blocking variable is discrete, establishing the categories for a randomized blocks design poses no problem since the natural categories themselves can be used. However, if the blocking variable is continuous, it must be divided into discrete categories in order to form the levels of the blocking dimension in the randomized blocks design. Two opposing considerations become relevant: degrees of freedom associated with the blocking variable are lost from error, and the larger the number of levels of the blocking dimension, the greater will be the homogeneity of subjects within levels. The degrees of freedom for mean squares in a randomized blocks design are found in the same way as for factorial designs. Thus, if one blocking variable is added to a CRD, the resulting design is formally equivalent to a two-dimensional factorial design. If we call the treatment dimension A and the blocking dimension B, the effects are A, B, and AB. If A exists at p levels and B at q levels, the degrees of freedom for treatment are $p - 1$, for blocks are $q - 1$, and for interaction are $(p - 1)(q - 1)$. Assuming n observations per cell, the error sum of squares has $pq(n - 1)$ degrees of freedom. In the corresponding CRD, treatment has $p - 1$ degrees of freedom; also, to make the total number of observations equal to that in the randomized blocks design, we must assign qn observations to each treatment. Thus, the error sum of squares has $p(qn - 1)$ degrees of freedom. Note that the combined degrees of freedom for blocks, interaction, and error from the randomized blocks design are

$$(q-1)+(p-1)(q-1)+pq(n-1)$$
$$= (q-1)+(pq-p-q+1)+(pqn-pq)$$
$$= pqn - p = p(qn-1)$$

The error term in the randomized blocks design has

$$(q-1)+(p-1)\,(q-1) = p(q-1)$$

fewer degrees of freedom than the error term in the corresponding completely randomized design. If we remember that q is the number of levels of the blocking variable, it is apparent that the loss of degrees of freedom from error is linearly related to q. Thus, if the researcher chooses q very large in order to assure highly homogeneous levels of the blocking dimension, he is, at the same time, sacrificing large numbers of degrees of freedom from the error term. Although not immediately apparent, there is an optimal number of levels of the blocking dimension for various sample sizes and various numbers of levels of A, the treatment variable. Utilizing the concept of *apparent imprecision*, Feldt (1958) has tabled some of these optimal conditions. The apparent imprecision of a randomized blocks design is defined as the ratio of the sampling error averaged over blocks to the theoretically minimum sampling error; this ratio is adjusted in terms of degrees of freedom. The values derived by Feldt, which are reproduced in Table 5-2, were chosen so as to minimize the apparent imprecision. Table 5-2 shows the optimal number of levels of a single blocking variable in terms of the total size of the experiment (that is, n_t is the total number of subjects in the experiment; it is assumed that each cell of the design has an equal number of observations, or as nearly equal as possible for the selected values of n_t), the correlation between the blocking and

table 5-2 *Optimal numbers of levels for a blocking variable*

		n_t					
r	levels of A	20	30	50	70	100	150
.2	2	2	3	4	5	7	9
	5	1	2	2	3	4	6
.4	2	3	4	6	9	13	17
	5	2	3	4	5	7	10
.6	2	4	6	9	13	17	25
	5	2	3	5	7	9	14
.8	2	5	7	12	17	23	25
	5	2	3	5	7	10	15

criterion variable (r), and the number of levels of the treatment dimension (i.e., either two or five as examples).

Three features of Table 5-2 are readily apparent. First, for experiments having a larger number of subjects, there is an increase in the optimal number of levels for the blocking variable. Second, as the correlation between the blocking and criterion variables increases, the optimal number of blocks increases. And third, greater numbers of levels for blocks are optimal for smaller numbers of levels of the treatment dimension.

The remainder of this chapter is concerned with some specific circumstances involving randomized blocks. We begin by considering the case in which each cell of the design contains only one observation. Next, we extend this to cases having more than one observation per cell, and then turn to a general consideration of appending one or more blocking dimensions to a basic factorial design. Finally, we discuss cases in which the available block sizes are too small to accommodate the complete factorial design; this problem can be circumvented by the use of *incomplete-block designs*, which, however, may involve confounding of effects.

designs with one replicate per cell

In this section we assume that the researcher is interested in utilizing a completely randomized design (or one-way analysis of variance) but also wishes to reduce experimental error by introducing a single blocking variable. Furthermore, complete matching of individuals across the levels of the treatment dimension is planned. That is, there is a supply of n_t subjects to be assigned to the p levels of the treatment variable. In order for matching to be effected, n_t must be a multiple of p (for example, if three levels of treatment exist, $n_t/3$ must be an integer; allowable values would include 6, 9, 12, etc.). Prior to forming subjects into treatment groups, the subjects are rank-ordered on the basis of their scores on the matching variable. The first p individuals in the ranking constitute the first block in the experiment; the second group of p individuals from the ranking constitute the second block; and so forth. Within each block of p individuals, the subjects are randomly assigned to a specific level of the treatment. In effect, this procedure produces p-tuples of matched subjects. For reference, the number of p-tuples is q (that is, $n_t/p = q$), and the blocking dimension will be called the B variable. The resulting design is formally equivalent to a two-dimensional factorial design with one observation per cell. However, the design differs from a factorial design in the important respect of not involving completely random assignment of subjects to treatment combinations. The assignment of a subject to a block (or p-tuple) is on the basis of his relative standing on the blocking variable, which cannot be assumed to represent a random process (e.g., prior achievement in science may constitute the blocking variable; it would not be reasonable to contend that this achievement is a random process); however, random assignment to treatment levels within blocks is basic to the design.

To illustrate the procedure for forming matched p-tuples, consider an experiment in which the treatment of interest involves three variations of procedure in teaching simple number conservation concepts to elementary school children. Although the experiment can be carried out in a single experimental session, administrative convenience dictates the use of intact classroom groups of first-graders. In this circumstance, the proper experimental unit is the mean performance of classroom groups. Assume, further, that the researcher has 15 classrooms which have been made available for the experiment. Thus, the task is to form five triples of classrooms based on a relevant blocking variable. Since the experimental tasks involve number concepts, a premeasure of numerical aptitude is obtained from students in all 15 classrooms. Since the experimental unit is the classroom, the mean score per classroom on the premeasure is utilized to rank the classrooms. These data are presented in Table 5-3. In order to form the triples of classrooms, we rank-order the classrooms by their mean aptitude scores. Scanning the list of means, we see that the three classrooms with the largest mean scores are 7, 2, and 6. Thus, these three classrooms form our first block. These must now be randomly assigned to the treatment levels. This might be done by simply writing their numbers on slips of paper and drawing from a bowl or by going to a table of random numbers and selecting one-digit numbers until 7, 2, and 6 turn up in some order. Returning to Table 5-3, we see that the second triple of classrooms would be 11, 4, and 13 since they constitute ranks 4 through 6. The classrooms comprising each of the five triples are 7, 2, and 6; 11, 4 and 13; 9, 12, and

table 5-3 *Premeasure means on a matching variable for 15 classrooms*

classroom	mean numerical aptitude score	classroom	mean numerical aptitude score
1	63.7	9	77.1
2	81.4	10	71.1
3	71.5	11	80.6
4	80.4	12	74.3
5	69.9	13	78.4
6	81.3	14	66.9
7	90.0	15	73.1
8	65.4		

table 5-4 *Notation for a randomized blocks design with one replicate per cell*

		TREATMENT (A)					
		1	2	3	$\cdots$	p	sums
BLOCKS (B)	1	Y_{11}	Y_{12}	Y_{13}	$\cdots$	Y_{1p}	$Y_{1.}$
	2	Y_{21}	Y_{22}	Y_{23}	$\cdots$	Y_{2p}	$Y_{2.}$
	$\cdots$	$\cdots$	$\cdots$	$\cdots$	$\cdots$	$\cdots$	$\cdots$
	q	Y_{q1}	Y_{q2}	Y_{q3}	$\cdots$	Y_{qp}	$Y_{q.}$
Sums		$Y_{.1}$	$Y_{.2}$	$Y_{.3}$	$\cdots$	$Y_{.p}$	$Y_{..}$

15; 3, 10, and 5; and 14, 8, and 1. In the event of identical, or tied, scores on the blocking variable, the researcher must randomly break the tie in assigning rank positions. Although this makes no difference if the tied scores fall completely within one p-tuple, it does assure an arbitrary assignment if the tied scores span more than one block.

In a randomized blocks design, the blocking variable plays no further role once the blocks have been formed. In the present example, the information from the premeasure of numerical aptitude is only of interest for the purpose of forming the five triples of classrooms.

The analysis of a randomized blocks design with one replicate per cell is identical with that for a factorial design. There is, of course, no within-cell-error term estimable from the data since there is only one score per cell. With the notation of Table 5-4, the sums of squares are

$$SS_A = SS_{\text{treat}} = \sum_{j=1}^{p} \frac{(Y_{.j})^2}{q} - \frac{(Y_{..})^2}{pq}$$

$$SS_B = SS_{\text{blocks}} = \sum_{i=1}^{q} \frac{(Y_{i.})^2}{p} - \frac{(Y_{..})^2}{pq}$$

$$SS_{AB} = \sum_{i=1}^{q} \sum_{j=1}^{p} Y_{ij}^2 - \frac{(Y_{..})^2}{pq} - SS_A - SS_B$$

$$SS_{\text{total}} = \sum_{i=1}^{q} \sum_{j=1}^{p} Y_{ij}^2 - \frac{(Y_{..})^2}{pq}$$

The analysis-of-variance summary table appears as in Table 5-5.

In a randomized blocks design, the null hypothesis of major interest is the one involving treatment effects. The blocking variable has been introduced because of a known relationship to the criterion variable. Thus, we expect that the blocks will reveal large differences. Indeed, if they do not, we have, apparently, been unsuccessful in forming homogeneous subgroups of experimental subjects by means of the blocking

table 5-5 *Analysis-of-variance summary table for a randomized blocks design with one replicate per cell*

source	df	sum of squares	mean square
(A) Treatment	$p-1$	SS_{treat}	$SS_{treat}/(p-1)$
(B) Blocks	$q-1$	SS_{blocks}	$SS_{blocks}/(q-1)$
(AB) Interaction	$(p-1)(q-1)$	SS_{AB}	$SS_{AB}/(p-1)(q-1)$
Total	$pq-1$	SS_{total}	

procedure. A major concern in randomized blocks designs is the nature of the dimensions, that is, whether they represent random or fixed effects. In most applications, the treatment dimension will be fixed since the researcher will apply those levels of the treatment which are of particular interest to him. The blocking dimension, however, is often a random-effects dimension. Consider the example which involved assigning 15 classrooms among blocks (Table 5-3). The cutting points which divide the blocks were not preset by the researcher; they depend upon the rank ordering of the classroom means on the premeasure of numerical aptitude. Although these cutting points were not, strictly speaking, determined by a random process, it is clear that they are largely arbitrary from the point of view of the researcher. Thus, we should consider the design to represent a mixed model with the treatment levels fixed but the blocks random. There are cases, of course, in which the levels for blocking would be predetermined, and the resulting design would represent a fixed-effects model. For example, if subjects were blocked by sex, this clearly would represent a fixed blocking dimension. The expected values for mean squares under the fixed-effects, random-effects, and mixed models are shown in Table 5-6 (the mixed design with treatments random and blocks fixed is not shown; it is rare in practice, but its structure can be easily inferred from the mixed model which is shown).

table 5-6 *Expected values of mean squares for the randomized blocks design with one replicate per cell*

mean square	fixed-effects model	random-effects model	mixed model (treatments fixed; blocks random)
Treatment (A)	$\sigma^2 + q\sigma_\alpha^2$	$\sigma^2 + \sigma_{\alpha\beta}^2 + q\sigma_\alpha^2$	$\sigma^2 + \sigma_{\alpha\beta}^2 + q\sigma_\alpha^2$
Blocks (B)	$\sigma^2 + p\sigma_\beta^2$	$\sigma^2 + \sigma_{\alpha\beta}^2 + p\sigma_\beta^2$	$\sigma^2 + p\sigma_\beta^2$
Interaction (AB)	$\sigma^2 + \sigma_{\alpha\beta}^2$	$\sigma^2 + \sigma_{\alpha\beta}^2$	$\sigma^2 + \sigma_{\alpha\beta}^2$

Since there is no within-cell variability of scores (that is, there is only one score per cell and, hence, no basis for estimating σ^2), the only available term which might serve as a denominator for F ratios is interaction. However, for the fixed-effects model, neither treatments nor blocks may be legitimately tested against interaction, as is apparent from the expected values for the mean squares. The use of interaction in the denominator of an F ratio would produce a distribution which has an expected value less than 1 when the null hypothesis is true. Thus, if the usual F values from tables were utilized for decisions concerning the null hypothesis, too few rejections would occur. On the other hand, if a null hypothesis were rejected, we could be assured that the actual level of significance is less than the nominal value selected for the test. In this sense the test would be "conservative" and might be utilized if the researcher were willing to sacrifice power.

The random-effects model has a straightforward analysis since the interaction mean square is relevant to test both the treatments and the blocks effects. In the mixed model, there is a valid test for treatments but not for blocks (although a conservative test could be set up using interaction). Since our major concern is with the treatment dimension, this lack of an exact test for blocks is not a serious drawback to the design. Also, when blocking is at all successful, the mean square for blocks will be large, and rejection will probably occur even with the conservative test.

The expected values of the mean squares for the various possible models reveal that the only models having completely adequate analyses are the random-effects model and the mixed model with fixed levels for the treatments. If both treatment levels and blocks are fixed, only a conservative test of either effect exists. Fortunately, in practice, the mixed model represented in Table 5-6 is the one commonly encountered when there is only one observation per cell in a randomized blocks design.

Although the computational procedures are quite simple, this design bears illustration in order to reemphasize the randomization requirements. Let us return to the hypothetical situation for which premeasure means are reported in Table 5-3. These means have been rank-ordered and the matched triples of classrooms formed by taking consecutive sets of three classrooms from the rank ordering. Assume that the treatment variable involves three ways of teaching conservation principles with respect to simple numerical concepts. These methods, coded 1, 2, and 3, are presented to the intact classroom groups via a standardized procedure, and the criterion measure involves a reasoning test, orally administered 10 minutes after completion of the instructional period. Table 5-7 presents the response data; each cell entry is the mean score for one classroom of student participants.

table 5-7 *Response data from a fictitious experiment involving a randomized blocks design with one replicate per cell*

		TREATMENT (A) 1	2	3	sums
BLOCKS (B)	1	21.4	24.2	20.0	65.6
	2	19.9	21.6	19.0	60.5
	3	19.6	19.4	16.3	55.3
	4	17.4	19.1	14.7	51.2
	5	18.1	18.3	14.5	50.9
Sums		96.4	102.6	84.5	283.5

By the reasoning presented earlier, the cutting points for the blocking variable are arbitrary from the experimenter's point of view, and we may consider the blocking dimension to be a random dimension. The treatment levels are, of course, fixed by the experimenter. The model is, then, a mixed model of the type shown in Table 5-6. The only hypothesis of real interest involves the treatment variable: $H_0: \alpha_1 = \alpha_2 = \alpha_3 = 0$. The hypothesis for the blocks, $H_0: \beta_1 = \cdots = \beta_5 = 0$, can be tested by utilizing a conservative procedure if this is of interest to the researcher. There is, of course, no possible test of an interaction effect in this design since the interaction term itself is serving in lieu of an error term.

Sums of squares can be computed most easily from marginal sums.

$$\begin{aligned} \text{SS}_A = \text{SS}_{\text{treat}} &= \frac{96.4^2}{5} + \frac{102.6^2}{5} + \frac{84.5^2}{5} - \frac{283.5^2}{15} \\ &= \frac{26{,}959.97}{5} - \frac{80{,}372.25}{15} \\ &= 5{,}391.99 - 5{,}358.15 \\ &= 33.84 \end{aligned}$$

$$\begin{aligned} \text{SS}_B = \text{SS}_{\text{blocks}} &= \frac{65.6^2}{3} + \cdots + \frac{50.9^2}{3} - \frac{283.5^2}{15} \\ &= \frac{16{,}233.95}{3} - 5{,}358.15 \\ &= 5{,}411.32 - 5{,}358.15 \\ &= 53.17 \end{aligned}$$

In order to find SS_{total}, and then SS_{AB} by subtraction, we need the total sum of squared scores. This is,

$$\sum_{i=1}^{5} \sum_{j=1}^{3} Y_{ij}^2 = 5{,}449.79$$

Then

$$SS_{total} = 5{,}449.79 - \frac{283.5^2}{15} = 5{,}449.79 - 5{,}358.15 = 91.64$$

By subtraction,

$$SS_{AB} = SS_{total} - SS_A - SS_B = 91.64 - 33.84 - 53.17 = 4.63$$

These results can be summarized in an analysis-of-variance summary table (Table 5-8). The F ratios were both computed by using the interaction mean square in the denominator (remember that this results in a conservative test for the blocks effect). The treatment effect is significant at beyond the .001 level, as is the blocks effect. Since there are three levels of treatment, additional analysis is required to locate specific differences. To illustrate the procedure in this application, let us assume that level 3 of the treatment was a control condition, and levels 1 and 2 were experimental variations. Further, assume that the researcher has *preplanned* the following contrasts involving the treatments: $(\bar{Y}_{.1} - \bar{Y}_{.2})$ and $\frac{1}{2}(\bar{Y}_{.1} + \bar{Y}_{.2}) - \bar{Y}_{.3}$. The patterns of coefficients for these contrasts are

$$1 \quad -1 \quad 0$$

$$\tfrac{1}{2} \quad \tfrac{1}{2} \quad -1$$

Each contrast satisfies the definition of a contrast since the sum of coefficients is equal to 0, and the two contrasts are orthogonal since the sum of their cross products is 0. Since MS_{AB} was used to test the effects of the treatments, this same term is used in lieu of an error mean square in computing F ratios for the 1 degree of freedom contrasts. From the column sums in Table 5-7, the means for the three levels of treatment are 19.28, 20.52, and 16.90. The sums of squares for the two

table 5-8 *Analysis-of-variance summary table for the fictitious randomized blocks design*

source	*df*	sum of squares	mean square	*F*	*p*
Treatment	2	33.84	16.92	29.23	$<.001$
Blocks	4	53.17	13.29	22.96	$<.001$
Interaction	8	4.63	.58		
Total	14	91.64			

contrasts are (note that $n = 5$ in this application since each column mean is based on 5 observations)

$$\mathrm{SS}_{C_1} = \frac{5(19.28 - 20.52)^2}{1^2 + 1^2} = \frac{7.6880}{2} = 3.84$$

$$\mathrm{SS}_{C_2} = \frac{5[(19.28 + 20.52)/2 - 16.90]^2}{\frac{1}{2}^2 + \frac{1}{2}^2 + 1^2} = \frac{5(3.00)^2}{1.5} = \frac{45.00}{1.5} = 30.00$$

Note that the sum of these two contrast sums of squares checks with the treatment sum of squares from the analysis-of-variance summary table (that is, $\mathrm{SS}_{C_1} + \mathrm{SS}_{C_2} = 3.84 + 30.00 = 33.84$, and $\mathrm{SS}_{\text{treat}} = 33.84$).

The tests of significance for the contrasts are $F = 3.84/.58 = 6.63$, which, with 1 and 8 degrees of freedom, is significant at the .05 but not at the .01 level of significance; and $F = 30.00/.58 = 51.72$, which is significant at well beyond the .001 level. In terms of the original sum of squares for treatment, the contrast between the two experimental treatment levels (1 and 2) accounts for about 11 percent, and the contrast of these two levels with the control level (3) accounts for about 89 percent.

Although an analysis similar to that for treatments could be carried out on the block means, this is generally of little interest. In the present instance, the means over the five levels of the blocking dimension (21.87, 20.17, 18.43, 17.07, and 16.97) are in exactly the predictable rank order, and the researcher would probably not wish to pursue additional analysis. Also, note that the mean square for blocks was highly significant even on the basis of a conservative test of significance.

During the analysis, no mention was made of testing the homogeneity-of-variance assumption. Because of the lack of any within-cell variability, there is no estimate of an error mean square and no separate within-cell-error terms to pool. Thus, it becomes irrelevant to test for homogeneity.

The discussion and analysis of the randomized blocks design with one observation per cell has focused on the case with one treatment variable and one blocking variable. The extension to designs with two or more treatment dimensions and/or two or more blocking dimensions is relatively simple if the analogy with factorial designs is kept in mind. As long as there is only one observation per cell, there will be no within-cell-error term. However, sums of squares for all main effects and interaction effects can be computed by any standard procedure (e.g., summary table method or method of orthogonal contrasts). In order to set up valid F ratios, the researcher must determine the expected values of all mean squares since there will ordinarily be a mixture of random and fixed effects. The computational procedures do not bear illustration since they are identical with those for factorial designs, to which the user may refer. However, some comments on expected values will prove useful. If all blocking variables are random and all treatment variables are fixed, there may or may not be appropriate terms to use for testing all the effects which are of interest. We shall consider two examples and urge the reader to try other combinations. Of primary importance is the determination of expected values prior to actually implementing a design; a design which appears perfectly reasonable may have no totally satisfactory analysis. In this instance, the researcher may wish to consider other possible designs. Both of our examples involve three-dimensional designs, and their expected values are displayed in Table 5-9.

The first example is essentially a two-dimensional factorial design with a blocking variable added to the design. Valid tests for both treatment variables are possible by utilizing the first-order interaction of the treatment with the blocking variable. Thus, the mean square for A is tested against the AC interaction, and the mean square for B is tested against the BC interaction. There is also a valid test of the AB interaction utilizing the second-order interaction ABC. The remaining effects, each of which involves the random blocking dimension, cannot be tested against entirely appropriate terms. A conservative test of the blocking variable C could be set up by utilizing either the AC, BC, or ABC interaction mean square. Since it is more likely in practice for the higher-order interaction to be small, the use of ABC would be the best choice.

The second example presented in Table 5-9 poses analytical difficulties. The design is basically a one-dimensional analysis of variance to which have been added two blocking dimensions. In this case, there

table 5-9 *Two examples of three-dimensional designs involving randomized blocks*

A and B fixed treatment variables; C a random blocking variable		A a fixed treatment variable; B and C random blocking variables	
mean square	expected value	mean square	expected value
A	$\sigma^2 + q\sigma_{\alpha\gamma}^2 + qr\sigma_\alpha^2$	A	$\sigma^2 + \sigma_{\alpha\beta\gamma}^2 + r\sigma_{\alpha\beta}^2 + q\sigma_{\alpha\gamma}^2 + qr\sigma_\alpha^2$
B	$\sigma^2 + p\sigma_{\beta\gamma}^2 + pr\sigma_\beta^2$	B	$\sigma^2 + p\sigma_{\beta\gamma}^2 + pr\sigma_\beta^2$
C	$\sigma^2 + pq\sigma_\gamma^2$	C	$\sigma^2 + p\sigma_{\beta\gamma}^2 + pq\sigma_\gamma^2$
AB	$\sigma^2 + \sigma_{\alpha\beta\gamma}^2 + r\sigma_{\alpha\beta}^2$	AB	$\sigma^2 + \sigma_{\alpha\beta\gamma}^2 + r\sigma_{\alpha\beta}^2$
AC	$\sigma^2 + q\sigma_{\alpha\gamma}^2$	AC	$\sigma^2 + \sigma_{\alpha\beta\gamma}^2 + q\sigma_{\alpha\gamma}^2$
BC	$\sigma^2 + p\sigma_{\beta\gamma}^2$	BC	$\sigma^2 + p\sigma_{\beta\gamma}^2$
ABC	$\sigma^2 + \sigma_{\alpha\beta\gamma}^2$	ABC	$\sigma^2 + \sigma_{\alpha\beta\gamma}^2$

are valid tests for the blocking variables but not for the treatment variables. That is, B and C can both be tested against the BC interaction term. Also, AB and AC can both be tested against the second-order interaction ABC. There is, however, no mean square with an expectation appropriate to test the main effect for A. It should be emphasized that the use of any of the interaction terms would *not* be a conservative procedure since the expectation for A contains additional terms not included in these interactions. A quasi mean square can be produced from AB, AC, and ABC which does have the proper terms in its expectation. That is, the sum $\text{MS}_{AB} + \text{MS}_{AC} - \text{MS}_{ABC}$ has as its expectation

$$(\sigma^2 + \sigma^2_{\alpha\beta\gamma} + r\sigma_{\alpha\beta}{}^2) + (\sigma^2 + \sigma^2_{\alpha\beta\gamma} + q\sigma_{\alpha\gamma}{}^2) - (\sigma^2 + \sigma^2_{\alpha\beta\gamma})$$
$$= \sigma^2 + \sigma^2_{\alpha\beta\gamma} + r\sigma_{\alpha\beta}{}^2 + q\sigma_{\alpha\gamma}{}^2$$

This expression differs from the expectation for A only in its lack of the term specific to A, $qr\sigma_\alpha{}^2$. Thus, the ratio $\text{MS}_A/(\text{MS}_{AB} + \text{MS}_{AC} - \text{MS}_{ABC})$ provides a quasi F ratio. The appropriate degrees of freedom for referring this quasi F ratio to a standard table of the distribution of F are presented in the chapter on factorial designs. The only remaining effect in the second example in Table 5-9 is the BC interaction. A conservative test of this term is possible by using the second-order interaction ABC. Note that in this example, the existence of a within-cells-error term would contribute little toward providing a more satisfactory analysis. Although there would be valid tests for both BC and ABC, it would still be necessary to utilize a quasi mean square to test the treatment variable A.

designs with more than one replicate per cell

The matching of subjects across experimental treatments results in as nearly comparable p-tuples of individuals as is possible on the basis of the blocking variable. However, in many research situations this degree of matching may be undesirable because of practical difficulty in obtaining complete data on the blocking variable or because of the heavy loss of degrees of freedom. In the one-replicate-per-cell design there is no direct estimate of experimental error, and this may cause difficulties in constructing valid tests of hypotheses. The category of designs considered in this section involves matching, but by groups of subjects rather than on the basis of individual subjects. Randomized blocks designs with more than one replicate per cell provide data for estimating experimental error and also provide some basis for testing interactions involving the blocking and treatment variable. Such interaction information may be highly desirable in many research settings.

Thus, if the treatment dimension is composed of variations in instructional methods and the blocking variable concerns levels of a relevant aptitude measure, the interaction of treatment and blocking variable is extremely important information since we should want to know whether or not the treatments operate in the same manner for students of different initial ability.

Formally, a randomized blocks with more than one replicate per cell is equivalent to a factorial design. There will be within-cell variability which can be pooled to estimate experimental error. All main effects and interactions can be estimated; however, if the blocking variable is a random dimension, some caution must be exercised in order to set up appropriate F ratios for all hypotheses of interest.

The formation of levels for a blocking variable typically occurs by one of two procedures. It may be that subjects are naturally labeled in convenient ways, and these labelings may provide the basis for blocking. For example, in some schools, students are "tracked" in terms of previous achievement or in terms of measured aptitude in specific academic areas. These administrative classifications often provide a convenient basis for blocking students in school experiments. Similarly, in many settings the researcher may expect that the response measure is sensitive to sex-linked differences. Blocking subjects by sex is an extremely common procedure in educational research. If natural classifications are not available, a process similar to that utilized for forming matched p-tuples of subjects where there is only one replicate per cell may be used. That is, all subjects are administered a premeasure of a relevant blocking variable. The total distribution is then divided by selecting cutting points to produce the desired number of levels of the blocking variable. Subjects within blocks are, of course, randomly assigned to levels of the treatment dimensions.

If it is assumed that there is one treatment dimension and one blocking dimension, the structural model for a score is

$$Y_{ijk} = \mu + \alpha_j + \beta_i + \alpha\beta_{ji} + \epsilon_{ijk}$$

where μ = grand-mean effect

α_j = effect of being at jth level of treatment

β_i = effect of being at ith level of blocking variable

$\alpha\beta_{ji}$ = effect unique to cell (ij) (that is, the interaction component)

ϵ_{ijk} = random error, assumed to be normally distributed with constant variance and expected value of 0

The sums of squares corresponding to each of the effects can be computed by the same procedures which are appropriate for factorial designs. Depending upon his preference, the researcher may utilize sets of orthogonal contrasts to carry out the analysis, or he may utilize the summary table procedures. A proper analysis depends upon the correct choice of model with respect to fixed and random effects. Most commonly, the design will be a fixed-effects model or a mixed model with the blocking variable representing a random effect. Table 5-10 shows expectations of mean squares for these two cases. For the fixed-effects model, the within-cells-error mean square is the appropriate denominator for all F ratios, and the analysis is very straightforward. For the mixed model with the treatment levels fixed but the block levels random, the treatment effect MS_A is tested against the interaction term MS_{AB}. Both the block effect and the interaction effect are tested against within-cell error.

Since the analysis of randomized blocks designs is identical with that for factorial designs, we shall not present an example of the computational steps. In practice, the researcher will begin with some basic experimental design, perhaps a CRD or a factorial design, and the blocking variable or variables will be appended to the basic design in order to increase the homogeneity of the experimental subjects. The resulting design is equivalent to a higher-order factorial design since there is complete crossing of the blocking variables with all other variables. When there is more than one blocking dimension to add to the factorial design, the researcher must form the blocking levels sequentially for the different blocking variables. For example, assume that an experiment is designed, and blocking by sex and initial anxiety level is desired. We begin by dividing the subjects into male and female groups. Then, *within* each sex group, the subjects are ranked on the basis of their anxiety test scores. The levels of the blocking variable of anxiety are then formed by dividing the subjects into the desired number of groups within each sex group. In this example, the levels of the anxiety variable are considered to constitute a random dimension, and the sex variable represents a fixed dimension.

In general, any of the analytical procedures discussed in connection with completely randomized designs or factorial designs can also be

table 5-10 *Expectations of mean squares for the randomized blocks design with more than one replicate per cell*

FIXED-EFFECTS MODEL (A, B FIXED)		MIXED MODEL (A FIXED, B RANDOM)	
mean square	expected value	mean square	expected value
A	$\sigma^2 + nq\sigma_\alpha^2$	A	$\sigma^2 + n\sigma_{\alpha\beta}^2 + nq\sigma_\alpha^2$
B	$\sigma^2 + np\sigma_\beta^2$	B	$\sigma^2 + np\sigma_\beta^2$
AB	$\sigma^2 + n\sigma_{\alpha\beta}^2$	AB	$\sigma^2 + n\sigma_{\alpha\beta}^2$
Error	σ^2	Error	σ^2

utilized in connection with randomized blocks. Thus, randomized blocks designs make an equi-variance assumption; this assumption can be tested by the Bartlett statistic, the Hartley F_{max} test, etc. Similarly, a priori contrasts among the block levels can be tested by means of sets of orthogonal contrasts; if the levels of the blocking dimensions represent points along a quantitative variable, then the techniques of orthogonal polynomials are also appropriate. Procedures such as the Newman-Keuls test for pairwise contrasts can be applied to a blocking variable. In randomized blocks designs, some care must be taken to assure that the correct error term is used in formulas associated with tests such as the Newman-Keuls. In general, in testing contrasts on a dimension, the appropriate error term is the mean square which was utilized to test the significance of the effect during the analysis of variance. Thus, if the AB interaction mean square were used in the denominator of an F ratio for testing the A effect, then this same mean square would be utilized in standard-error formulas associated with contrast procedures in testing, say, pairwise contrasts along the A dimension.

incomplete-block designs

When a blocking dimension is added to a completely randomized design to form a randomized blocks design, the experiment is, essentially, a replication of the completely randomized design for each level of the blocking dimension. That is, within each blocking level, all levels of the treatment variable are represented. Similarly, if a blocking dimension is added to a factorial design, the resulting randomized blocks design may be conceptualized as a replication of the factorial design at each level of the blocking dimension. In practical situations, when the number of treatment combinations is relatively large prior to addition of the blocking dimension, it may be difficult or impossible to obtain the required number of subjects at each blocking level. Instead of the complete randomized blocks design, the researcher may turn to the use of designs in which a complete replication of the basic design is not attempted for each blocking level. Rather, in each block, only certain treatment levels, or combinations of treatment levels, are administered to groups of experimental subjects. Such designs may be constructed with varying amounts of balance; the goal is to reduce the size of the blocks while retaining estimation of the experimental effects which are of major interest. When incomplete blocks are used in conjunction with factorial designs, there is, at least, partial confounding of some of the interaction effects.

Incomplete-block designs are rarely found in actual use in educational or psychological research. The explanation for this is twofold:

First, when the basic design is a CRD, the use of incomplete blocks is dictated only for large numbers of levels of the treatment dimension. Since the typical experiment in the behavioral sciences seldom has more than six or eight levels of a treatment, there is no need to avoid the use of the full randomized blocks design. Second, when incomplete blocks are used in connection with factorial designs, the resulting confounding of interaction effects is viewed by most researchers as an undesirable feature since little is known about the nature of these interaction effects independent of the experiment being conducted. Despite these practical limitations on the apparent usefulness of incomplete-block designs, we do undertake the exposition of some exemplary incomplete designs in this section. The rationale for the inclusion of these designs is that they are popular in fields outside the behavioral sciences, and the student may have to delve into these fields in the course of his reading, and that these designs are likely to become more popular in education and psychology as these fields become more highly developed.

The design of incomplete-block experiments with the CRD as the basic design is a quite different procedure from that used when the basic design is a factorial design, and they are treated separately in this section.

BALANCED INCOMPLETE-BLOCK DESIGNS

Balanced incomplete-block designs (or BIB designs) may be utilized to reduce the required block size when a blocking dimension is to be added to a completely randomized design, or one-way analysis of variance. The "balance" in these designs means that each possible different pair of treatment levels occurs equally often within a single block. For notational purposes, we introduce the following symbols:

p = number of levels of treatment dimension
b = number of different blocks (i.e., number of levels of blocking dimension)
k = block size (i.e., number of treatment levels actually applied per blocking level)
r = number of replications (i.e., number of times each treatment level occurs in entire experiment)
λ = number of repetitions of each treatment pair within a single block

It is apparent that pr must equal bk. That is, the total number of treatment groups can be found in two ways. If each of the p treatment levels occurs r times, the total number of groups must be pr. Similarly, if each of b blocks contains k treatment levels, the total number of groups must be bk. Thus, $pr = bk$. Furthermore, by simple rearrangement, $b = pr/k$.

A BIB design must satisfy this condition, although for certain choices of constants it may be actually impossible to construct a BIB design. For example, if p is five treatment levels and we wish to arrange these in blocks of size 3, the expression becomes $b = 5r/3$. Since only integral solutions are permissible (that is, b must be a whole number), the first possibility is $r = 3$, which gives $b = 5$. It is, however, impossible to construct a BIB design to these specifications. The next possible solution is $r = 6$, which gives $b = 10$, and the appropriate design can be constructed (see Table 5-11). Note that each pair of treatment levels occurs exactly three times within a single block. For example, levels 1 and 2 occur together in blocks 1, 2, and 4, and levels 3 and 4 occur together in blocks 5, 7, and 10.

Within the general category of balanced incomplete-block designs, there are a number of specific designs which have their own names due to their special properties. We shall present a sample of these and then turn to the problem of analyzing BIB designs in general.

If p is a perfect square (that is, $p = k^2$) and k is a prime number, a class of designs known as *balanced simple lattices* may be constructed. In these designs, the block size is also k (that is, k retains the same definition attributed to it for BIB designs). In order to achieve balance in the sense of each different pair of treatment levels occurring together in a block the same number of times, there must be $k(k + 1)$ blocks in the design. For values of p equal to 9, 16, 25, 49, 64, and 81, Cochran and Cox (1957) present balanced simple-lattice designs with $\lambda = 1$ (that is, each pair of treatments occurs only once in the overall design). There is no possible design for $p = 36$, and no examples have been produced for $p = 100$ or larger. A distinguishing feature of balanced simple lattices is that the blocks can be grouped together to form distinct replications of the treatments. That is, we can form sets of the blocks in such a way that within each set each different treatment level occurs exactly once. For $\lambda = 1$, there will be $k + 1$ distinct replicates. It is instructive to study the simplest design with $p = 9$ (that is, $k^2 = p = 9$, and $k = 3$). This design is reproduced in Table 5-12. Since $k = 3$, there is a total of $3 \cdot 4$, or

table 5-11 *Balanced incomplete-block design* ($p = 5, b = 10, k = 3, r = 6, \lambda = 3$)

block	treatment levels	block	treatment levels
1	1, 2, 3	6	1, 3, 5
2	1, 2, 5	7	2, 3, 4
3	1, 4, 5	8	2, 3, 5
4	1, 2, 4	9	2, 4, 5
5	1, 3, 4	10	3, 4, 5

table 5-12 *Balanced simple lattice* ($p = 9$, $b = 12$, $k = 3$, $r = 4$, $\lambda = 1$)

block	rep 1	block	rep 2	block	rep 3	block	rep 4
1	1, 2, 3	4	1, 4, 7	7	1, 5, 9	10	1, 8, 6
2	4, 5, 6	5	2, 5, 8	8	7, 2, 6	11	4, 2, 9
3	7, 8, 9	6	3, 6, 9	9	4, 8, 3	12	7, 5, 3

12, blocks in the design. Note that each pair of treatment levels occurs exactly once somewhere in the design. For example, 1 and 2 and 1 and 3 occur together in block 1, and 1 and 4 and 1 and 7 occur together in block 4, and so forth. Also, by judicious choices of blocks, it is possible to arrange the blocks in distinct replications, as has been done in Table 5-12. Within a single replication, each of the nine treatment levels occurs exactly once. Thus, the first three blocks represent the application of all nine treatment levels; if the blocks are ignored, this constitutes one administration of the entire completely randomized design. Similarly, blocks 4 through 6 represent a second replicate of the CRD, and so on for blocks 7 through 9 and blocks 10 through 12. If we ignore the blocks completely, the design becomes a completely randomized design based on four observations per treatment group. Of course, from an analytical point of view the blocks cannot be ignored.

A balanced simple lattice is balanced with respect to the occurrence of pairs of treatment levels in the rows, or blocks, of the design. An additional degree of balance can be introduced by requiring that each pair of treatment levels also occurs an equal number of times in the columns of the design. Such designs are known as *balanced lattice-square designs*. Since, once again, p must equal k^2, the required number of replicates in order to achieve this balance is $k + 1$. For $k = 3$ and $p = k^2 = 9$, the lattice-square design is shown in Table 5-13. Within blocks, the condition of each pair of treatment levels occurring once and only once is met, as was true for simple balanced lattices. In addition, each pair also occurs once and only once in a column of the design.

Lattice designs exist only when p, the number of treatment levels, is a perfect square and when $k = \sqrt{p}$ is a prime number. A class of BIB

table 5-13 *Balanced lattice-square design* ($p = 9$, $b = 12$, $k = 3$, $r = 4$, $\lambda = 1$)

block	rep 1	block	rep 2	block	rep 3	block	rep 4
1	1, 2, 3	4	1, 4, 7	7	1, 6, 8	10	1, 9, 5
2	4, 5, 6	5	2, 5, 8	8	9, 2, 4	11	6, 2, 7
3	7, 8, 9	6	3, 6, 9	9	5, 7, 3	12	8, 4, 3

designs which share some of the balance characteristics of lattices without imposing limits on the number of treatment levels is *Youden squares.* These designs, despite their name, are actually rectangular arrangements. As is required in any BIB design, each pair of treatment levels occurs an equal number of times in the rows (blocks) of a Youden square. In addition, each treatment level (and, hence, each pair of treatment levels) occurs exactly once in each column of a Youden square. Thus, each column represents a separate replicate of the design, and the number of replicates is the same as the block size. From the foregoing, the following restrictions are apparent: $p = b$ and $k = r$. These designs are called Youden *squares* since they can be constructed from cyclic Latin squares. In our discussion of Latin squares, we showed that a standard Latin square could always be constructed by a process of cyclic permutation. By selecting appropriate numbers of columns from such Latin squares, Youden squares can be formed. For convenience, we replace the usual letters in the cells of a cyclic Latin square by numbers to conform to the labeling of levels in a CRD. For example, if $p = 6$, a Youden square can be constructed with $p = b = 6$ and $k = r = 5$. We begin with a 6×6 Latin square generated by cyclic permutation:

1	2	3	4	5	6
2	3	4	5	6	1
3	4	5	6	1	2
4	5	6	1	2	3
5	6	1	2	3	4
6	1	2	3	4	5

Either the first or the last column may be deleted to yield a Youden square. If the final column is chosen for deletion and the rows correspond to blocks, the Youden square appears as in Table 5-14. Note that each pair of treatment levels occurs four times within blocks. Thus, levels 1 and 2 occur together in blocks 1, 4, 5, and 6. For six treatment

table 5-14 *Youden-square design* ($p = b = 6$, $k = r = 5$, $\lambda = 4$)

block	rep 1	rep 2	rep 3	rep 4	rep 5
1	1	2	3	4	5
2	2	3	4	5	6
3	3	4	5	6	1
4	4	5	6	1	2
5	5	6	1	2	3
6	6	1	2	3	4

levels, the block size (and number of replicates) must equal 5. The student may wish to try to construct a design with a smaller block size to convince himself of its impossibility.

Relatively complete indexes to the types of designs described above, as well as a variety of others, may be found in Cochran and Cox (1957). Chapter 10 of that reference treats simple lattices, Chapter 11 treats balanced incomplete-block designs in general, Chapter 12 treats lattice squares, and Youden squares are discussed in Chapter 13 under the heading of incomplete Latin squares. At the end of each of these chapters, the authors provide indexes of available plans and also provide numerous sample plans.

The analysis of balanced incomplete-block designs follows a general plan regardless of the specific type of design. However, the analysis may be simplified or extended for certain of the designs. Complete treatments of the analyses of all the types of designs enumerated above are found in Cochran and Cox (1957) and in Winer (1962). Here, we present the analysis in general terms for balanced incomplete-block designs. For designs in which the blocks can be arranged in distinct replicates, the analysis includes the replicates as a source of variation, and the analysis differs somewhat from that illustrated here.

We assume a strictly additive model for a score; there is assumed to be no interaction between the treatment and blocks. This model for the jth score in block i is

$$Y_{ij} = \mu + \alpha_j + \beta_i + \epsilon_{ij}$$

where μ = usual grand-mean effect
α_j = effect due to jth level of treatment
β_i = effect due to being in block i
ϵ_{ij} = experimental error, which is assumed to be normally and homogeneously distributed for all treatment levels

Throughout the analysis as developed here it is assumed that there is one observation per treatment level in a block (that is, $n = 1$). If the entire design is replicated to yield a total of $n = 2$ or more subjects per treatment level in a block, the analysis is changed only slightly, but treatment of this extension is left to the outside references cited above. From an analytical point of view, the task is to obtain a sum of squares for treatment which is uncontaminated by block differences.

The fact of this contamination is best seen by studying an actual design. The BIB design presented in Table 5-11 contains three repetitions of each pair of treatment levels. For example, treatment levels 1 and 2 occur together in blocks 1, 2, and 4. In addition, treatment level 1 occurs in blocks 3, 5, and 6, and level 2 occurs in blocks 7, 8, and 9. Thus, a contrast of level 1 and level 2 involves some interblock comparisons as well as the treatment comparison. For example, if blocks 3, 5, and 6 have a large positive effect on all scores in these blocks and blocks 7, 8, and 9 have no effect, or a negative effect, a comparison of levels 1 and 2 of the treatment will be affected. Of course, that part of the comparison which is intrablock (that is, from blocks 1, 2, and 4) is not contaminated by block differences. Since the sum of squares for treatment must reflect treatment-level comparisons only, the effects of the blocks must be removed. Similarly, a comparison of block levels is contaminated by treatment, and to obtain a sum of squares for blocks which reflects only block effects, some adjustment must be carried out. The nature of the required adjustment can be most clearly understood by deriving unbiased estimates of the treatment effects (that is, of the α_j terms in the structural model for a score). We place the usual restrictions on the treatment and block effects: $\Sigma_{j=1}^{p} \alpha_j = 0$ and $\Sigma_{i=1}^{b} \beta_i = 0$. We also assume a fixed-effects model. Each treatment level is replicated r times. If we sum the r observations for the jth treatment, $\Sigma_{(i)} Y_{ji}$, we estimate $r\mu + r\alpha_j + \Sigma_{(j)} \beta_i$, where the use of the summation limits (i) and (j) is intended to imply summation over the appropriate blocks [that is, $\Sigma_{(i)} Y_{ji}$ is the sum for the jth level of the treatment dimension, where this sum is taken only from those blocks containing these treatment levels; $\Sigma_{(j)} \beta_i$ is the sum of the blocks which contain the jth level of the treatment]. Note that the treatment sum is contaminated by components representing block effects. We can eliminate this block contamination by a roundabout procedure. The sum of all scores in blocks which contain the jth level of the treatment is

$$\sum Y_{\beta(j)} = kr\mu + r\alpha_j + \lambda\sum\alpha'_j + k\sum_{(j)} \beta_i$$

where $\Sigma\alpha'_j$ is the sum of treatment effects for treatment levels other than the jth level. Focusing on the middle terms on the right-hand side of the expression: $r\alpha_j + \lambda\Sigma\alpha'_j = r\alpha_j - \lambda\alpha_j + \lambda\alpha_j + \lambda\Sigma\alpha'_j$. But $\lambda(\Sigma\alpha_j) = \lambda\alpha_j + \lambda\Sigma\alpha'_j = 0$ [since $\Sigma\alpha_j = 0$, then $\lambda(\Sigma\alpha_j) = 0$]. Thus, $r\alpha_j + \lambda\Sigma\alpha'_j = (r - \lambda)\alpha_j$ and $\Sigma Y_{\beta(j)} = kr\mu + (r - \lambda)\alpha_j + k\,\Sigma_{(j)} \beta_i$. Both $\Sigma_{(i)} Y_{ji}$ and $\Sigma Y_{\beta(j)}$ contain block effects. However, by combining these two expressions, we can eliminate these effects:

$$k\Big(\sum_{(i)} Y_{ji}\Big) - \sum Y_{\beta(j)}$$
$$= kr\mu + kr\alpha_j + k\sum_{(j)} \beta_i - kr\mu - (r - \lambda)\alpha_j - k\sum_{(j)} \beta_i = (kr - r + \lambda)\alpha_j$$

We have thus shown that it is possible to combine treatment and block information to estimate treatment effects. With further manipulation it is possible to find expressions for treatment means and treatment totals which are free of block effects. With our previous notation, the sum of observations for the jth level of the treatment is $\Sigma_{(i)} Y_{ji}$. For simplicity, the adjusted total will be denoted T'_j. Then

$$T'_j = \sum_{(i)} Y_{ji} + \theta W_j$$

where

$$\theta = \frac{(b-1)(E_b - E_e)}{p(k-1)(b-1)E_b + (p-k)(b-p)E_e}$$

$$W_j = (p-k)\sum_{(i)} Y_{ji} - (p-1)\sum Y_{\beta(j)} + (k-1)G$$

$$G = \sum_{i=1}^{b}\sum_{j=1}^{p} Y_{ij}$$

$$E_b = \text{SS}_{\text{blocks(adj)}}$$

$$\text{SS}_{\text{blocks(adj)}} = \text{SS}_{\text{blocks(unadj)}} + \text{SS}_{\text{treat(adj for blocks)}} - \text{SS}_{\text{treat(unadj)}}$$

$$E_e = \text{SS}_{\text{total}} - \text{SS}_{\text{treat(unadj)}} - \text{SS}_{\text{blocks(adj)}}$$

$$\text{SS}_{\text{treat(unadj)}} = \frac{\sum_{j=1}^{p}\left(\sum_{(i)} Y_{ji}\right)^2}{r} - \frac{G^2}{pr}$$

$$\text{SS}_{\text{treat(adj for blocks)}} = \frac{\sum_{j=1}^{p}\left[k\left(\sum_{(i)} Y_{ji}\right) - \sum Y_{\beta(j)}\right]^2}{k(kr - r + \lambda)}$$

$$\text{SS}_{\text{blocks(unadj)}} = \frac{\sum_{i=1}^{b} B_i^2}{k} - \frac{G^2}{pr}$$

$$B_i = \text{sum of scores in } i\text{th block}$$

The treatment sums T'_j reflect both intrablock and interblock information concerning the treatments. A sum of squares reflecting these sources is

$$\text{SS}_{\text{treat(adj)}} = \frac{\sum_{j=1}^{p} (T'_j)^2}{r} - \frac{G^2}{pr}$$

This sum of squares has $p - 1$ degrees of freedom. The effective error term for testing the mean square based on adjusted treatments is $E' = E_e[1 + (p - k)\theta]$. An overall F test for treatment takes the form

$$F = \frac{\text{MS}_{\text{treat(adj)}}}{E'}$$

where

$$\text{MS}_{\text{treat(adj)}} = \frac{\text{SS}_{\text{treat(adj)}}}{p - 1}$$

The degrees of freedom for this ratio are $p - 1$ and $rp - p - b + 1$. This test is only approximate, even when the usual assumptions of the analysis of variance are met, since sampling variability in the value of θ is not taken into account. For reasonably large numbers of levels of the blocking dimension (say 10 or more) the approximation is quite satisfactory when the ordinary F distributions are used to test the significance of the computed F. An exact test which, however, ignores interblock information on the treatment effects can be set up using $\text{SS}_{\text{treat(adj for blocks)}}$. The appropriate ratio is

$$F = \frac{\text{MS}_{\text{treat(adj for blocks)}}}{E_e}$$

where

$$\text{MS}_{\text{treat(adj for blocks)}} = \frac{\text{SS}_{\text{treat(adj for blocks)}}}{p - 1}$$

The degrees of freedom are the same as above. This test includes only intrablock treatment information (i.e., is based only on contrasts involving pairs of treatment levels occurring together in the same block). If the analysis is based on the adjusted sums T'_j, so that both interblock and intrablock treatment information is utilized, additional analysis (e.g., orthogonal contrasts and Newman-Keuls pairwise contrasts) would use E' as the appropriate error term.

The steps involved in adjusting the treatment sums are best laid out in tabular form. The design originally presented in Table 5-11 will be used to illustrate the analysis. We assume that in each block, each treatment level that appears is applied to only one experimental subject. Hypothetical data for the design of Table 5-11 are presented in Table 5-15. The treatment-level designation is shown in parentheses just prior to each score. Table 5-16 presents a summary table for treatment sums, block sums (for blocks containing the given treatment

table 5-15 *Example of a balanced incomplete-block design* ($p = 5, b = 10, k = 3, r = 6, \lambda = 3$)

block	treatment levels and scores						block	treatment levels and scores					
1	(1)	27	(2)	37	(3)	19	6	(1)	22	(3)	12	(5)	11
2	(1)	35	(2)	42	(5)	20	7	(2)	51	(3)	30	(4)	57
3	(1)	29	(4)	50	(5)	17	8	(2)	46	(3)	22	(5)	20
4	(1)	17	(2)	24	(4)	35	9	(2)	37	(4)	41	(5)	11
5	(1)	22	(3)	9	(4)	39	10	(3)	19	(4)	38	(5)	14

level), and the adjusted treatment totals. In order to compute the T'_j, the correction factor θ must be known. For the example, this involves a number of intermediate steps:

$$\text{SS}_{\text{total}} = 27^2 + 37^2 + \cdots + 14^2 - \frac{G^2}{30} = 29{,}325 - \frac{853^2}{30} = 29{,}325 \ -24{,}253.63$$
$$= 5{,}071.37$$

$$\text{SS}_{\text{treat (unadj)}} = \frac{152^2 + \cdots + 93^2}{6} - \frac{G^2}{30} = 27{,}973.83 - 24{,}253.63 = 3{,}720.20$$

$$\text{SS}_{\text{blocks (unadj)}} = \frac{83^2 + \cdots + 71^2}{3} - \frac{G^2}{30} = 25{,}988.33 - 24{,}253.63 = 1{,}734.70$$

$$\text{SS}_{\text{treat (adj for blocks)}} = \frac{(3 \cdot 152 - 467)^2 + \cdots + (3 \cdot 93 - 486)^2}{45} = 3{,}253.64$$

$$E_b = \text{SS}_{\text{blocks (adj)}} = \text{SS}_{\text{blocks (unadj)}} + \text{SS}_{\text{treat (adj for blocks)}} - \text{SS}_{\text{treat (unadj)}}$$
$$= 1{,}734.70 + 3{,}253.64 - 3{,}720.20 = 1{,}268.14$$

table 5-16 *Summary table for* BIB *analysis*

treatment level	$\sum_{(i)} Y_{ji}$	$\sum Y_{\beta(j)}$	W_j	T'_j
1	152	467	142	165.37
2	237	571	−104	227.21
3	111	495	−52	106.10
4	260	540	66	266.21
5	93	486	−52	88.10
Sums	$853 = G$	$2{,}559 = 3G$	0	852.99

$$E_e = \text{SS}_{\text{total}} - \text{SS}_{\text{treat(unadj)}} - \text{SS}_{\text{blocks(adj)}}$$
$$= 5{,}071.37 - 3{,}720.20 - 1{,}268.14 = 83.03$$

Then

$$\theta = \frac{9(E_b - E_e)}{90E_b - 10E_e} = \frac{10{,}665.99}{113{,}302.30} = .09414$$

In the final column of Table 5-16, the adjusted treatment totals T'_j were computed as $\Sigma_{(i)} Y_{ji} + \theta W_j$. Now the sum of squares for the adjusted treatments can be found:

$$\text{SS}_{\text{treat(adj)}} = \frac{165.37^2 + \cdots + 88.10^2}{6} - \frac{G^2}{30} = 28{,}143.03 - 24{,}253.63$$
$$= 3{,}889.40$$

The effective-error term for testing the treatment effect is

$$E' = E_e(1 + 2\theta) = 83.03(1 + .1883) = 98.66$$

Then

$$\text{MS}_{\text{treat(adj)}} = \frac{\text{SS}_{\text{treat(adj)}}}{4} = \frac{3{,}889.40}{4} = 972.35$$

and

$$F = \frac{972.35}{98.66} = 9.86$$

With 4 and 16 degrees of freedom, this value is significant at beyond the .01 level.

Assume that on the basis of theoretical considerations, the following contrasts among the adjusted sample means were relevant to this example:

		LEVEL		
1	2	3	4	5
1	1	0	−1	−1
1	−1	0	0	0
0	0	0	1	−1
1	1	−4	1	1

The adjusted means can be found from the adjusted treatment sums by dividing by 6 (the number of replicates). Then $\bar{Y}'_1 = 27.56$, $\bar{Y}'_2 = 37.87$, $\bar{Y}'_3 = 17.68$, $\bar{Y}'_4 = 44.37$, $\bar{Y}'_5 = 14.68$. The values of the contrasts are

$$C_1 = 27.56 + 37.87 - 44.37 - 14.68 = 6.38$$

$$C_2 = 27.56 - 37.87 = -10.31$$

$$C_3 = 44.37 - 14.68 = 29.69$$

$$C_4 = 27.56 + 37.87 - 4(17.68) + 44.37 + 14.68 = 53.76$$

The associated sums of squares are

$$\text{SS}_{C_1} = \frac{6(6.38)^2}{1^2 + 1^2 + 1^2 + 1^2} = \frac{244.2264}{4} = 61.0566$$

$$\text{SS}_{C_2} = \frac{6(-10.31)^2}{1^2 + 1^2} = \frac{637.1526}{2} = 318.5763$$

$$\text{SS}_{C_3} = \frac{6(29.69)^2}{1^2 + 1^2} = \frac{5,288.9766}{2} = 2,644.4883$$

$$\text{SS}_{C_4} = \frac{6(53.76)^2}{1^2 + 1^2 + 4^2 + 1^2 + 1^2} = \frac{17,340.8256}{20} = 867.0413$$

Each contrast sum of squares can be tested against the effective-error mean square E'; the resulting F ratios are

$$F_1 = \frac{61.0566}{98.66} = .62 \qquad F_2 = \frac{318.5763}{98.66} = 3.23$$

$$F_3 = \frac{2,644.4883}{98.66} = 26.80 \qquad F_4 = \frac{867.0413}{98.66} = 8.79$$

Each of these F ratios has 1 and 16 degrees of freedom; thus, the contrasts based on C_3 and C_4 are significant, and C_1 and C_2 are nonsignificant. In terms of adjusted treatment means, we can conclude that treatment level 4 produces a higher mean than treatment level 5 and the average of treatment levels 1, 2, 4, and 5 produces a higher mean than treatment level 3. However, there is no evidence for a difference in effect of treatment levels 1 and 2, or for the average of 1 and 2 against the average of 4 and 5.

The analysis for BIB designs presented here is relevant to cases in which there is one observation per treatment level in a block and the blocks are not arranged in distinct replications. The principles can be extended to the other cases, but we leave this to the outside sources which were referenced earlier. In addition to balanced incomplete-block designs, there is a class of experimental designs known as *partially balanced incomplete-block* designs which do not have all the balance features of BIB designs. Discussion of these designs can be found in Winer (1962) and Cochran and Cox (1957).

FACTORIAL DESIGNS IN INCOMPLETE BLOCKS

The incomplete-block designs considered up to this point have all been for the case in which the basic design is a completely randomized design (i.e., there is only one treatment dimension). We now turn to the more general case of constructing incomplete blocks when the basic design is a factorial design. The principles discussed in relation to incomplete factorial designs will be utilized. In essence, factorial designs in incomplete blocks involve either partial or complete confounding of one or more higher-order interaction effects.

As was true for BIB designs, we shall present only some relevant details and illustrations of factorial designs in incomplete blocks and leave more extensive treatment to outside sources.

A researcher would consider the use of incomplete blocks only if the available experimental material were insufficiently homogeneous or too limited in supply to permit a complete replication of the factorial design for each level of the blocking variable. This might arise in an educational setting if an experiment were too large to conduct with available classes of students in one school. If schools are considered as blocks, the complete experimental plan may involve classes of students from two or more schools without the complete design's being conducted in a single school. Thus, certain treatment combinations would not occur in certain of the schools; if the choices of treatment combinations to apply in the schools are made judiciously, there will still be intrablock (i.e., within-school) information for estimating main effects and some of the interactions. However, other interactions will be either completely or partially confounded with interblock differences, and this must be taken into account in the analysis of the specific design.

In the simplest case, incomplete blocks can be formed by choosing to confound the highest-order interaction effect with block effects. If the basic design is a 2^3 factorial, an incomplete-block design, with two blocks of size 4 each, can be constructed by confounding the ABC interaction with blocks. The same principles for choosing the specific treatment combinations to use in a block apply as for choosing treatment combinations to form a $\frac{1}{2}$ replicate of the factorial design. Thus, one block would contain all cells (i.e., treatment combinations) which enter positively into the ABC interaction, and the second block would contain all the negative terms. The plan, in terms of cells, would be:

block 1	block 2
(111)	(211)
(221)	(121)
(212)	(112)
(122)	(222)

In the 2^3 factorial design, the ABC interaction has 1 degree of freedom and is based on the contrast of cells (111), (221), (212), and (122) with cells (211), (121), (112), and (222). The incomplete-block design in two blocks of size 4 has simply allocated the two halves of this interaction to the two blocks. Thus, if a sum of squares is computed for the block difference, this is equivalent to the ABC interaction; therefore, blocks and the ABC interaction are confounded. The analysis of this particular design is very simple since all sums of squares are computed in the same way as for a complete factorial design. The sum of squares for the ABC interaction, however, has the sum of squares for blocks as an alias.

An incomplete-block design of the type just presented might be replicated in one of two different ways. First, within each block, a total of n subjects might be administered each of the treatment combinations. In this case, the analysis is identical with the complete factorial design (except for the confounding of blocks with the ABC interaction), and there will be a within-cells-error term with $8(n-1)$ degrees of freedom. Second, the entire experiment may be replicated by using two different blocks for each replication:

block 1	block 2	block 3	block 4	···	block $b-1$	block b
(111)	(211)	(111)	(211)		(111)	(211)
(221)	(121)	(221)	(121)		(221)	(121)
(212)	(112)	(212)	(112)		(212)	(112)
(122)	(222)	(122)	(222)		(122)	(222)

In this case, there are $b-1$ degrees of freedom for blocks (i.e., assuming a total of b levels of the blocking dimension). The degrees of freedom are allocated as:

source	df
Blocks	$b-1$
A, B, C	3
AB, AC, BC	3
Residual	$3b-6$

The term labeled "residual" is made up of interaction components involving the blocks and the treatments. If the blocks represent a random dimension and if the block-by-treatment interaction is negligible, the residual sum of squares can be converted to a residual mean square and used as the error term for testing the main and interaction effects from the factorial design.

The design presented for a 2^3 factorial design in blocks of size 4 provides no information on the ABC interaction; it is completely confounded with interblock differences. It is, however, possible to construct a balanced design with some intrablock information on all interaction effects including the ABC term. Such a design requires a minimum of eight blocks and involves four replicates of the 2^3 factorial design. In one replicate (i.e., pair of blocks), the ABC interaction is confounded with blocks; in one of the remaining replicates, each of the first-order interactions is confounded with blocks (i.e., in one replicate, AB is confounded with blocks; in another, AC is confounded with blocks; and in the third, BC is confounded with blocks). The appropriate cells to confound with blocks can always be found by studying the contrast for the particular interaction effect. The design appears as in Table 5-17.

In the first replicate, the ABC interaction has been confounded with blocks. In the second replicate, however, there is intrablock information on the ABC interaction, but the AB interaction is confounded with blocks. Similarly, in replicates 3 and 4, there is intrablock informa-

table 5-17 2^3 *factorial design in blocks of size* 4 *with partial information on all interactions*

	REPLICATE 1		REPLICATE 2		REPLICATE 3		REPLICATE 4	
	block 1	block 2	block 3	block 4	block 5	block 6	block 7	block 8
	(111)	(211)	(111)	(121)	(111)	(112)	(111)	(112)
	(221)	(121)	(112)	(122)	(121)	(122)	(122)	(121)
	(212)	(112)	(221)	(211)	(212)	(211)	(211)	(212)
	(122)	(222)	(222)	(212)	(222)	(221)	(222)	(221)
Blocks confounded with	ABC		AB		AC		BC	

tion on ABC, but, in each case, a first-order interaction is confounded with blocks. There is, therefore, information in six of the eight blocks with respect to the ABC interaction. This condition is described by saying that the "relative intrablock information on the ABC interaction is $\frac{6}{8}$ or $\frac{3}{4}$." Similarly, each of the first-order interactions can be estimated from six of the eight blocks, and each is based on $\frac{3}{4}$ intrablock information. For analysis, all sums of squares are found in the same manner as for complete factorial designs; however, in computing sums of squares for interaction effects, data from replicates in which an interaction is confounded are not included. For the design shown in Table 5-17, the breakdown of degrees of freedom is:

source	df
Blocks	7
A, B, C	3
AB, AC, BC	3
ABC	1
Residual	17
Total	31

The residual term contains interaction effects among the replicates and the main effects and among the replicates and the interaction effects (first-order and second-order). If these interactions can be assumed to be negligible, the residual term provides a relevant error term for testing all main and interaction effects.

From the examples given, the student can easily generalize the notion of confounding with blocks in 2^k-series factorial designs. It is always possible to completely confound the highest-order interaction with interblock differences; on the other hand, partial intrablock information can be obtained on all interactions by confounding lower-order interactions for some of the replicates of the design. Incomplete blocks in factorial designs can be extended to cases other than the 2^k-series factorial designs. Designs are available for 3^k-series designs in blocks of sizes which are multiples of 3. Also, plans for certain mixed designs such as $3 \times 2 \times 2$, $3 \times 3 \times 2$, and $2 \times 2 \times 4$ have been tabled. For examples of these designs and a more complete treatment of the procedures for constructing factorial designs in incomplete blocks, the reader should consult Chapter 6 of Cochran and Cox (1957) and Chapter 8 of Winer (1962).

The analysis of factorial designs in incomplete blocks is very straightforward when the design is balanced in the sense that all interactions at a given level are estimated with the same amount of intrablock information. In practice, the same formulas may be utilized as are appropriate for complete factorial designs. The number of cases entering into the interaction sums must be kept clearly in mind when there is partial information, and appropriate adjustments must be made in the formulas for the complete factorial. To illustrate these procedures, consider the data in Table 5-18. This design is replicates 1, 3, and 4 from Table 5-17. Replicate 2 has been omitted; there is intrablock information on all main effects. Also, the AB interaction is never confounded with blocks, and it can be estimated from all three replicates. There is, however, only $\frac{2}{3}$ relative information on the AC, BC, and ABC interactions. Although this design is not balanced, it presents no analytical problems. The setting for the experiment in Table 5-18 involves administration of three treatments; since the design could not be completely replicated within a single school, a total of six schools were utilized to obtain at least partial information on all interaction effects. Thus, a pair of schools (where the schools are considered to be the blocks) constitutes a single replicate of the basic 2^3 factorial design. Sums of squares for main effects can be computed directly from contrasts based on cell means, where the cell means are found by averaging over the three replicates. These cell means are

$$\bar{Y}_{111} = \frac{14 + 17 + 10}{3} = \frac{41}{3} = 13.67 \qquad \bar{Y}_{121} = \frac{43}{3} = 14.33$$

$$\bar{Y}_{112} = \frac{48}{3} = 16.00 \qquad \bar{Y}_{122} = \frac{39}{3} = 13.00$$

$$\bar{Y}_{211} = \frac{57}{3} = 19.00 \qquad \bar{Y}_{212} = \frac{79}{3} = 26.33$$

table 5-18 *Illustrative 2^3 factorial design and data in blocks of size 4 with ABC, AC, and BC partially confounded*

ABC CONFOUNDED replicate 1				AC CONFOUNDED replicate 2				BC CONFOUNDED replicate 3			
school 1		school 2		school 3		school 4		school 5		school 6	
(111)	14	(211)	21	(111)	17	(112)	15	(111)	10	(112)	16
(221)	23	(121)	13	(121)	16	(122)	13	(122)	11	(121)	14
(212)	26	(112)	17	(212)	29	(211)	19	(211)	17	(212)	24
(122)	15	(222)	31	(222)	34	(221)	21	(222)	27	(221)	20

$$\bar{Y}_{221} = \frac{64}{3} = 21.33 \qquad \bar{Y}_{222} = \frac{92}{3} = 30.67$$

Also, the grand mean is

$$\bar{Y}_{...} = \frac{463}{24} = 19.29$$

The contrast for the A dimension is

$$\bar{Y}_{111} + \bar{Y}_{121} + \bar{Y}_{112} + \bar{Y}_{122} - \bar{Y}_{211} - \bar{Y}_{212} - \bar{Y}_{221} - \bar{Y}_{222} = -40.33$$

Thus,

$$\mathrm{SS}_A = \frac{n(-40.33)^2}{1^2 + 1^2 + 1^2 + 1^2 + 1^2 + 1^2 + 1^2 + 1^2} = \frac{3(1{,}626.5089)}{8}$$
$$= \frac{4{,}879.5267}{8} = 609.94$$

Similarly, the contrast for the B dimension is $\bar{Y}_{111} + \bar{Y}_{112} + \bar{Y}_{211} + \bar{Y}_{212} - \bar{Y}_{121} - \bar{Y}_{122} - \bar{Y}_{221} - \bar{Y}_{222}$, and the sum of squares is

$$\mathrm{SS}_B = \frac{3(-4.33)^2}{8} = \frac{56.2467}{8} = 7.03$$

The contrast for the C dimension is $\bar{Y}_{111} + \bar{Y}_{221} + \bar{Y}_{211} + \bar{Y}_{121} - \bar{Y}_{212} - \bar{Y}_{122} - \bar{Y}_{112} - \bar{Y}_{222}$, and the sum of squares is

$$\mathrm{SS}_C = \frac{3(-17.67)^2}{8} = \frac{936.6867}{8} = 117.09$$

Also, suitable contrasts can be set up to yield the interaction sums of squares. However, with the exception of AB, which is based on all observations, only 16 of the 24 cells enter into each of the interaction terms. Starting with the AB interaction, the contrast is $\bar{Y}_{111} + \bar{Y}_{112} + \bar{Y}_{221} + \bar{Y}_{222} - \bar{Y}_{121} - \bar{Y}_{122} - \bar{Y}_{211} - \bar{Y}_{212}$, and the sum of squares is

$$\mathrm{SS}_{AB} = \frac{3(9.01)^2}{8} = \frac{243.5403}{8} = 30.44$$

For AC the means entering into the contrast must be based only on replicates 1 and 3. These are

$$\bar{Y}'_{111} = \frac{14 + 10}{2} = \frac{24}{2} = 12.00 \qquad \bar{Y}'_{121} = \frac{27}{2} = 13.5$$

$$\bar{Y}'_{112} = \frac{33}{2} = 16.5 \qquad \bar{Y}'_{122} = \frac{26}{2} = 13.00$$

$$\bar{Y}'_{211} = \frac{38}{2} = 19.00 \qquad \bar{Y}'_{221} = \frac{43}{2} = 21.5$$

$$\bar{Y}'_{212} = \frac{50}{2} = 25.00 \qquad \bar{Y}'_{222} = \frac{58}{2} = 29.00$$

The contrast is $\bar{Y}'_{111} + \bar{Y}'_{121} + \bar{Y}'_{212} + \bar{Y}'_{222} - \bar{Y}'_{211} - \bar{Y}'_{221} - \bar{Y}'_{112} - \bar{Y}'_{122}$, and the sum of squares is

$$\mathrm{SS}_{AC} = \frac{2(9.50)^2}{8} = \frac{180.50}{8} = 22.56$$

A similar procedure must be used for the BC interaction. In this case, the means are based on the sums of replicates 1 and 2:

$$\bar{Y}''_{111} = \frac{31}{2} = 15.5 \qquad \bar{Y}''_{121} = \frac{29}{2} = 14.5 \qquad \bar{Y}''_{112} = \frac{32}{2} = 16.00$$

$$\bar{Y}''_{122} = \frac{28}{2} = 14.00 \qquad \bar{Y}''_{211} = \frac{40}{2} = 20.00 \qquad \bar{Y}''_{221} = \frac{44}{2} = 22.00$$

$$\bar{Y}''_{212} = \frac{55}{2} = 27.5 \qquad \bar{Y}''_{222} = \frac{65}{2} = 32.5$$

The contrast is $\bar{Y}''_{111} + \bar{Y}''_{211} + \bar{Y}''_{122} + \bar{Y}''_{222} - \bar{Y}''_{121} - \bar{Y}''_{221} - \bar{Y}''_{112} - \bar{Y}''_{212}$, and the sum of squares is

$$\mathrm{SS}_{BC} = \frac{2(2.00)^2}{8} = \frac{8.00}{8} = 1.00$$

Finally, the ABC interaction must be based on information from replicates 2 and 3. The cell means are

$$\bar{Y}'''_{111} = \frac{27}{2} = 13.5 \qquad \bar{Y}'''_{121} = \frac{30}{2} = 15.0 \qquad \bar{Y}'''_{112} = \frac{31}{2} = 15.5$$

$$\bar{Y}'''_{122} = \frac{24}{2} = 12.0 \qquad \bar{Y}'''_{211} = \frac{36}{2} = 18.0 \qquad \bar{Y}'''_{221} = \frac{41}{2} = 20.5$$

$$\bar{Y}'''_{212} = \frac{53}{2} = 26.5 \qquad \bar{Y}'''_{222} = \frac{61}{2} = 30.5$$

The contrast is $\bar{Y}'''_{111} + \bar{Y}'''_{122} + \bar{Y}'''_{221} + \bar{Y}'''_{212} - \bar{Y}'''_{121} - \bar{Y}'''_{112} - \bar{Y}'''_{211} - \bar{Y}'''_{222}$, and the sum of squares is

$$SS_{ABC} = \frac{2(-6.50)^2}{8} = \frac{84.50}{8} = 10.56$$

Since there are six levels of the blocking variable (schools), there are also 5 degrees of freedom for blocks; this sum of squares is based on the block sums (e.g., for block 1, the sum is $14 + 23 + 26 + 15 = 78$). Then

$$SS_{\text{blocks}} = \frac{78^2 + 82^2 + 96^2 + 68^2 + 65^2 + 74^2}{4} - \frac{463^2}{24} = 155.21$$

The residual term must be found by subtraction. The total sum of squared scores is

$$\sum_{i=1}^{2}\sum_{j=1}^{2}\sum_{k=1}^{2}\sum_{r=1}^{3} Y^2_{ijkr} = 9{,}891$$

and the total sum of squares is

$$SS_{\text{total}} = 9{,}891 - \frac{463^2}{24} = 958.96$$

Then

$$\begin{aligned} SS_{\text{res}} &= SS_{\text{total}} - SS_{\text{blocks}} - SS_A - SS_B - SS_C \\ &\qquad - SS_{AB} - SS_{AC} - SS_{BC} - SS_{ABC} \\ &= 958.96 - 155.21 - 609.94 - 7.03 - 117.09 - 30.44 \\ &\qquad - 22.56 - 1.00 - 10.56 = 5.13 \end{aligned}$$

These results are summarized in the analysis-of-variance table 5-19.

For these illustrative data, all effects, main and interaction, are significant with the exception of the BC interaction. If it is desired, the mean square for blocks can also be tested against the residual term. Furthermore, the sum of squares for blocks can be partitioned into several components. The three replicates of the design each contain two blocks. Thus, differences among the replicates are part of the block differences. Also, the contrast of block 1 with block 2 (i.e., school 1 with school 2) is confounded with the ABC interaction, block 3 and 4 differences are confounded with the AC interaction, and block 5 and 6

table 5-19 *Analysis-of-variance summary table for the 2^3 factorial design in blocks of size 4 with partial confounding*

source	df	sum of squares	mean square	F	p
Blocks	5	155.21	31.04		
A	1	609.94	609.94	1,297.95	<.01
B	1	7.03	7.03	14.96	<.01
C	1	117.09	117.09	249.17	<.01
AB	1	30.44	30.44	64.78	<.01
AC'	1	22.56	22.56	48.01	<.01
BC'	1	1.00	1.00	2.13	ns
ABC'	1	10.56	10.56	22.47	<.01
Residual	11	5.14	.47		
Total	23				

differences are confounded with the BC interaction. The 5 degrees of freedom for blocks can be decomposed as:

2 df for among-replicates differences
1 df for ABC interaction (confounded with block 1 versus block 2)
1 df for AC interaction (confounded with block 3 versus block 4)
1 df for BC interaction (confounded with block 5 versus block 6)

Although it may be of interest to separate out the difference due to replicates, confounding prevents any useful interpretation of the remaining three components.

exemplary applications of randomized blocks designs

example 1 Goldhaber (1967) utilized a three-dimensional design to investigate comprehension of speeded speech (or compressed speech) under varying conditions of rate and difficulty level of material. The basic design was a 4×3 factorial with sex of experimental subjects constituting the third dimension. Thus, the total design comprised 24 groups of subjects (that is, 12 groups of males and 12 groups of females); utilizing 10 subjects per cell, the researcher randomly selected 240 students from undergraduate speech classes at a large state university.

Speeded speech involves, essentially, removing silent or blank segments from tape recordings of ordinary speech. By controlling the amount of silent material removed, the rate of presentation of the material can be varied. In his study, Goldhaber produced tapes at 175, 325, 375, and 425 words per minute. The basic rate of unspeeded speech was 175 words per minute. The material being presented was concerned with the history of broadcasting, and the difficulty level (e.g., vocabulary level and complexity of sentence structure) of the material was varied over three levels: eighth grade, college freshman, and graduate level. Achievement was measured by means of a 20-item multiple-choice examination which was administered to all subjects immediately upon completion of the experimental sessions. The design scheme was:

		SEX OF SUBJECT					
		male			female		
DIFFICULTY:		8th grade	fresh	grad	8th grade	fresh	grad
RATE	175	G_1	G_2	G_3	G_{13}	G_{14}	G_{15}
	325	G_4	G_5	G_6	G_{16}	G_{17}	G_{18}
	375	G_7	G_8	G_9	G_{19}	G_{20}	G_{21}
	425	G_{10}	G_{11}	G_{12}	G_{22}	G_{23}	G_{24}

In addition to the usual hypotheses concerning main and interaction effects, the author was interested in the functional relationship (or trend) between comprehension and rate of speech and between comprehension and difficulty level of the message being presented. Thus, trend analysis was performed on the main dimensions of rate and difficulty level. The analysis-of-variance summary was:

source	*df*	sum of squares	mean square	*F*	*p*
A Sex of subject	1	98 82	98 82	22.79	$<.01$
B Difficulty	2	32.26	16.13	3.28	$<.05$
C Rate	3	644.93	214.98	48.11	$<.01$
AB	2	3.86	1 93	.43	$>.05$
AC	3	4.32	1.44	.32	$>.05$
BC	6	27.04	4.51	1.01	$>.05$
ABC	6	22.11	3.68	.83	$>.05$
Error	216	962.40	4.46		
Total	239	1,795.73			

Note that since all three dimensions of the design represent fixed factors, the appropriate denominator in all F ratios is the MS_{error}. From the analysis, the main effects are each significant, while none of the interaction effects approaches significance. From inspection of the marginal means, it was found that the higher level of comprehension occurred for male students. The additional analysis performed on the rate and difficulty dimensions utilized the techniques of orthogonal polynomials. The difficulty levels had been chosen to represent, as nearly as possible, equally spaced points; thus, the contrast coefficients $-1, 0, +1$ and $+1, -2, +1$ were used to separate linear- and quadratic-trend components from the sum of squares associated with difficulty levels. The results were:

source	*df*	mean square	*F*	*p*
Linear trend	1	.23	.05	$>.05$
Quadratic trend	1	32.03	7.24	$<.01$
Total (difficulty)	2	32.26		

It is interesting to note that the linear-trend component is nonsignificant (and accounts for less than 1 percent of the sum of squares for difficulty), while the quadratic component is highly significant (and accounts for over 99 percent of the sum of squares for difficulty). The plot of the trend is shown in Figure 5-1. The trend, which is distinctly parabolic, reveals that the highest performance occurred with the difficulty level set at a college-freshman level (which is, also, the approximate level of the students involved in the experiment). Turning to the results for rate of presentation, a problem arises since the rates are not equally spaced and, therefore, the usual tabled coefficients for trend components cannot be used. The author followed the method outlined by Peng (1967, pages 57–60). The results were:

source	*df*	mean square	*F*	*p*
Linear trend	1	610.30	136.98	$<.01$
Quadratic trend	1	34.28	7.69	$<.01$
Cubic trend	1	.35	.08	$>.05$
Total (rate)	3	644.93		

In this instance, both the linear component and the quadratic components reach significance, and no higher-order component is significant. The plot of the trend is shown in Figure 5-2. It is apparent that there is a decline in comprehension associated with increasing rates of presenta-

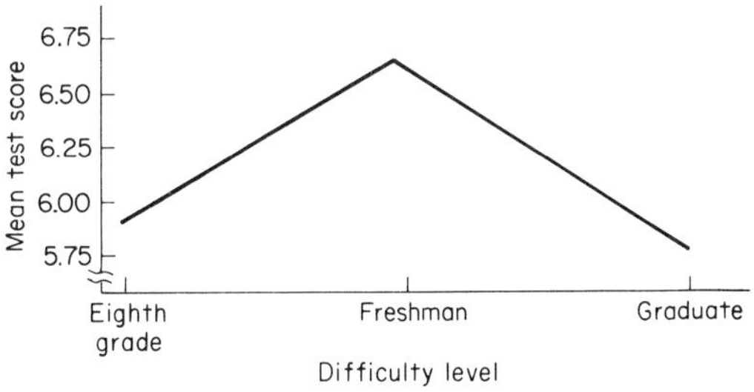

figure 5-1 *Graph of achievement and difficulty level of message.*

tion; however, the decline is less for the higher rates than between 175 and 325 words per minute. This difference in rate of decline in performance accounts for the significant quadratic-trend component.

example 2 Merrill and Stolurow (1966) report an investigation involving the use of hierarchically organized learning sequences. The basic design was a CRD at six levels; in addition, subjects were blocked on quantitative aptitude and on verbal aptitude. The treatment dimension consisted of variations in the instruction given subjects concerning problem-solving procedures for an imaginary science (the "Xenograde system"). All instruction was programmed and presented by means of computer-based teaching machines (known as SOCRATES). The levels of the treatment were:

1. Presentation of question frames only; two attempts at a correct response were allowed, and feedback concerning correctness of response was given.
2. Presentation of five sets of hierarchically arranged summary statements (each based on one principle of the imaginary science) followed by the same presentation of question frames as in level 1.
3. Presentation of question frames as in level 1 with general review frames which were relevant to the problem posed in the question frame.

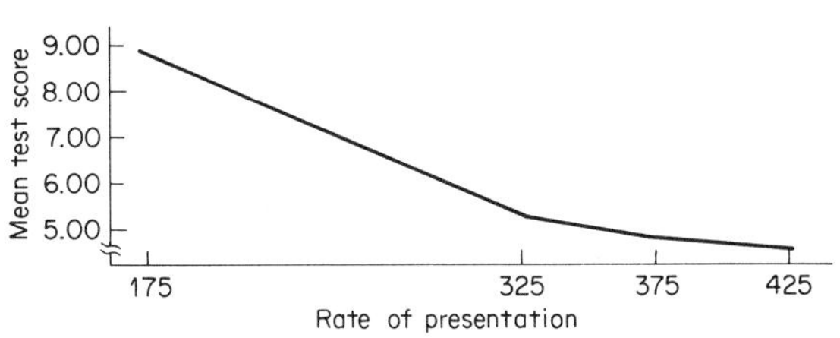

figure 5-2 *Graph of achievement and rate of presentation of message.*

4. Presentation of question frames as in level 1 with a step-by-step presentation of the solution to the problem posed in the question frame.

5. Presentation of question frames as in level 1 with general review frames if an initial error was made and the specific step-by-step solution frames if a second error was made in response.

6. Same presentation as in level 5 except that hierarchically arranged summary statements were presented first as in level 2.

The levels for the two blocking variables quantitative aptitude and verbal aptitude were selected by randomly sampling a total of 675 male freshmen at a large state university. The distribution of the quantitative and verbal aptitude scores was doubly stratified to yield six cells; verbal aptitude was divided into thirds, and quantitative aptitude was divided into halves. From each cell, 12 subjects were selected, and two were randomly assigned to each treatment level. Thus, the overall design contained 36 cells (i.e., six levels of treatment by three levels of verbal aptitude by two levels of quantitative aptitude). With $n = 2$ per cell, the experiment required 72 subjects. Although the authors summarized results for several variables (e.g., total time spent on the program and time per frame), we restrict our attention here to the results from a 68-item achievement test administered 1 week after the experimental sessions. Analysis of variance was applied to these scores; also, specific, preplanned contrasts were tested for the six levels of the treatment dimension. The results from the analysis of variance were:

source	*df*	sum of squares	mean square	*F*	*p*
A Treatment	5	3,245.0	649.0	3.83	<.01
B Verbal aptitude	2	603.0	301.5	1.78	>.05
C Quantitative aptitude	1	325.0	325.0	1.92	>.05
AB	10	755.0	75.5	.45	>.05
AC	5	108.0	21.6	.13	>.05
BC	2	205.0	102.5	.61	>.05
ABC	10	1,001.0	100.1	.59	>.05
Error	35	5,929.0	169.4		
Total	70†	12,171.0			

†Since the total design called for 72 subjects, there would be, normally, 71 total degrees of freedom; however, one cell contained only 1 subject instead of 2; the authors elected to count this score twice in the analysis but to keep the correct degrees of freedom for error.

Although the main effect associated with treatments is highly significant, neither of the blocking variables suggests differential effects, and all interactions are nonsignificant. As mentioned above, the authors had preplanned a set of orthogonal contrasts among the treatment levels:

contrast	TREATMENT LEVEL 1	2	3	4	5	6
1	+1	−1	0	0	+1	−1
2	+1	0	−1	+1	−1	0
3	+1	0	+1	−1	−1	0

The three contrasts account for 3 of the 5 degrees of freedom among the treatment levels; the remaining contrasts were of no interest to the authors, and their combined sum of squares can be found by subtraction. The results of this analysis were:

source	*df*	sum of squares	mean square	*F*	*p*
Contrast 1	1	918.8	918.8	5.42	$<.05$
Contrast 2	1	140.1	140.1	.83	$>.05$
Contrast 3	1	1,496.3	1,496.3	8.83	$<.01$
Contrasts 4 + 5	2	689.8	344.9	1.66	$>.05$
Total (treatment)	5	3,245.0			

(*Note:* As an exercise, the student may wish to check the above calculations; the means for the six levels of the treatment were 23.58, 35.25, 27.59, 35.33, 38.17, and 44.00; since the means are reported to only two decimal places, the sums of squares for the contrasts based on these means will vary slightly from those reported above.) The first contrast, which compared those groups receiving the hierarchically arranged summary statements initially with those groups not receiving it initially, was significant, and inspection of the means revealed that superior performance was manifested by the groups receiving the initial review. Also, the third contrast, which compared those groups receiving the specific step-by-step review with those not receiving this review, was highly significant, and the means reveal that the groups receiving the specific step-by-step review were superior in performance. The second contrast, involving a comparison of groups receiving and not receiving the general review statements, was nonsignificant.

chapter six

nested designs

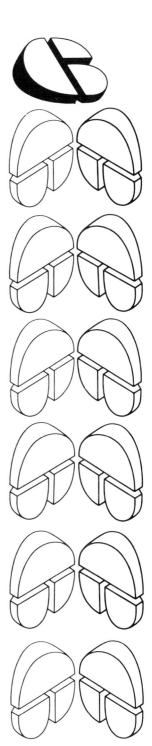

introduction

In behavioral science research, it is not unusual for factors to enter into an experiment which are not completely crossed with the treatment dimensions. Furthermore, it is usually reasonable to assume that these factors will have a unique effect on the outcome of the experiment and, therefore, should be taken into account in the design of the experiment. For example, in a laboratory learning experiment, it may be necessary to have the treatments applied by a number of research assistants (e.g., graduate students who are assisting a professor in the conduct of an experiment). It is likely that no two such assistants are identical in their handling of subjects, their administration of the treatment conditions, etc.; thus, the possible effects of these research assistants should be formally taken into account in the structural model for a score and in the design of the experiment. In a different situation, teaching methods may be studied by having a number of randomly formed groups of students taught by different teaching approaches. Further, two or more teachers might be utilized to teach separate groups of students receiving the same treatment level. Once again, it is reasonable to assume that the teacher will contribute a unique effect to the outcome of the experiment, and this should be taken into account.

In circumstances of the type described above, factors which are not crossed with the treatment dimensions are *nested* within the treatment levels. The concept of nesting can be best understood by studying the design scheme for a teaching methods study. In Table 6-1 three levels of the treatment dimension are shown; each of these levels is a variation in the method of presenting subject-matter content to students. At each treatment level, three teachers are used to instruct separate, random groups of experimental subjects. Note that it is a different group of three teachers at each level of the treatment dimension. Thus, a total of nine teachers are required in the experiment. In order to conduct the experiment as a factorial design, with teachers as one dimension of the design, it would be necessary for each teacher to teach three groups of experimental subjects (i.e., to teach by each of the three methods which

table 6-1 *Example of nesting teachers within treatment levels*

	TREATMENT (A)								
	1			2			3		
TEACHERS (B):	1	2	3	4	5	6	7	8	9
	G_1	G_2	G_3	G_4	G_5	G_6	G_7	G_8	G_9

represent the levels of the treatment dimension). In an actual situation, this may be impractical because of the additional time required to carry out the experiment, because of special training required to ready the teachers to apply the methods, etc. By use of nesting, the experiment can be carried out simultaneously on the nine groups of experimental subjects.

In nested designs, it is impossible to obtain information on interaction effects involving a nested treatment variable and the other treatment dimensions in which it is nested. From Table 6-1, it can be seen that each teacher is involved in only one type of treatment; thus, the differential effects of teachers for different treatment levels cannot be assessed. In many research situations, this would not be a serious shortcoming since the interactions involving nested variables are rarely of intrinsic interest to the researcher. If they are, however, the use of a factorial design is dictated.

Nested treatment variables can be included in higher-dimensional designs (either factorial designs or randomized blocks designs). When there are two or more treatment dimensions besides the nested variable, it is possible for the nesting to be either complete or partial. In complete nesting, the levels of the nested variable are nested within the *cells* of the higher-dimensional design. In partial nesting, the nested variable is crossed (in factorial fashion) with some of the treatment dimensions, but is nested within others. The first case, complete nesting, is illustrated in Table 6-2. Within each cell of the 2 × 2 factorial design, three levels of a nested factor are shown. A total of 12 levels of the nested variable appear in the experiment (e.g., if the "levels" for nesting were teachers, a total of 12 teachers would be utilized in the experiment). A design with complete nesting is also referred to as a *hierarchical design*. In Table 6-3, we represent a 2 × 2 factorial design with the nested factor nested only within the A treatment dimension and crossed with the B treatment dimension. This design is also described

table 6-2 *Completely nested design*

			TREATMENT A				
			1			2	
	nested factor:	1	2	3	4	5	6
TREATMENT B 1		G_1	G_2	G_3	G_4	G_5	G_6
	nested factor:	7	8	9	10	11	12
TREATMENT B 2		G_7	G_8	G_9	G_{10}	G_{11}	G_{12}

table 6-3 *Partially nested design*

		TREATMENT A					
		1			2		
nested factor:		1	2	3	4	5	6
TREATMENT B	1	G_1	G_2	G_3	G_4	G_5	G_6
	2	G_7	G_8	G_9	G_{10}	G_{11}	G_{12}

as a *partially hierarchical design*. In the first example, no interaction information involving the nested factor would be available (of course, main and interaction effects for A and B, the treatments, would be available). In the second example, the interaction of the nested factor and treatment B could be estimated since these two dimensions are crossed. However, no interaction of treatment A and the nested factor could be estimated in the design. Note that in Table 6-3, both levels of treatment dimension B are associated with each level of the partially nested variable (e.g., groups 1 and 7 both receive level 1 of the nested variable, but group 1 is at level 1 of B while group 7 is at level 2 of B; thus, there is crossing of B and the nested variable). However, the nested factor is clearly nested within the levels of A (for example, levels 1, 2, and 3 of the nested factor occur in conjunction with level 1 of A, while levels 4, 5, and 6 of the nested factor occur with level 2 of A; thus, there is no crossing of the nested factor and the A treatment).

In writing the structural model for a score in a nested design, it is necessary to employ some notational convention to indicate clearly the nature of the nesting which is built into the experimental design. The system adopted here involves placing appropriate subscripts in parentheses to indicate the variables in which nesting takes place. Consider the design presented in Table 6-1. The basic design is a completely randomized design with three levels of the treatment. The nested factor, teachers, occurs within the levels of this treatment dimension. In addition to a grand-mean effect and experimental error, the only other effects are due to treatment and to teachers (there is no interaction estimable within the framework of the design). If we call the treatment dimension A and the nested factor B, the structural model for a score is

$$Y_{ijk} = \mu + \alpha_j + \beta_{i(j)} + \epsilon_{ijk}$$

where μ = usual grand-mean effect
α_j = effect of being at jth level of treatment dimension
$\beta_{i(j)}$ = effect of being at ith level of nested factor (within jth level of treatment dimension)
ϵ_{ijk} = experimental-error component which is unique to kth individual within cell (ij) of design

The notation $\beta_{i(j)}$ clearly denotes the fact that B is nested within A. For convenience in writing formulas, the subscript for nested factors will be taken to represent the position of the level of the nested factor within a given level of the main treatment dimension. For the present example, $\beta_{2(1)}$ means the effect of the second level of the nested factor within level 1 of the treatment dimension (i.e., the effect of B within group 2); also, $\beta_{2(2)}$ means the effect of the second level of the nested factor within the second level of the treatment (i.e., the effect of B within group 5). The designs considered in this chapter all assume that the nested factor exists at the same number of levels for each level of the dimensions within which it is nested. Thus, our convention for denoting the nested factor cannot lead to any confusion.

Consider, next, the design represented in Table 6-2. This is a completely nested design, but the basic design is a 2×2 factorial. The structural model for a score in this design is

$$Y_{ijkm} = \mu + \alpha_k + \beta_j + \alpha\beta_{kj} + \gamma_{i(jk)} + \epsilon_{ijkm}$$

where C is the nested factor. Note that all effects from the 2×2 factorial design occur in the model (that is, α_k, β_j, and $\alpha\beta_{kj}$); in addition, the term $\gamma_{i(jk)}$ indicates the ith level of C, the nested factor, within cell (jk) of the design. In group 8, for example, this component would be $\gamma_{2(21)}$ since group 8 is the second group within cell (21) of the design.

Finally, we turn to the partially nested design in Table 6-3. If we let C represent the nested factor, the structural model for a score is

$$Y_{ijkm} = \mu + \alpha_k + \beta_j + \alpha\beta_{kj} + \gamma_{i(k)} + \beta\gamma_{ji(k)} + \epsilon_{ijkm}$$

Again, all the components associated with the basic 2×2 factorial design occur (that is, α_k, β_j, and $\alpha\beta_{kj}$). The main effect due to nesting is $\gamma_{i(k)}$, which indicates that the ith level of the nested factor is nested within the kth level of treatment dimension A (but not nested within B since no j subscript occurs within parentheses). Since C, the nested factor, is crossed with treatment B, the interaction of B and C can be estimated within the design. The term $\beta\gamma_{ji(k)}$ reflects this source of variation among scores. Note that the nesting of C within A is also taken into account in writing this term; the estimation of the BC interaction occurs within levels of the A dimension because of the nesting of C within A.

The concept of nesting can be extended to include two or more nested factors in a basic design of any dimensionality. The nesting of any or all the nested factors can be complete or partial, and nested variables can be nested within other nested variables. To illustrate the possible complexity of the situation, consider the design in Table 6-4. The basic design is a 2×2 factorial. In addition, both factors C and D

table 6-4 *Nested design with two nested factors*

		TREATMENT A											
		1						2					
	factor C:	1		2		3		4		5		6	
	factor D:	1	2	3	4	5	6	13	14	15	16	17	18
TREATMENT B	1	G_1	G_2	G_3	G_4	G_5	G_6	G_{13}	G_{14}	G_{15}	G_{16}	G_{17}	G_{18}
	factor D:	7	8	9	10	11	12	19	20	21	22	23	24
	2	G_7	G_8	G_9	G_{10}	G_{11}	G_{12}	G_{19}	G_{20}	G_{21}	G_{22}	G_{23}	G_{24}

are nested. Factor C is nested within A but crossed with B; factor D is nested within C and also within B. A total of 24 groups of experimental subjects are required for the experiment. The effects which can be estimated are (1) main effects due to A, B, and C nested within A, and D nested within A, B, and C (note that factor D is not crossed with any of the other dimensions); (2) interaction effects due to AB and BC. Factor D cannot enter into any interaction since it is not crossed with any other dimension. Similarly, the AC interaction is not estimable since C is nested within A. The appropriate structural model for a score in the design of Table 6-4 is

$$Y_{ijkmr} = \mu + \alpha_m + \beta_k + \gamma_{j(m)} + \delta_{i(jkm)} + \alpha\beta_{mk} + \beta\gamma_{kj(m)} + \epsilon_{ijkmr}$$

Note that the use of subscripts within parentheses clearly indicates the nature of the nesting which is built into the structure of the experimental design.

The remainder of this chapter is devoted to the analysis and interpretation of designs which include nested factors. We begin with the simplest case, that is, one nested factor added to a completely randomized design. Then a variety of examples of designs with complete and/or partial nesting of treatments are discussed. Throughout, consideration must be given to the nature of the dimensions from the point of view of representing fixed or random variables. The method for deriving expectations of mean squares which was presented in Chapter 3 will be extended to include designs having nested factors.

nesting in one-dimensional designs

In this section, we assume that the basic design is a CRD which exists at p different levels. Within each level of the treatment dimension, there

table 6-5 *Score notation for a nested design ($p = 4$ and $q = 3$)*

	TREATMENT (A)											
	1			2			3			4		
nested factor (B):	1(1)	2(2)	3(3)	1(4)	2(5)	3(6)	1(7)	2(8)	3(9)	1(10)	2(11)	3(12)
	Y_{111}	Y_{211}	Y_{311}	Y_{121}	Y_{221}	Y_{321}	Y_{131}	Y_{231}	Y_{331}	Y_{141}	Y_{241}	Y_{341}
	Y_{112}	Y_{212}	Y_{312}	Y_{122}	Y_{222}	Y_{322}	Y_{132}	Y_{232}	Y_{332}	Y_{142}	Y_{242}	Y_{342}
	...	...	...	...	...	...	...	...	...	...	...	...
	Y_{11n}	Y_{21n}	Y_{31n}	Y_{12n}	Y_{22n}	Y_{32n}	Y_{13n}	Y_{23n}	Y_{33n}	Y_{14n}	Y_{24n}	Y_{34n}

are nested q levels of the nested factor. There is, therefore, a total of pq groups of experimental subjects in the experiment. The data from such a design can be represented as in Table 6-5. In order to denote the nesting condition, each level of the nested factor has two codes; the first stands for the position of the group within the given treatment level, and the second represents the level of the nested factor. For example, $B_{2(2)}$ is the second group under treatment level 1 and is also the second level of the nested factor; $B_{2(5)}$ is the second group within treatment level 2 and is, overall, the fifth level of the nested variable. In writing formulas, the notation within parentheses will be omitted; it is, however, useful for reminding us of the nested nature of the experimental design.

From the point of view of score notation, the system is the same as for a two-dimensional factorial design with n replicates per cell. That is, the score Y_{ijk} represents the kth score within cell (ij) of the design. The usual "dot" notation is utilized to represent summation; the cell and marginal sums are represented in the same manner as for two-dimensional factorial designs. That is, a cell sum of scores would be, in general, $Y_{ij.}$. This sum is also equivalent to the sum for just one level of the nested factor [for example, $Y_{11.}$ is the sum of scores in cell (11) of the design and is also the same as the sum of scores for the first level of the nested factor]. Similarly, $Y_{.j.}$ is the sum of scores for level j of the A treatment dimension, and $Y_{...}$ is the grand sum of the scores. Note that the sum $Y_{i..}$ is of no relevance since this would represent a sum of p different levels of the nested factor (that is, $Y_{1..}$ would be the sum of scores from the p different levels of the nested factor which happen to appear first under each level of the A treatment dimension).

The structural model for a score, as given earlier, is

$$Y_{ijk} = \mu + \alpha_j + \beta_{i(j)} + \epsilon_{ijk}$$

There are two null hypotheses of interest. Of primary concern is the treatment variable A, and the null hypothesis is $H_0: \alpha_1 = \alpha_2 = \cdots = \alpha_p$. Similarly, for the nested factor B we can write $H_0: \beta_{1(1)} = \beta_{2(1)} = \cdots = \beta_{q(p)}$.

The analysis of a one-dimensional nested design is relatively straightforward. With n observations per cell, there is, of course, a within-cell-error term. Thus, the sums of squares which can be computed are associated with treatment (that is, A), with the nested factor (that is, B within A), and with experimental error. The sum of squares associated with treatment is computed from the treatment sums (i.e., the sums for each treatment level, summing over the nested factor). Thus

$$\mathrm{SS}_A = \sum_{j=1}^{p} \frac{(Y_{\cdot j \cdot})^2}{qn} - \frac{(Y_{\dots})^2}{pqn}$$

The denominator of the left-hand term is qn since each level of A is summed over q levels of B, the nested factor, and each level of B contains n observations. The sum of squares for B, the nested factor, is computed from information *within* the levels of A since B is nested within A. That is, this sum of squares reflects the variability of the q levels of B within a level of A (and, of course, summed over the levels of A). For example, within level 1 of A, there are q levels of B. Part of the sum of squares for B is the variability of these q levels of B. Similarly, within level 2 of A there are q levels of B, and the variability of these levels of B constitutes another component of the sum of squares for B. Thus, the sum of squares for B (within A) is a combination of sources of variability each of which is confined to one level of A. Computationally, the sum of squares for B within level 1 of A is

$$\mathrm{SS}_{B(A_1)} = \sum_{i=1}^{q} \frac{(Y_{i1\cdot})^2}{n} - \frac{(Y_{\cdot 1 \cdot})^2}{qn}$$

At level 2 of A, the formula is identical except that the subscript 1 is replaced by 2. The sum of squares for B (within A) is, then,

$$\mathrm{SS}_{B(A)} = \mathrm{SS}_{B(A_1)} + \mathrm{SS}_{B(A_2)} + \cdots + \mathrm{SS}_{B(A_p)}$$

From the definition of the sum of squares for B within A, it is apparent that the degrees of freedom for this source are also computed from within the levels of A. For each level of A, B has $q - 1$ degrees of freedom; thus, overall, the degrees of freedom for $\mathrm{SS}_{B(A)}$ are $p(q - 1)$.

The within-cell-error term can be found either from within-cell variability or by subtraction, after computing a total sum of squares. By the former procedure

$$\mathrm{SS}_{\text{error}} = \sum_{i=1}^{q} \sum_{j=1}^{p} \sum_{k=1}^{n} Y_{ijk}^2 - \sum_{i=1}^{q} \sum_{j=1}^{p} \frac{(Y_{ij\cdot})^2}{n}$$

table 6-6 *Analysis-of-variance summary table for a one-dimensional nested design*

source	df	sum of squares	mean square
Treatment (A)	$p-1$	SS_A	$SS_A/(p-1)$
Nested factor (B)	$p(q-1)$	$SS_{B(A)}$	$SS_{B(A)}/p(q-1)$
Error	$pq(n-1)$	SS_{error}	$SS_{error}/pq(n-1)$

Utilizing the subtractive procedure,

$$SS_{total} = \sum_{i=1}^{q} \sum_{j=1}^{p} \sum_{k=1}^{n} Y_{ijk}^2 - \frac{(Y_{...})^2}{pqn}$$

$$SS_{error} = SS_{total} - SS_A - SS_{B(A)}$$

These two procedures will yield identical results, and both can be computed in order to provide a check on the accuracy of the results. The degrees of freedom for the error sum of squares are, as would be expected, $pq(n-1)$. In effect, each of the pq cells of the experimental design has $n-1$ degrees of freedom. The results, through computation of mean squares, can be displayed as in Table 6-6.

As has been true throughout our discussions of analysis-of-variance designs, setting up F ratios depends upon the nature of the dimensions entering into the design. That is, we must specify whether each factor is fixed or random before proper tests can be ascertained. For the one-dimensional nested design the situation is relatively simple. If both the treatment dimension and the nested dimension represent fixed effects, MS_{error} will provide the denominator for setting up both F ratios. If B is random and A is fixed (the more common mixed model), then the nested factor is tested against error, but the treatment effect is tested against the mean square for the nested factor (i.e., the two F ratios are $MS_A/MS_{B(A)}$ and $MS_{B(A)}/MS_{error}$). For the mixed model with A random and B fixed, MS_A is tested against error, and $MS_{B(A)}$ is tested against MS_A. Finally, for the random-effects model, there is no appropriate denominator for either MS_A or $MS_{B(A)}$.

Since the computation of the sum of squares for the nested factor represents a departure from our usual procedures, this nested design merits illustration. The data in Table 6-7 illustrate an experiment in which the treatment variable exists at just two levels; however, within each treatment level, four levels of the nested factor are applied. In an educational context, the treatment may represent two methods of instructing junior high school social studies, say lecture versus discussion. The nested factor represents different teachers who instruct randomly formed groups of experimental students (note that if the

table 6-7 *Illustrative one-dimensional nested design*

	TREATMENT (A)							
	1 (lecture)				2 (discussion)			
TEACHERS (B):	1	2	3	4	5	6	7	8
	28	34	24	28	30	30	35	30
	21	30	21	24	25	21	32	28
	17	19	18	33	22	20	27	37
	14	27	27	30	27	18	29	30
	19	26	24	27	25	22	25	31
Sums	99	136	114	142	129	111	148	156
Treatment sums			491				544	

groups of students were intact classroom groups, this design would not be appropriate; rather, classroom means would provide the basic data and there would be four observations per cell in a CRD; furthermore, if the intact classroom groups were not randomly assigned to the treatments, no valid analysis would be possible by analysis-of-variance techniques). For an example, a sample size of five per cell was chosen; the scores represent posttest achievement after completion of instruction in the social studies unit.

The sums of squares can be found from group- and treatment-level sums. For treatments, we have

$$SS_{\text{treat}} = \frac{491^2 + 544^2}{20} - \frac{1{,}035^2}{40} = 26{,}850.85 - 26{,}780.62 = 70.23$$

$$SS_{\text{teach (treat)}} = SS_{B(A_1)} + SS_{B(A_2)}$$

and

$$SS_{B(A_1)} = \frac{99^2 + 136^2 + 114^2 + 142^2}{5} - \frac{491^2}{20} = 12{,}291.4 - 12{,}054.05 = 237.35$$

$$SS_{B(A_2)} = \frac{129^2 + 111^2 + 148^2 + 156^2}{5} - \frac{544^2}{20} = 15{,}040.4 - 14{,}796.80 = 243.60$$

Thus,

$$SS_{\text{teach (treat)}} = 237.35 + 243.60 = 480.95$$

The error term can be found by subtraction after computing the total sum of squares.

table 6-8 *Analysis-of-variance summary table for the illustrative one-dimensional nested design*

source	df	sum of squares	mean square
Treatment	1	70.23	70.23
Teachers within treatment	6	480.95	80.16
Error	32	555.20	17.35
Total	39	1,106.38	

$$SS_{\text{total}} = 27{,}887 - \frac{1{,}035^2}{40} = 27{,}887 - 26{,}780.62 = 1{,}106.38$$

Then

$$\begin{aligned} SS_{\text{error}} = SS_{\text{total}} - SS_{\text{treat}} - SS_{\text{teach(treat)}} &= 1{,}106.38 - 70.23 - 480.95 \\ &= 555.20 \end{aligned}$$

These results are summarized in Table 6-8. Constructing F ratios depends upon the nature of the treatment and nested variables. Clearly, the treatment variable represents a fixed dimension; if the teachers were randomly assigned between the two instructional methods, we should consider teachers as representing a random factor. Then treatment would be tested against the nested teacher effect, and the nested effect would be tested against experimental error. Assuming this to be the appropriate model for the present example, we have $F = 70.23/80.16 = .88$ and $F = 80.16/17.35 = 4.62$. The first ratio, for the treatment effect, is clearly nonsignificant since it does not exceed 1.0. The second ratio, for the nested teacher effect, has 6 and 32 degrees of freedom and is significant at beyond the .01 level. Thus, we conclude that the two instructional methods do not result in differential achievement among students; however, different teachers within one instructional method do differ significantly in the amount of achievement displayed by their students. The latter conclusion is not very surprising and has been often documented in actual research settings.

The analysis carried out on the one-dimensional nested design utilized the treatment and group sums. An alternative procedure involves constructing sets of orthogonal contrasts to absorb the degrees of freedom in the design. Since there are eight groups of subjects, there are 7 degrees of freedom among these groups. As is apparent from Table 6-8, these are divided into 1 degree of freedom for treatment and 6 degrees of freedom for the nested factor. Contrasts which correspond to these 7 degrees of freedom are:

	1	2	3	4	5	6	7	8
A	1	1	1	1	−1	−1	−1	−1
B_1	1	−1	0	0	0	0	0	0
B_2	1	1	−2	0	0	0	0	0
B_3	1	1	1	−3	0	0	0	0
B_4	0	0	0	0	1	−1	0	0
B_5	0	0	0	0	1	1	−2	0
B_6	0	0	0	0	1	1	1	−3

The contrast for the treatment A involves comparing groups 1 through 4 with groups 5 through 8. The 6 degrees of freedom for the nested factor B are taken up by contrasts which are entirely within the levels of A. By utilizing patterned contrasts, the components B_1 through B_3 consume 3 degrees of freedom by contrasting the levels of B *within* the first level of A. Similarly, contrast components B_4 through B_6 involve comparisons among the levels of B strictly *within* the second level of A. The sum of the six components B_1 through B_6 constitutes the B-within-A effect. In Table 6-9 are shown the means of the eight groups, and the contrasts and associated sums of squares have been computed. The sum of squares for A agrees within a small rounding error with that obtained earlier and displayed in Table 6-8. Also, the sum $B_1 + B_2 + B_3 + B_4 + B_5 + B_6$ is 480.95, which also agrees with the previously computed value. From the orthogonal contrasts, it is apparent that the major teacher differences which contributed to the significant B-within-A effect involved group 1 versus 2 and groups 1 through 3 versus 4 within level 1 of the treatment and groups 5 and 6 versus 7 and groups 5 through 7 versus 8 within level 2 of the treatment. In completing the

table 6-9 *Orthogonal-contrast procedure applied to a one-dimensional nested design*

GROUP:	1	2	3	4	5	6	7	8	$\sum_{h=1}^{8} c_{gh}\overline{Y}_h = C_g$	$nC_g^2/\sum_{h=1}^{8} c_{gh}^2$
MEAN:	19.8	27.2	22.8	28.4	25.8	22.2	29.6	31.2		
effect										
A	1	1	1	1	−1	−1	−1	−1	−10.6	70.22
B_1	1	−1	0	0	0	0	0	0	−7.4	136.90
B_2	1	1	−2	0	0	0	0	0	1.4	1.63
B_3	1	1	1	−3	0	0	0	0	−15.4	98.82
B_4	0	0	0	0	1	−1	0	0	3.6	32.40
B_5	0	0	0	0	1	1	−2	0	−11.2	104.53
B_6	0	0	0	0	1	1	1	−3	−16.0	106.67

analysis via the method of orthogonal contrasts, the total and error sums of squares would be found in the usual manner.

The orthogonal-contrast procedure for analyzing nested designs can be used in general; patterned contrasts can always be defined *within* the levels of the treatment variable in which nesting takes place, and the degrees of freedom associated with the nested factor can be completely accounted for in this manner.

If the major treatment variable exists at more than two levels, additional analysis is required to locate specific differences when a significant overall F is obtained from the analysis of variance. Any of the multiple-comparison procedures (e.g., Duncan's test or the Newman-Keuls test) can be utilized for this purpose; however, the appropriate error term in these tests is the same term used in the denominator of the F ratio for the treatment effect. Thus, in the case of a mixed model with treatments fixed and the nested factor random, the B-within-A effect (i.e., the nested effect) would be the appropriate error term. Multiple-comparison procedures can also be utilized for the nested factor if it is significant; this is, however, usually of little interest to the researcher.

Also, in designs with unequal sample sizes (or in cases in which a test of the equi-variance assumption is of interest in its own right) a homogeneity-of-variance test should be conducted on the group variances. Any of the standard procedures can be utilized (e.g., the Hartley $F_{\max}$ test or Bartlett's procedure).

complete nesting in one-dimensional designs

When the basic design is a CRD, more than one dimension of nesting can be added to the design. In this section, we study the case in which two nested factors are present and the design is completely hierarchical; that is, the second nested factor is completely nested within the first nested factor. Data from such a design would appear as in Table 6-10, which is for a treatment variable existing at three levels and each nested factor existing at two levels. Labeling the nested factors as B and C, respectively, note that the levels of C are completely nested within the levels of B. Because of the complete nesting of variables, there is no interaction estimable within the framework of the experimental design. There are, however, main effects attributable to A, B within A, and C within B within A. The model for a score is

$$Y_{ijkm} = \mu + \alpha_k + \beta_{j(k)} + \gamma_{i(jk)} + \epsilon_{ijkm}$$

The parenthetical subscripts indicate the nature of the nesting; that is, B is nested within the levels of A, and C is nested within both A and B. The appropriate null hypotheses are

table 6-10 *Score notation for a nested design with two nested factors completely nested* $(p = 3, q = 2, r = 2)$

	TREATMENT A											
	1				2				3			
FACTOR B:	1(1)		2(2)		1(3)		2(4)		1(5)		2(6)	
FACTOR C:	1(1)	2(2)	1(3)	2(4)	1(5)	2(6)	1(7)	2(8)	1(9)	2(10)	1(11)	2(12)
	Y_{1111}	Y_{2111}	Y_{1211}	Y_{2211}	Y_{1121}	Y_{2121}	Y_{1221}	Y_{2221}	Y_{1131}	Y_{2131}	Y_{1231}	Y_{2231}
	Y_{1112}	Y_{2112}	Y_{1212}	Y_{2212}	Y_{1122}	Y_{2122}	Y_{1222}	Y_{2222}	Y_{1132}	Y_{2132}	Y_{1232}	Y_{2232}
	⋯	⋯	⋯	⋯	⋯	⋯	⋯	⋯	⋯	⋯	⋯	⋯
	Y_{111n}	Y_{211n}	Y_{121n}	Y_{221n}	Y_{112n}	Y_{212n}	Y_{122n}	Y_{222n}	Y_{113n}	Y_{213n}	Y_{123n}	Y_{223n}

$$H_0: \alpha_1 = \alpha_2 = \cdots = \alpha_p \qquad H_0: \beta_{1(1)} = \beta_{2(1)} = \cdots = \beta_{q(p)}$$

$$H_0: \gamma_{1(11)} = \gamma_{2(11)} = \cdots = \gamma_{r(pq)}$$

assuming that A exists at p levels, B at q levels, and C at r levels. With n observations per cell, formulas can be written for the sums of squares which follow the general procedures introduced in the case of one nested factor. For both B and C, the variability which is analyzed is completely within the nesting factors (that is, A is the *nesting* factor for B, and A and B are nesting factors for C). Utilizing group sums, we have

$$\mathrm{SS}_A = \sum_{k=1}^{p} \frac{(Y_{..k.})^2}{qrn} - \frac{(Y_{....})^2}{pqrn}$$

$$\mathrm{SS}_{B(A)} = \mathrm{SS}_{B(A_1)} + \mathrm{SS}_{B(A_2)} + \cdots + \mathrm{SS}_{B(A_p)}$$

where

$$\mathrm{SS}_{B(A_1)} = \sum_{j=1}^{q} \frac{(Y_{.j1.})^2}{rn} - \frac{(Y_{..1.})^2}{qrn}$$

and $\mathrm{SS}_{B(A_2)}$ through $\mathrm{SS}_{B(A_p)}$ are defined similarly (i.e., in each case the sum of squares is computed entirely within one level of A).

$$\mathrm{SS}_{C(AB)} = \mathrm{SS}_{C(AB_{11})} + \mathrm{SS}_{C(AB_{12})} + \cdots + \mathrm{SS}_{C(AB_{pq})}$$

where

$$\mathrm{SS}_{C(AB_{11})} = \sum_{i=1}^{r} \frac{(Y_{i11.})^2}{n} - \frac{(Y_{.11.})^2}{rn}$$

and $SS_{C(AB12)}$ through $SS_{C(ABpq)}$ are similarly defined (note that the middle two subscripts are constant within each of the terms constituting $SS_{C(AB)}$). Finally, either the error sum of squares can be computed directly from within-cell information, or the total sum of squares can be computed first and the error term found by subtraction.

As an alternative to the computational approach based on group sums, sets of orthogonal contrasts can be built to correspond to the various effects. Once again, the use of patterned contrasts is most expedient. For the example presented in Table 6-10, there are 12 cells (or groups), and contrasts corresponding to the effects are:

effect	1	2	3	4	5	6	7	8	9	10	11	12
A_1	1	1	1	1	−1	−1	−1	−1	0	0	0	0
A_2	1	1	1	1	1	1	1	1	−2	−2	−2	−2
B_1	1	1	−1	−1	0	0	0	0	0	0	0	0
B_2	0	0	0	0	1	1	−1	−1	0	0	0	0
B_3	0	0	0	0	0	0	0	0	1	1	−1	−1
C_1	1	−1	0	0	0	0	0	0	0	0	0	0
C_2	0	0	1	−1	0	0	0	0	0	0	0	0
C_3	0	0	0	0	1	−1	0	0	0	0	0	0
C_4	0	0	0	0	0	0	1	−1	0	0	0	0
C_5	0	0	0	0	0	0	0	0	1	−1	0	0
C_6	0	0	0	0	0	0	0	0	0	0	1	−1

The A contrasts involve level 1 versus level 2 and level 3 versus the sum of levels 1 and 2. For B within A, the contrasts involve the four groups included under each level of A; two of these groups represent the first level of B, and two represent the second level of B. Similarly, within each level of B (within A) there are two groups, one at the first level of C and one at the second level of C. For the total of 12 groups, there are 11 degrees of freedom among groups, and these have been completely accounted for by the set of 11 orthogonal contrasts. In the present example, each of the contrasts would be reasonably simple to compute (especially those for C which each involve only two groups and those for B which each involve only 4 of the 12 groups). In practice, the method based on group sums and the orthogonal-contrast procedure yield results which are identical within rounding errors, and a choice between them can be left to the personal preferences of the research analyst.

To exemplify this design, consider an extension of the experiment proposed as an illustration of the case of one nested variable. In that

table 6-11 *Illustrative completely nested design with two nested factors*

	TREATMENT (A)							
	1 (lecture)				2 (discussion)			
TEACHERS (B):	1		2		3		4	
CLASSES (C):	1	2	3	4	5	6	7	8
	26	28	35	27	32	24	36	31
	23	22	31	30	26	34	33	29
	19	18	19	27	22	20	25	39
	29	13	25	33	28	19	30	32
	25	19	27	25	25	26	25	33
Sums	122	100	137	142	133	123	149	164
Group means	24.4	20.0	27.4	28.4	26.6	24.6	29.8	32.8
Teacher sums	222		279		256		313	
Teacher means	22.2		27.9		25.6		31.3	
Treatment sums	501				569			
Treatment means	25.05				28.45			

example (see Table 6-7), the treatment existed at two levels, which represented two different instructional procedures, and the nested factor was teachers (four per treatment level). Assume, additionally, that each teacher taught two classes of students but both classes were taught by the same method. We should hypothesize that membership in a particular class could have an effect on a student's performance in the experiment; thus, teachers are nested within methods, and classes are nested within both teachers and methods. The data in Table 6-11 are for a design of this type; however, the number of teachers per level of the treatment has been set at two; thus, a total of eight groups of experimental subjects constitute the experiment.

Since each of the dimensions exists at only two levels, the method of orthogonal contrasts is especially convenient as a form of analysis for this example. The contrast matrix and associated sums of squares are displayed in Table 6-12. In order to obtain the sums of squares for B and C (that is, teachers and classes), it is necessary to pool the 1 degree of freedom components. Thus,

$$\text{SS}_{B(A)} = 162.45 + 162.45 = 324.90$$

and

$$\text{SS}_{C(AB)} = 48.40 + 2.50 + 10.00 + 22.50 = 83.40$$

table 6-12 *Contrast matrix and sums of squares for the illustrative completely nested design*

GROUP:	1	2	3	4	5	6	7	8	$\sum_{h=1}^{8} c_{gh}\bar{Y}_h = C_g$	$\frac{nC_g^2}{\sum_{h=1}^{8} c_{gh}^2}$
MEAN:	24.4	20.0	27.4	28.4	26.6	24.6	29.8	32.8		
A_1	1	1	1	1	1	−1	−1	−1	−13.6	115.60
B_1	1	1	−1	−1	0	0	0	0	−11.4	162.45
B_2	0	0	0	0	1	1	−1	−1	−11.4	162.45
C_1	1	−1	0	0	0	0	0	0	4.4	48.40
C_2	0	0	1	−1	0	0	0	0	−1.0	2.50
C_3	0	0	0	0	1	−1	0	0	2.0	10.00
C_4	0	0	0	0	0	0	1	−1	−3.0	22.50

The error sum of squares can be computed by subtraction after finding the total sum of squares. From the original data, the sum of squared scores is

$$\sum_{i=1}^{r}\sum_{j=1}^{q}\sum_{k=1}^{p}\sum_{m=1}^{n} Y_{ijkm}^2 = 29{,}860$$

and the total sum of scores is $Y_{....} = 1{,}070$. Then

$$\text{SS}_{\text{total}} = 29{,}860 - \frac{1{,}070^2}{40} = 29{,}860 - 28{,}622.50 = 1{,}237.50$$

By subtraction,

$$\text{SS}_{\text{error}} = 1{,}237.50 - 115.60 - 324.90 - 83.40 = 713.60$$

These results are summarized in Table 6-13 along with mean squares. If we assume that classes were randomly assigned to teachers and that teachers were randomly assigned to treatment levels, the appropriate model is mixed, with A fixed and B and C random. In this case, A is tested against B, B is tested against C, and C is tested against within-cell error. (A general procedure for finding expectations in nested designs is presented in the next section of this chapter.) The appropriate F ratios are then

$$F = \frac{115.60}{162.45} = .72 \qquad F = \frac{162.45}{20.85} = 7.79 \qquad F = \frac{20.85}{22.30} = .93$$

table 6-13 *Analysis-of-variance summary table for illustrative completely nested design*

source	df	sum of squares	mean square
Treatment (A)	1	115.60	115.60
Teachers (B within A)	2	324.90	162.45
Classes (C within A and B)	4	83.40	20.85
Error	32	713.60	22.30
Total	39	1,237.50	

The ratios for the treatment effect and for classes are obviously nonsignificant since they do not exceed 1. With 2 and 4 degrees of freedom, the ratio for teachers reaches significance at the .05 level but not at the .01 level. From the contrast sums of squares in Table 6-12 it is apparent that the variability due to teachers is evenly divided between the two treatment levels.

From the two examples of nested designs which have been presented, it can be seen that when the nested factor (or factors) is a random variable, the treatment effect, which is of primary interest to the researcher, is tested by means of an F ratio which has relatively few degrees of freedom in its denominator. In the first example (see Table 6-8), this ratio had only 6 degrees of freedom for its denominator, and, as a result, the test has little power. Similarly, in the second example (see Table 6-13), the denominator had only 2 degrees of freedom, and an extremely large F is required for significance. Note that within-cell error enters only into the tests for nested factors, and it is these tests which are of little real interest to the researcher. Thus, in the construction of experiments involving nested factors, the researcher should be aware of the terms entering into F ratios prior to settling on details of the design. In the first example, the degrees of freedom for the nested factor (which is the denominator in the F ratio for the treatment effect) are $p(q-1)$. Since p, the number of levels of the treatment dimension, is ordinarily fixed by other considerations, the researcher can increase his degrees of freedom for the F test only by increasing q, the number of levels, per level of A, of the nested factor. Similarly, in the second example, the degrees of freedom for the B nested factor are $p(q-1)$, and the same principle applies. In effect, the number of observations per cell does not enter into consideration with respect to the degrees of freedom for tests of the treatment variable. Thus, to effectively increase the power of the experiment, it is the number of levels of the nested factor which must be increased, and the sample size per group is irrelevant. Of course, this reasoning applies only when the nested factor is random; if the model for the design is completely fixed, the within-cell-error term is used in all F ratios, and increasing power depends upon increasing the sample size per group.

expectations of mean squares in nested designs

In Chapter 3, a procedure (the Cornfield and Tukey algorithm) was presented for determining expected values of mean squares from factorial designs. That technique can be extended to apply to designs which have either partial or complete nesting of factors.

The first steps of the augmented procedure are identical with those presented earlier. That is, we begin by writing down the structural model for a score in the design of interest. It is especially important that the nesting of factors be indicated by the parenthetical notation introduced earlier. Next, we construct a table using the main and interaction effects as row headings and the different subscripts (including those which are nested and those which are not nested) as column headings. The major innovation in the procedure occurs in step 3. Recalling the definitions of D_p, D_q, etc., from Chapter 3, we see that step 3 becomes:

3. The cell entries in column k are found from:

a. Enter D_p for each effect which contains a *k which is not nested* in its subscripts.

b. Enter 1 *for each effect which contains a k which is nested.*

c. Enter p for each effect which does not contain k in its subscripts.

Steps 4 through 9 are the same as presented earlier. That is, all columns are filled in by the procedure of step 4, the error term is filled in by the procedure of step 5, and expectations are found by the weighting procedure described in steps 6 through 9.

To illustrate the procedure for a nested design, consider the completely hierarchical design involving one major treatment dimension and two nested variables. The structural model for a score is

$$Y_{ijkm} = \mu + \alpha_k + \beta_{j(k)} + \gamma_{i(jk)} + \epsilon_{ijkm}$$

The special operators are $D_p = 1 - p/P$ for the A treatment, which has p levels; $D_q = 1 - q/Q$ for the nested B factor, which has q levels per level of A; and $D_r = 1 - r/R$ for the nested C factor, which has r levels per cell of the design. Table 6-14 has been set up with the three effects (for A, B, and C) as rows and the different subscripts (i, j, k, and m) as columns. From step 3, the entries in the column headed "k" (for the A dimension) are either D_p, 1, or p. D_p is entered in the one instance in which a k occurs and is not nested; 1 appears for both of the other two effects since they each contain a nested k (that is, a k in parentheses);

table 6-14 *Expected values of mean squares for a completely nested design with two nested factors and one treatment variable*

effect	k	j	i	m	E(MS)
α_k	D_p	q	r	n	$\sigma^2 + nD_r\sigma_\gamma^2 + nrD_q\sigma_\beta^2 + nqr\sigma_\alpha^2$
$\beta_{j(k)}$	1	D_q	r	n	$\sigma^2 + nD_r\sigma_\gamma^2 + nr\sigma_\beta^2$
$\gamma_{i(jk)}$	1	1	D_r	n	$\sigma^2 + n\sigma_\gamma^2$
ϵ_{ijkm}	1	1	1	1	σ^2

p does not appear since all three effects contain k among their subscripts. For the column headed "j," the entries are either D_q, 1, or q. D_q occurs in the one case of an unnested j, 1 appears in the one case of a nested j, and q occurs in the one case in which j does not appear among the subscripts. Similarly, for the i subscript we enter D_r in the one case containing an unnested i, enter r in the two cases not containing an i, and have no instance of a nested i among the subscripts on the effects. The entries for m are all n's since m appears among the subscripts for none of the effects. Next, the error effect is added as a row, and 1s are entered for each column. At this point, the left side of the table is completed. To obtain the expected values, we find the weighted sum of effects (omitting the column or columns whose subscripts correspond to those for the effect in question), where the weights are the column entries which have just been determined. Thus, for the A effect, we omit column k and find the weighted sum by using the other column entries. The student can easily verify the results which are shown.

To make the example specific, assume that A represents a fixed treatment dimension, and B and C both represent random nested factors. Under these circumstances, $D_p = 0$, while $D_q = 1$ and $D_r = 1$. Substituting these values gives the expectations for the mean squares as

$$E(\text{MS}_A) = \sigma^2 + n\sigma_\gamma^2 + nr\sigma_\beta^2 + nqr\sigma_\alpha^2$$

$$E(\text{MS}_{B(A)}) = \sigma^2 + n\sigma_\gamma^2 + nr\sigma_\beta^2$$

$$E(\text{MS}_{C(AB)}) = \sigma^2 + n\sigma_\gamma^2$$

Note that $E(\text{MS}_A)$ and $E(\text{MS}_{B(A)})$ differ only in that the former contains an effect specific to A. Thus, the ratio $\text{MS}_A/\text{MS}_{B(A)}$ has an expected value of 1 when the null hypothesis concerning A is correct but an expectation larger than 1 when this null hypothesis is incorrect. Similarly, appropriate ratios for the B and C effects are $\text{MS}_{B(A)}/\text{MS}_{C(AB)}$ and $\text{MS}_{C(AB)}/\text{MS}_{\text{error}}$. These are, of course, the same tests presented *ex cathedra* when an example of this design was analyzed in the preceding section of this chapter.

The full-blown version of the Cornfield and Tukey algorithm which has now been presented and illustrated has very wide applicability to experimental designs. The algorithm yields expected values of mean squares for any factorial or nested design, and the nesting may be complete or partial in any basic experimental design. Additional illustration of the procedure will be presented in conjunction with examples of more complex nested designs.

relationship between a one-dimensional nested design and a two-dimensional factorial design

In some circumstances, the nested design with one treatment variable and one nested factor may be considered as an alternative design to a two-dimensional factorial design. Consider the case in which the nested factor represents teachers; at each level of the treatment, q different teachers each apply the treatment to random groups of experimental subjects. For $p = 3$ and $q = 3$, the design is:

	TREATMENT (A)								
	1			2			3		
TEACHERS (B):	1(1)	2(2)	3(3)	1(4)	2(5)	3(6)	1(7)	2(8)	3(9)
	G_1	G_2	G_3	G_4	G_5	G_6	G_7	G_8	G_9

If, instead of nesting the teachers, only three teachers were used in the experiment, but each teacher applied each treatment level to a different group of subjects, the design would be two-dimensional:

		TREATMENT (A)		
		1	2	3
	1	G_1	G_4	G_7
TEACHERS (B)	2	G_2	G_5	G_8
	3	G_3	G_6	G_9

The experiment still contains nine groups of experimental subjects; however, there is crossing of teachers and treatment levels. Thus, sums of squares could be computed corresponding to A, B, and AB. For the nested design, the only available sums of squares are for A and B within A. It is interesting to inquire into the relationships between these two types of analysis. Note that for the nested design, the effect due to A is based on contrasts involving groups 1, 2, 3 versus 4, 5, 6 versus 7, 8, 9 (that is, if the orthogonal-contrast scheme of computing sums of squares were utilized, there would be two 1-degree-of-freedom contrasts set up corresponding to the A effects; in terms of patterned contrasts, these would involve groups $1 + 2 + 3$ versus $4 + 5 + 6$ and groups $1 + 2 + 3 + 4 + 5 + 6$ versus $7 + 8 + 9$). Similarly, the analysis of the two-dimensional design would involve contrasting exactly the same groups in computing a sum of squares for the A effect. If identical scores were assumed as data in the two designs, the calculated values of the sum of squares for A would be identical in the nested and two-dimensional designs. Again, with identical scores, it is obvious that SS_{total} would be the same in the two designs. Now, in the nested design, the error term is computed from within-group information; also, in the two-dimensional design the error term is computed within groups. Since we have the same nine groups in both designs and since we are assuming identical scores, it is apparent that SS_{error} would be identical in the two designs. Also, for the nested design, $SS_{total} = SS_A + SS_{B(A)} + SS_{error}$; and for the two-dimensional design, $SS_{total} = SS_A + SS_B + SS_{AB} + SS_{error}$. By the preceding assumption of identical scores, the terms SS_{total}, SS_A, and SS_{error} are identical in the two expressions for SS_{total}. Thus, it is apparent that $SS_{B(A)}$ in the nested design is equivalent to $SS_B + SS_{AB}$ from the two-dimensional design. This formulation supplies some additional interpretation of the sum of squares due to the nested factor. In effect, the sum of squares associated with the nested factor is reflecting both an effect due to the nested factor and an effect due to the interaction of the nested factor and the treatment factor. These two sources of variability are completely confounded within the framework of the nested design; however, in interpreting the outcome of a design which involves nesting it is useful to realize the nature of components entering into the nested effect.

relationship between a one-dimensional nested design and a CRD based on group means

In the previous section, we documented a relationship between one-dimensional nested designs and two-dimensional factorial designs. There is another interesting interpretation of nested designs when the nested factor represents a random variable. Consider a one-dimensional nested design with one nested factor. If the nested factor is random, the appropriate F ratio for testing the treatment effect is

$MS_A/MS_{B(A)}$. If the analysis is performed on group means, an equivalent F ratio results. As presented earlier, the nested design for $p = 3$ and $q = 3$ is

	TREATMENT (A)								
	1			2			3		
TEACHERS (B):	1(1)	2(2)	3(3)	1(4)	2(5)	3(6)	1(7)	2(8)	3(9)
	G_1	G_2	G_3	G_4	G_5	G_6	G_7	G_8	G_9

Each of the groups G_1 through G_9 has a mean; call this $\bar{G}_{ij}$ for the ith group within the jth level of A. The CRD utilizing these group means as the basic data is:

TREATMENT (A)		
1	2	3
$\bar{G}_{11}$	$\bar{G}_{12}$	$\bar{G}_{13}$
$\bar{G}_{21}$	$\bar{G}_{22}$	$\bar{G}_{23}$
$\bar{G}_{31}$	$\bar{G}_{32}$	$\bar{G}_{33}$

In effect, the CRD is a one-way analysis of variance with three observations per cell. In the analysis of this design, the means $\bar{G}_{ij}$ are treated as individual scores. Thus, a sum of squares for treatment SS_A can be computed from the column sums, and a sum of squares for error SS_{error} can be computed from within-column information. The appropriate F ratio for testing the null hypothesis concerning treatment effects is MS_A/MS_{error}. The ratio MS_A/MS_{error} computed from the CRD with group means as scores will be identical with the ratio $MS_A/MS_{B(A)}$ computed from the nested design. Note, also, that the degrees of freedom for the F ratio will be, in general, $p - 1$ and $p(q - 1)$ for both cases. Although the ratios are equal, the actual sums of squares and mean squares will not be the same. In fact, MS_A computed from the nested design will be n times as large as that computed from the CRD with group means as scores. However, $MS_{B(A)}$ computed from the nested design will also be n times as large as MS_{error} computed from the CRD; thus, the F ratios are identical. Establishing these relationships between the mean squares from the nested and completely randomized designs can be accomplished by writing out the computational formulas for all the

table 6-15 *Nested design means presented as a* CRD

	TREATMENT (A) 1 (lecture)	2 (discussion)
	19.8	25.8
	27.2	22.2
	22.8	29.6
	28.4	31.2
Sums	98.2	108.8

sums of squares and then substituting appropriately. The interested student should be able to carry through the algebra without difficulty.

To illustrate the equivalence of the nested design and CRD based on group means, we shall utilize the data presented in Table 6-7. The design involves two treatment levels and four levels of the nested factor within each treatment level. Table 6-15 presents the data in the form of a CRD. Under each treatment level, we show the group means as single observations. Since $q = 4$ in the nested design, there are four observations per level of A in the CRD.

The sum of squares for A, based on the column sums, is

$$\text{SS}_A = \frac{98.2^2 + 108.8^2}{4} - \frac{207.0^2}{8} = 5{,}370.17 - 5{,}356.12 = 14.05$$

The sum of squares for experimental error can be found by subtraction after computing a total sum of squares. From the data,

$$\sum_{i=1}^{4} \sum_{j=1}^{2} Y_{ij}^2 = 5{,}466.36 \qquad \text{and} \qquad \sum_{i=1}^{4} \sum_{j=1}^{2} Y_{ij} = 207.0$$

then

$$\text{SS}_{\text{total}} = 5{,}466.36 - \frac{207.0^2}{8} = 110.24$$

By subtraction

$$\text{SS}_{\text{error}} = 110.24 - 14.05 = 96.19$$

The analysis-of-variance summary table is presented in Table 6-16. For the nested design presented in Table 6-7, the calculated F ratio ($\text{MS}_A/\text{MS}_{B(A)}$) was .88, and this is the same as the value resulting from analysis of the group means in a CRD. Note also the relationships be-

table 6-16 *Analysis-of-variance summary table for a* CRD *based on group means from a nested design*

source	df	sum of squares	mean square	F
Treatment (A)	1	14.05	14.05	.88
Error	6	96.19	16.03	
Total	7	110.24		

tween the mean squares in Table 6-16 and those presented in Table 6-8 for the nested design. From the nested design, $MS_A = 70.23$; from the CRD, $MS_A = 14.05$. Since the nested design was based on $n = 5$ observations per group, we find that $5(14.05) = 70.25$, which is within rounding error of the mean square from the nested design. Also, from the nested design, $MS_{B(A)} = 80.16$; and from the CRD, $MS_{\text{error}} = 16.03$. Further, $5(16.03) = 80.15$, which is within rounding error of the mean square from the nested design.

complete nesting in factorial designs

We now turn to the consideration of additional nested designs. In particular, one or more nested factors can be added to a basic factorial design if a research situation warrants this approach. When nesting is complete for all the nested factors, the principles developed with respect to one-dimensional designs easily generalize to factorial designs. Basically, the main and interaction effects for the factorial design are computed by ignoring the presence of the nested factors. Then sums of squares for the nested factors are computed entirely from within-cell information. The method of analysis based on orthogonal contrasts is especially attractive in the present setting since it avoids building cumbersome computational formulas based on group sums for each different design. In general, the orthogonal-contrast matrix is relatively easy to construct if patterned contrasts based on successive differences are utilized throughout. Then the sums of squares for the 1 degree of freedom components can be computed from familiar formulas and combined to form appropriate sums of squares for main, interaction, and nested effects. In considering the use of a particular nested design, the researcher should specify the structural model for a score and apply the Cornfield and Tukey algorithm in order to determine expected values of mean squares. This should be done prior to a final choice of design to ensure that an appropriate analysis is possible.

To illustrate the construction of an appropriate contrast matrix and the ensuing analysis, consider a basic 2×3 factorial design to which

table 6-17 *Two nested factors in a* 2 $\times$ 3 *factorial design*

	TREATMENT A											
	1				2				3			
factor C:	1(1)		2(2)		1(3)		2(4)		1(5)		2(6)	
factor D:	1(1)	2(2)	1(3)	2(4)	1(5)	2(6)	1(7)	2(8)	1(9)	2(10)	1(11)	2(12)
1	G_1	G_2	G_3	G_4	G_5	G_6	G_7	G_8	G_9	G_{10}	G_{11}	G_{12}
TREATMENT B												
factor C:	1(7)		2(8)		1(9)		2(10)		1(11)		2(12)	
factor D:	1(13)	2(14)	1(15)	2(16)	1(17)	2(18)	1(19)	2(20)	1(21)	2(22)	1(23)	2(24)
2	G_{13}	G_{14}	G_{15}	G_{16}	G_{17}	G_{18}	G_{19}	G_{20}	G_{21}	G_{22}	G_{23}	G_{24}

two completely nested factors have been added. The scheme for such a design is presented in Table 6-17. If we set each of the nested factors C and D at two levels, the total experiment requires $2 \cdot 3 \cdot 2 \cdot 2$ or 24 groups of experimental subjects. In an educational context, the basic design might represent any two manipulatable aspects of instruction while the nested factors represent, respectively, schools within treatment combinations and teachers within schools within treatment combinations. Thus, in cell (11) of the design, two schools occur, and in each of these schools two different teachers apply the treatment combination to independent, random groups of experimental subjects. Since C and D are completely nested, they do not enter into any interactions either between themselves or with the treatment variables. Of course, all effects from the basic 2 $\times$ 3 factorial design can be estimated, as well as effects due to C within AB and D within ABC. The structural model for a score is

$$Y_{ijkmh} = \mu + \alpha_m + \beta_k + \alpha\beta_{mk} + \gamma_{j(km)} + \delta_{i(jkm)} + \epsilon_{ijkmh}$$

If we exclude the grand-mean effect μ, the first three components represent the effects from the 2 $\times$ 3 factorial design. The component $\gamma_{j(km)}$ reflects the influence of receiving the jth level of the C nested factor within cell (km), and the component $\delta_{i(jkm)}$ reflects the influence of receiving the ith level of the D nested factor within the jth level of C and in cell (km). Since the design contains a total of 24 groups, there will be 23 degrees of freedom to divide among treatment effects and effects due to nested factors. The basic factorial design absorbs 2, 1, and 2 degrees of freedom, respectively, for A, B, and AB. The C-within-AB nested effect has 1 degree of freedom per cell of the design (since there are two levels of C per cell). Thus, C within AB absorbs an additional 6 degrees of freedom. Also, D within ABC accounts for 1 degree of freedom per ABC combination, and there are 12 of these. The breakdown of degrees of freedom is:

source	df
A	2
B	1
AB	2
$C(AB)$	6
$D(ABC)$	12
	23

If we assume, further, that there is a total of n observations per cell of the design, the total degrees of freedom are $24n - 1$, and the degrees of freedom for error are $24(n - 1)$.

For the nested design of Table 6-17, expected values for mean squares based on the various main, interaction, and nested effects can be derived by means of the Cornfield and Tukey algorithm. The structural model for a score has already been specified. The table of effects and subscripts is shown as Table 6-18. Since the design contains four different dimensions, we use the operators $D_p = 1 - p/P$, $D_q = 1 - q/Q$, $D_r = 1 - r/R$, and $D_s = 1 - s/S$, assuming that A, B, C, and D exist at p, q, r, and s levels, respectively. Note that for a four-dimensional design, there is just one random model (i.e., all dimensions random) and just one fixed model (i.e., all dimensions fixed), but there are 14 different mixed models which can occur (that is, four ways in which just one dimension is random, six ways in which just two dimensions are

table 6-18 *Expected values of mean squares for a design with two nested factors added to a basic* 2 × 3 *factorial design*

effect	m	k	j	i	h	E(MS)
α_m	D_p	q	r	s	n	$\sigma^2 + nD\sigma_\delta^2 + snD_{sr}\sigma_\gamma^2 + rsnD_q\sigma_{\alpha\beta}^2 + qrsn\sigma_\alpha^2$
β_k	p	D_q	r	s	n	$\sigma^2 + nD_s\sigma_\delta^2 + snD_r\sigma_\gamma^2 + rsnD_p\sigma_{\alpha\beta}^2 + prsn\sigma_\beta^2$
$\alpha\beta_{mk}$	D_p	D_q	r	s	n	$\sigma^2 + nD_s\sigma_\delta^2 + snD_r\sigma_\gamma^2 + rsn\sigma_{\alpha\beta}^2$
$\gamma_{j(km)}$	1	1	D_r	s	n	$\sigma^2 + nD_s\sigma_\delta^2 + sn\sigma_\gamma^2$
$\delta_{i(jkm)}$	1	1	1	D_s	n	$\sigma^2 + n\sigma_\delta^2$
ϵ_{ijkmh}	1	1	1	1	1	σ^2

random, and four ways in which just three dimensions are random). Also, in Table 6-18, the column entries under the subscripts have been entered according to the rules presented earlier. For example, under the subscript m, which corresponds to the A treatment dimension, we have placed D_p in the two rows which contain unnested m's among their subscripts, we have placed 1s in the two rows having nested m's among their subscripts, and we have placed p in the one row not containing an m among its subscripts. After all rows for effects had been completed, the error term was appended and 1s were placed in all columns. The expected values were obtained by the weighting process described earlier.

To make the model specific, assume that both nested factors represent random variables, while both dimensions of the basic factorial design represent fixed variables. Then $D_p = 0$, $D_q = 0$, $D_r = 1$, and $D_s = 1$. The expected values become:

source	E(MS)
A	$\sigma^2 + n\sigma_\delta^2 + sn\sigma_\gamma^2 + qrsn\sigma_\alpha^2$
B	$\sigma^2 + n\sigma_\delta^2 + sn\sigma_\gamma^2 + prsn\sigma_\beta^2$
AB	$\sigma^2 + n\sigma_\delta^2 + sn\sigma_\gamma^2 + rsn\sigma_{\alpha\beta}^2$
$C(AB)$	$\sigma^2 + n\sigma_\delta^2 + sn\sigma_\gamma^2$
$D(ABC)$	$\sigma^2 + n\sigma_\delta^2$
Error	σ^2

It is apparent that all three effects from the basic factorial design, A, B, and AB, are appropriately tested against the C-within-AB term. Then, the C-within-AB term is tested against D within ABC, and D within ABC is tested against experimental error. If we recall our original specification that $p = 3$, $q = 2$, $r = 2$, and $s = 2$, each of the effects from the basic factorial design is tested in an F ratio which has only 6 degrees of freedom for its denominator. To increase these degrees of freedom, and hence the power of these tests, it is necessary to increase the number of levels of the C nested factor. For example, nesting three, rather than two, levels of C results in 12 degrees of freedom for C within AB, and nesting five levels would yield 24 degrees of freedom.

Let us now take a different specification for the same design. Assume that, in addition to C and D, the treatment variable B is also a random component. Then $D_p = 0$, $D_q = 1$, $D_r = 1$, and $D_s = 1$. The expected values are:

source	E(MS)
A	$\sigma^2 + n\sigma_\delta^2 + sn\sigma_\gamma^2 + rsn\sigma_{\alpha\beta}^2 + qrsn\sigma_\alpha^2$
B	$\sigma^2 + n\sigma_\delta^2 + sn\sigma_\gamma^2 + prsn\sigma_\beta^2$
AB	$\sigma^2 + n\sigma_\delta^2 + sn\sigma_\gamma^2 + rsn\sigma_{\alpha\beta}^2$
$C(AB)$	$\sigma^2 + n\sigma_\delta^2 + sn\sigma_\gamma^2$
$D(ABC)$	$\sigma^2 + n\sigma_\delta^2$
Error	σ^2

For this mixed model, the A treatment effect is tested against the AB interaction, B and AB are tested against C within AB, C within AB is tested against D within ABC, and D within ABC is tested against experimental error. Note that the one fixed dimension, A, picks up in its expectations terms involving all three of the random dimensions. This model suffers greatly from lack of degrees of freedom for the test of the A treatment dimension. With $p = 3$, $q = 2$, $r = 2$, and $s = 2$, there are only 2 degrees of freedom for the denominator of the F ratio used to test A. To increase the power of this comparison, additional levels of the treatment variable B must be added to the design, and this would often be an impractical approach. However, if the basic design were really a randomized blocks design with two levels of the blocking variable (which is likely if B is a random dimension), then it might be possible to form different blocking units and increase q.

Having investigated the nature of the expected values for a 2×3 factorial design with two completely nested factors, we turn to a consideration of the computations involved in finding sums of squares. By utilizing the orthogonal-contrast procedure, a matrix of coefficients can be set up as in Table 6-19. Note that a pair of patterned contrasts will absorb the 2 degrees of freedom associated with the A treatment dimension. Since B occurs at only two levels, a single contrast will generate the appropriate sum of squares. Also, both of the nested factors, C and D, occur at only two levels; thus, their corresponding contrasts are especially simple to set up. If actual data were available, the computations required to find sums of squares for each of the 1 degree of freedom components would be based on the formulas

$$C_g = \sum_{h=1}^{24} c_{gh}\overline{Y}_h \qquad \text{and} \qquad \text{SS}_{C_g} = \frac{nC_g^2}{\sum_{h=1}^{24} c_{gh}^2}$$

where, in accordance with our previous usage, c_{gh} is the contrast coefficient in the gth contrast corresponding to cell h of the design, and $\overline{Y}_h$

table 6-19 *Matrix of contrast coefficients for a 2 × 3 factorial design with two completely nested factors*

effect	1111	2111	1211	2211	1112	2112	1212	2212	1113	2113	1213	2213	1121	2121	1221	2221	1122	2122	1222	2222	1123	2123	1223	2223
A_1	+1	+1	+1	+1	−1	−1	−1	−1	0	0	0	0	+1	+1	+1	+1	−1	−1	−1	−1	0	0	0	0
A_2	+1	+1	+1	+1	+1	+1	+1	+1	−2	−2	−2	−2	+1	+1	+1	+1	+1	+1	+1	+1	−2	−2	−2	−2
B	+1	+1	+1	+1	+1	+1	+1	+1	+1	+1	+1	+1	−1	−1	−1	−1	−1	−1	−1	−1	−1	−1	−1	−1
AB_1	+1	+1	+1	+1	−1	−1	−1	−1	0	0	0	0	−1	−1	−1	−1	+1	+1	+1	+1	0	0	0	0
AB_2	+1	+1	+1	+1	+1	+1	+1	+1	−2	−2	−2	−2	−1	−1	−1	−1	−1	−1	−1	−1	+2	+2	+2	+2
$C(AB)_1$	+1	+1	−1	−1	0	0	0	0	0	0	0	0	0	0	0	0	0	0	0	0	0	0	0	0
$C(AB)_2$	0	0	0	0	+1	+1	−1	−1	0	0	0	0	0	0	0	0	0	0	0	0	0	0	0	0
$C(AB)_3$	0	0	0	0	0	0	0	0	+1	+1	−1	−1	0	0	0	0	0	0	0	0	0	0	0	0
$C(AB)_4$	0	0	0	0	0	0	0	0	0	0	0	0	+1	+1	−1	−1	0	0	0	0	0	0	0	0
$C(AB)_5$	0	0	0	0	0	0	0	0	0	0	0	0	0	0	0	0	+1	+1	−1	−1	0	0	0	0
$C(AB)_6$	0	0	0	0	0	0	0	0	0	0	0	0	0	0	0	0	0	0	0	0	+1	+1	−1	−1
$D(ABC)_1$	+1	−1	0	0	0	0	0	0	0	0	0	0	0	0	0	0	0	0	0	0	0	0	0	0
$D(ABC)_2$	0	0	+1	−1	0	0	0	0	0	0	0	0	0	0	0	0	0	0	0	0	0	0	0	0
$D(ABC)_3$	0	0	0	0	+1	−1	0	0	0	0	0	0	0	0	0	0	0	0	0	0	0	0	0	0
$D(ABC)_4$	0	0	0	0	0	0	+1	−1	0	0	0	0	0	0	0	0	0	0	0	0	0	0	0	0
$D(ABC)_5$	0	0	0	0	0	0	0	0	+1	−1	0	0	0	0	0	0	0	0	0	0	0	0	0	0
$D(ABC)_6$	0	0	0	0	0	0	0	0	0	0	+1	−1	0	0	0	0	0	0	0	0	0	0	0	0
$D(ABC)_7$	0	0	0	0	0	0	0	0	0	0	0	0	+1	−1	0	0	0	0	0	0	0	0	0	0
$D(ABC)_8$	0	0	0	0	0	0	0	0	0	0	0	0	0	0	+1	−1	0	0	0	0	0	0	0	0
$D(ABC)_9$	0	0	0	0	0	0	0	0	0	0	0	0	0	0	0	0	+1	−1	0	0	0	0	0	0
$D(ABC)_{10}$	0	0	0	0	0	0	0	0	0	0	0	0	0	0	0	0	0	0	+1	−1	0	0	0	0
$D(ABC)_{11}$	0	0	0	0	0	0	0	0	0	0	0	0	0	0	0	0	0	0	0	0	+1	−1	0	0
$D(ABC)_{12}$	0	0	0	0	0	0	0	0	0	0	0	0	0	0	0	0	0	0	0	0	0	0	+1	−1

is the mean of cell h. Of course, the 1 degree of freedom components are pooled, as necessary, to yield the sums of squares for the main, interaction, and nested effects.

partial nesting in factorial designs

It is not necessary that a nested factor be nested within cells of a basic factorial design. In some applications, a nested factor can be conveniently crossed with one or more of the treatment dimensions but must be nested within others. Partial nesting may arise in educational research settings as a result of administrative difficulties which make either complete nesting or a complete factorial design involving the nested factor inconvenient. Consider an experiment in the area of programmed instruction. Assume that a researcher is interested in studying response mode and step size as it affects achievement in junior high school general science when programmed instruction is the teaching medium. If two response modes (e.g., multiple-choice and completion) and two step sizes (e.g., large and small) were selected as the treatment levels, the basic design would be a 2 × 2 factorial. Assume that randomly selected groups of students are available from six different junior high schools and that the entire experiment is to be accomplished within one class period in a given school. It would be impractical to attempt to replicate the 2 × 2 design within each school since the groups employing different response modes need different instructions relative to the use of the materials. However, students utilizing the large-and small-step versions of the program could easily be instructed at the same time. Thus, an appropriate design would be the basic 2 × 2 factorial design with schools nested within the response modes but crossed with step size. The scheme for this design is displayed in Table 6-20. Note that schools are nested within the A treatment; thus, students in schools 1, 2, and 3 receive only the program form with multiple-choice response format, and students in schools 4, 5, and 6 re-

table 6-20 *Partial nesting in a 2 × 2 factorial design*

	TREATMENT (A) — Response Format: 1 (mult. choice)			2 (completion)		
SCHOOLS (C):	1(1)	2(2)	3(3)	1(4)	2(5)	3(6)
TREATMENT (B) 1 (Small-step)	G_1	G_2	G_3	G_4	G_5	G_6
TREATMENT (B) 2 (Large-step)	G_7	G_8	G_9	G_{10}	G_{11}	G_{12}

ceive only the program with completion response format. However, the students within each school are divided between two groups, one of which takes the small-step version of the program, while the second group completes the large-step version. Thus, schools are nested under A but crossed with B.

The appropriate structural model for a score in the design of Table 6-20 is

$$Y_{ijkm} = \mu + \alpha_k + \beta_j + \alpha\beta_{kj} + \gamma_{i(k)} + \beta\gamma_{ji(k)} + \epsilon_{ijkm}$$

There are terms corresponding to the effects from the basic 2×2 factorial design: A, B, and AB. In addition, a term occurs representing the C-within-A nested factor; also, since C is crossed with B, an interaction effect due to BC within A can be estimated from the data. The allocation of degrees of freedom among these terms is of interest. Since 12 groups of subjects are involved in the total experiment, there are 11 degrees of freedom to distribute among the main, interaction, and nested effects; the breakdown is:

source	df
A	1
B	1
AB	1
$C(A)$	4
$BC(A)$	4
	11

The degrees of freedom for the three effects associated with the basic factorial design (that is, A, B, and AB) are derived in the usual fashion. The component $C(A)$ has 4 degrees of freedom since C exists at three levels within each level of A; thus, per level of A, C has 2 degrees of freedom. In general, if we assume that A exists at p levels and C at r levels, this component has $r(p-1)$ degrees of freedom. The interaction term $BC(A)$ also has 4 degrees of freedom, which can be found by multiplying the degrees of freedom associated with B by those associated with $C(A)$.

By using the Cornfield and Tukey algorithm, the expected values for mean squares for this design have been derived in Table 6-21. As for previous examples, it is instructive to make specific assumptions concerning the natures of the treatment and nested variables in order to study the appropriate F tests. Probably the most common design

table 6-21 *Expected values of mean squares for a 2 × 2 factorial design with one partially nested factor*

effect	k	j	i	m	E(MS)
α_k	D_p	q	r	n	$\sigma^2 + nD_rD_q\sigma_{\beta\gamma}^2 + qnD_r\sigma_\gamma^2 + rnD_q\sigma_{\alpha\beta}^2 + qrn\sigma_\alpha^2$
β_j	p	D_q	r	n	$\sigma^2 + nD_r\sigma_{\beta\gamma}^2 + rnD_p\sigma_{\alpha\beta}^2 + prn\sigma_\beta^2$
$\alpha\beta_{kj}$	D_p	D_q	r	n	$\sigma^2 + nD_r\sigma_{\beta\gamma}^2 + rn\sigma_{\alpha\beta}^2$
$\gamma_{i(k)}$	1	q	D_r	n	$\sigma^2 + nD_q\sigma_{\beta\gamma}^2 + qn\sigma_\gamma^2$
$\beta\gamma_{ji(k)}$	1	D_q	D_r	n	$\sigma^2 + n\sigma_{\beta\gamma}^2$
ϵ_{ijkm}	1	1	1	1	σ^2

would be mixed, with both A and B fixed but C random. This would arise in the case of the illustrative design (Table 6-20) if the schools were randomly divided between the two types of response format. For this mixed model, the values of the operators are $D_p = 0$, $D_q = 0$, and $D_r = 1$. Then, the expectations become:

source	E(MS)
A	$\sigma^2 + qn\sigma_\gamma^2 + qrn\sigma_\alpha^2$
B	$\sigma^2 + n\sigma_{\beta\gamma}^2 + prn\sigma_\beta^2$
AB	$\sigma^2 + n\sigma_{\beta\gamma}^2 + rn\sigma_{\alpha\beta}^2$
$C(A)$	$\sigma^2 + qn\sigma_\gamma^2$
$BC(A)$	$\sigma^2 + n\sigma_{\beta\gamma}^2$
Error	σ^2

In order to set up F ratios with expected values of 1 when the respective null hypothesis is correct, we must use the ratio $MS_A/MS_{C(A)}$ to test the A effect, the ratio $MS_B/MS_{BC(A)}$ to test the B effect, and the ratio $MS_{AB}/MS_{BC(A)}$ to test the AB interaction effect; and both $C(A)$ and $BC(A)$ are tested against within-cell experimental error.

The analysis of partially nested designs can be accomplished by the method of orthogonal contrasts. The illustrative design in Table 6-20 contains a total of 12 groups of experimental subjects. A contrast matrix which specifies components corresponding to the various main, interaction, and nested effects is shown in Table 6-22. The contrasts for the effects from the basic factorial design are defined in an obvious manner. The $C(A)$ effect has four components, and each is defined by ignoring the B treatment classification in the design table. For example, the contrast involving level 1 of C versus level 2 of C within level 1 of A compares

table 6-22 *Matrix of contrast coefficients for a 2 × 2 factorial design with one partially nested factor*

effect	(111)	(121)	(131)	(112)	(122)	(132)	(211)	(221)	(231)	(212)	(222)	(232)
A	+1	+1	+1	−1	−1	−1	+1	+1	+1	−1	−1	−1
B	+1	+1	+1	+1	+1	+1	−1	−1	−1	−1	−1	−1
AB	+1	+1	+1	−1	−1	−1	−1	−1	−1	+1	+1	+1
$C(A)_1$	+1	−1	0	0	0	0	+1	−1	0	0	0	0
$C(A)_2$	+1	+1	−2	0	0	0	+1	+1	−2	0	0	0
$C(A)_3$	0	0	0	+1	−1	0	0	0	0	+1	−1	0
$C(A)_4$	0	0	0	+1	+1	−2	0	0	0	+1	+1	−2
$BC(A)_1$	+1	−1	0	0	0	0	−1	+1	0	0	0	0
$BC(A)_2$	+1	+1	−2	0	0	0	−1	−1	+2	0	0	0
$BC(A)_3$	0	0	0	+1	−1	0	0	0	0	−1	+1	0
$BC(A)_4$	0	0	0	+1	+1	−2	0	0	0	−1	−1	+2

groups $1 + 7$ with groups $2 + 8$. Groups 1 and 2 are at level 1 of B, while groups 7 and 8 are at level 2 of B. The contrast coefficients for the $BC(A)$ components are found by the usual procedure for interaction terms; that is, we multiply together corresponding coefficients from the main effects for B and $C(A)$ to determine the coefficients for $BC(A)$.

Once the matrix of coefficients is determined, the computation of 1 degree of freedom sums of squares utilizes the familiar formulas

$$C_g = \sum_{h=1}^{12} c_{gh}\bar{Y}_h \qquad \text{and} \qquad \text{SS}_{C_g} = \frac{nC_g^2}{\sum_{h=1}^{12} c_{gh}^2}$$

where c_{gh} is the coefficient in the gth contrast corresponding to cell h in the design, and $\bar{Y}_h$ is the mean of the scores in cell h. Combining the appropriate 1 degree of freedom components yields the desired sums of squares for the main, interaction, and nested effects.

As an alternative to the use of a set of orthogonal contrasts, the analyst may wish to carry out computations for nested and partially nested designs in terms of group sums. Although specific formulas will not be developed (this is left as an exercise for the student), the general approach to such an analysis will be indicated. Consider, once again, the design in Table 6-20. The basic design is a 2 × 2 factorial. The effects associated with the 2 × 2 factorial design can be computed by standard formulas if the nested factor is ignored. In terms of the groups entering into the experiment, the analysis is based on the following type of table:

		A: 1	A: 2
B	1	$G_1 + G_2 + G_3$	$G_4 + G_5 + G_6$
	2	$G_7 + G_8 + G_9$	$G_{10} + G_{11} + G_{12}$

In effect, the scores from groups 1 through 3 and so forth are combined and the design is analyzed as if each cell simply contained rn observations (assuming n observations per cell and r levels of the nested factor per level of A). The computational procedures for the basic factorial design presented in Chapter 3 can be used. The nested component $C(A)$ is then found from the variation among the levels of C within the levels of A. In the present example, each level of C within each level of A is applied to two different groups (i.e., one at each of the two levels of B). Thus, the analysis is carried out as if B did not exist, and the data table is:

	A: 1			A: 2		
C:	1(1)	2(2)	3(3)	1(4)	2(5)	3(6)
	$G_1 + G_7$	$G_2 + G_8$	$G_3 + G_9$	$G_4 + G_{10}$	$G_5 + G_{11}$	$G_6 + G_{12}$

The $C(A)$ component has two subparts: $SS_{C(A_1)}$ and $SS_{C(A_2)}$. Each of these is computed from just the information in a single level of the A treatment dimension. Thus, $SS_{C(A_1)}$ is computed from the sums $G_1 + G_7$, $G_2 + G_8$, and $G_3 + G_9$. Each of these sums is based on qn observations (assuming n observations per group and q levels of the B treatment dimension). A formula for $SS_{C(A_1)}$ can be constructed by using the model of the SS_{treat} from a one-way analysis of variance. Similarly, $SS_{C(A_2)}$ can be computed from the sums $G_4 + G_{10}, G_5 + G_{11}$, and $G_6 + G_{12}$.

Once sums of squares for A, B, AB, and $C(A)$ have been computed, the only remaining terms are $BC(A)$ and experimental error. The $BC(A)$ term is best found by subtraction. Begin by computing an "among-groups" sum of squares; this is equivalent to performing a one-way analysis of variance on the 12 groups, ignoring the A, B, and C

classifications. That is, consider the 12 groups as constituting a CRD, and compute the SS_{treat} for the groups; call this term SS_{ag}. Then

$$SS_{BC(A)} = SS_{ag} - SS_A - SS_B - SS_{AB} - SS_{C(A)}$$

Finally, the sum of squares associated with experimental error can be found by subtraction after computing SS_{total} from the scores.

In the last two sections, we have presented some illustrative designs which involve nesting (complete or partial) when the basic design is factorial. Because of the large number of different ways in which complete, partial, or a combination of complete and partial nesting can occur in a design, it is impossible to provide really complete coverage of these possibilities. Rather, we have emphasized the principles involved in constructing and analyzing these designs, so that when they are encountered in practice the student should have no difficulty in generalizing to the new cases.

designs with unequal numbers of observations

Throughout this chapter we have worked on the assumption that each group involved in a nested design was based on the same number of scores, n. Under most circumstances, the introduction of unequal numbers of observations results in an unbalanced design which poses computational difficulties. However, if proportionality holds (in the sense discussed in Chapter 3), the formulas based on group sums will yield correct solutions if careful accounting is made of the proper sample sizes to enter into these formulas.

In cases in which the sample sizes are not only unequal but also disproportionate, the method of unweighted means can be easily adapted to nested designs. In effect, the design is treated as if the sample size per group were 1; the value comprising this n of 1 is the mean of the scores in the group. Either the method of orthogonal contrasts or the method based on group sums can be applied to determine sums of squares for all main, interaction, and nested effects. There is, of course, at this stage of the analysis no within-cell-error term which can be computed. However, within-cell error can be computed from the original scores (i.e., based on groups of unequal size). As indicated in Chapter 3, the harmonic mean of the sample sizes is used to adjust the sums of squares computed from the data table of means. Once this adjustment is carried out, setting up F ratios follows the principles already developed in this chapter. The method of unweighted means is appropriate as an analytical approach when the original plan of the experiment was to have equal sample sizes, but because of events such as

table 6-23 *One-dimensional nested design with unequal sample sizes*

	TREATMENT (A)							
	1 (lecture)				2 (discussion)			
TEACHERS (B):	1	2	3	4	5	6	7	8
	28	34	21	28	30	30	35	28
	17	30	27	24	25	21	32	31
	14	19	24	33	22	20	27	
	19	27		30	27	18	29	
		26		27	25		25	
Sample size	4	5	3	5	5	4	5	2
Sums	78	136	72	142	129	89	148	59
Means	19.5	27.2	24.0	28.4	25.8	22.25	29.6	29.5

subject mortality or loss of data, less than complete data become available for analysis.

To illustrate the general procedure involved in applying the method of unweighted means to nested designs, we return to the example of a one-dimensional nested design treated earlier in this chapter (Table 6-7). The entries in Table 6-23 have been generated by randomly deleting one score from group 1, two scores from group 3, one score from group 6, and three scores from group 8. For this unbalanced design, the harmonic mean of the sample sizes is

$$n_h = \frac{8}{1/4 + 1/5 + 1/3 + 1/5 + 1/5 + 1/4 + 1/5 + 1/2} = \frac{8}{2.1333} = 3.7501$$

The unweighted-means analysis begins by assuming that each cell contains one observation, the cell mean. Thus, the data would look like:

	TREATMENT (A)							
	1 (lecture)				2 (discussion)			
TEACHERS (B):	1	2	3	4	5	6	7	8
	19.5	27.2	24.0	28.4	25.8	22.25	29.6	29.5

Letting $n = 1$, we can set up a matrix of contrast coefficients and compute the sums of squares. This step is summarized in Table 6-24. The 6 degrees of freedom associated with $B(A)$ must be combined to

table 6-24 *Matrix of contrast coefficients and sums of squares for a one-dimensional nested design — unweighted-means analysis*

Teachers:	1	2	3	4	5	6	7	8	$C_g = \sum_{h=1}^{8} c_{gh}\bar{Y}_h$	$SS'_{C_g} = nC_g^2 / \sum_{h=1}^{8} c_{gh}^2$
Mean:	19.5	27.2	24.0	28.4	25.8	22.25	29.6	29.5		
effect										
A	+1	+1	+1	+1	−1	−1	−1	−1	−8.05	8.10
$B(A)_1$	+1	−1	0	0	0	0	0	0	−7.70	29.64
$B(A)_2$	+1	+1	−2	0	0	0	0	0	−1.30	.28
$B(A)_3$	+1	+1	+1	−3	0	0	0	0	−14.50	17.52
$B(A)_4$	0	0	0	0	+1	−1	0	0	3.55	6.30
$B(A)_5$	0	0	0	0	+1	+1	−2	0	−11.15	20.72
$B(A)_6$	0	0	0	0	+1	+1	+1	−3	−10.85	9.81

form the appropriate sum of squares. From this step we have $SS'_A = 8.10$ and $SS'_{B(A)} = 84.27$. These can be adjusted to the average cell size by multiplying by the harmonic mean of the cell sizes, n_h. Then $SS_A = n_h SS'_A = 30.38$, and $SS_{B(A)} = n_h SS'_{B(A)} = 316.02$. The within-cell-error term is found from the original data by pooling sums of squares for each of the eight groups. The sums of squares per group are $SS_1 = 109.00$, $SS_2 = 122.80$, $SS_3 = 18.00$, $SS_4 = 45.20$, $SS_5 = 34.80$, $SS_6 = 84.75$, $SS_7 = 63.20$, and $SS_8 = 4.50$. The sum of these eight terms is $SS_{error} = 482.25$.

The result of the unweighted-means analysis is presented in Table 6-25. If we assume that treatments represent a fixed dimension and that teachers represent a random dimension, then the appropriate F ratio for treatments is $MS_A/MS_{B(A)}$. It is obvious that this ratio does not reach significance since it does not exceed 1 in the present example. The teacher effect can be tested against within-cell error. This ratio is $MS_{B(A)}/MS_{error} = 52.67/19.29 = 2.73$, which is significant at the .05 level with 6 and 25 degrees of freedom.

table 6-25 *Analysis-of-variance summary table for the unweighted-means analysis of a one-dimensional nested design*

source	*df*	sum of squares	mean square
Methods (A)	1	30.38	30.38
Teachers $B(A)$	6	316.02	52.67
Error	25	482.25	19.29

split-plot designs

Within the field of agricultural research (and related areas), a class of designs known as *split-plots* is often encountered. These designs are equivalent to partially nested designs, and in this section we shall demonstrate their similarity. Since many of the "classic" reference works on experimental design were written for agricultural researchers (e.g., Cochran and Cox, 1957; Federer, 1955), the educational researcher who may wish to utilize these references will benefit from understanding the terminology and application of split-plot designs within agricultural settings.

A typical split-plot design may appear as

		plot 1	plot 2	plot 3	plot 4	plot 5	plot 6
				TREATMENT A			
		(A_2)	(A_1)	(A_1)	(A_3)	(A_2)	(A_3)
	1	SP_1	SP_3	SP_1	SP_2	SP_3	SP_2
TREATMENT B	2	SP_3	SP_2	SP_2	SP_1	SP_2	SP_3
	3	SP_2	SP_1	SP_3	SP_3	SP_1	SP_1

The columns correspond to plots of land which are to be used in an agricultural experiment; within each plot, the land has been subdivided into three subplots (that is, SP_1, SP_2, and SP_3 within each plot). The design paradigm corresponds to a mapping of the land being used in the experiment. There are two treatment dimensions, A and B. The levels of A are randomly assigned to the plots, and the levels of B are randomly assigned to the subplots within the levels of A (note that in the design we have, for convenience, shown the levels of B in their natural order, but have rearranged the subplots to correspond to their assignment to the levels of B after randomization). The sums of squares and associated degrees of freedom for such a design would be (assuming p levels of A, q levels of B, and r plots per level of A):

source	*df*	
Between plots	$pr-1$	5
A	$p-1$	2
Plots within A	$p(r-1)$	3
Within plots	$pr(q-1)$	12
B	$q-1$	2
AB	$(p-1)(q-1)$	4
B (plots within A)	$p(q-1)(r-1)$	6

That this split-plot design is equivalent to a partially nested design can be seen from a restructuring of the design scheme:

		A					
		1		2		3	
PLOTS (C):		2	3	1	5	4	6
	1	SP_3	SP_1	SP_1	SP_3	SP_2	SP_2
B	2	SP_2	SP_2	SP_3	SP_2	SP_1	SP_3
	3	SP_1	SP_3	SP_2	SP_1	SP_3	SP_1

In effect, the plots are nested within the levels of A (that is, plots 2 and 3 occur only in conjunction with level 1 of A, plots 1 and 5 only in conjunction with level 2 of A, and plots 4 and 6 only in conjunction with level 3 of A); however, the plots are crossedwith B since each level of B occurs within each plot. The subplots merely correspond to groups of subjects within the context of nested designs. The sums of squares and associated degrees of freedom using nested-design terminology are:

source	df	
A	$p-1$	2
B	$q-1$	2
$C(A)$	$p(r-1)$	3
AB	$(p-1)(q-1)$	4
$BC(A)$	$p(r-1)(q-1)$	6

Since split-plot designs can be rewritten as partially nested designs, there is no new analytical approach required for these designs. Rather, all the procedures developed for nested designs are directly applicable.

exemplary applications of nested designs

The research literature in education contains relatively few examples of the use of nesting in experimental designs. Two factors are likely contributors to this state of affairs. First, many experimenters are unfamiliar with the principles underlying nesting and, as a result, in planning experiments they do not take into account the possibility of including nested factors. Second, those researchers who are familiar with nesting may still hesitate to employ such designs since, when the nested factor is random, a powerful experiment requires a large number of levels of the nested dimension.

example 1 Herman, Potterfield, Dayton, and Amershek (1969) report a study which investigated outcomes, in terms of achievement and interest, from organizing fifth-grade social studies instruction in two ways: teacher-centered and pupil-centered. The experiment involved 6 weeks of instruction, and a total of 18 classrooms participated. In nine of the classrooms, teachers organized the social studies material in terms of their own goals and purposes; they lectured and gave illustrations, while pupils remained in a relatively passive role. In the other nine classrooms, pupils and teacher worked together to set goals and purposes; pupil reports were used in addition to lectures, and, overall, students were actively involved in determining their own activities.

The basic design for the study involved nesting teachers (and their classrooms) within treatments. However, in addition, the groups of pupils were classified in terms of mean intelligence level into above-average, average, and below-average categories. In the experimental design, classroom groups within intelligence levels were randomly assigned to the treatments (i.e., teacher-centered or pupil-centered organization). Also, within classrooms, both male and female pupils participated in the experiment; since the mixture of males and females was not equal for all groups and since there are known sex differences in school achievement of boys and girls, pupils within classrooms were blocked on sex. Thus, the final design involved one treatment dimension, one nested factor, and two blocking variables. The scheme for the design is:

		TREATMENT					
		teacher-centered			pupil-centered		
Classroom		1	2	3	4	5	6
ABOVE AVERAGE	M						
	F						
Classroom		7	8	9	10	11	12
AVERAGE	M						
	F						
Classroom		13	14	15	16	17	18
BELOW AVERAGE	M						
	F						

Note that classrooms are nested within treatment levels and within intelligence levels but that the remaining factors are crossed in factorial fashion. Thus, the structural model for a score is

$$Y_{ijkms} = \mu + \alpha_m + \beta_k + \gamma_j + \delta_{i(mk)} + \alpha\beta_{mk} + \alpha\gamma_{mj} + \beta\gamma_{kj} + \gamma\delta_{ji(mk)} + \alpha\beta\gamma_{mkj} + \epsilon_{ijkms}$$

where $\alpha_m = m$th level of treatment
$\beta_k = k$th level of intelligence
$\gamma_j = j$th level of sex
$\delta_{i(mk)} = i$th classroom within cell (mk) of design

Since classrooms were randomly assigned within the treatment/intelligence cells, the nested factor represents a random dimension. Thus, expectations for mean squares from this design are (assuming p levels of treatment, q levels of intelligence, r levels of sex, and s classrooms nested per treatment/intelligence level cell):

source	E(MS)
A Treatment	$\sigma^2 + rn\sigma_\delta^2 + qrsn\sigma_\alpha^2$
B Levels (intelligence)	$\sigma^2 + rn\sigma_\delta^2 + prsn\sigma_\beta^2$
C Sex	$\sigma^2 + n\sigma_{\gamma\delta}^2 + pqsn\sigma_\gamma^2$
D Classrooms (nested)	$\sigma^2 + rn\sigma_\delta^2$
AB	$\sigma^2 + rn\sigma_\delta^2 + rsn\sigma_{\alpha\beta}^2$
AC	$\sigma^2 + n\sigma_{\gamma\delta}^2 + qsn\sigma_{\alpha\gamma}^2$
BC	$\sigma^2 + n\sigma_{\gamma\delta}^2 + psn\sigma_{\beta\gamma}^2$
ABC	$\sigma^2 + n\sigma_{\gamma\delta}^2 + sn\sigma_{\alpha\beta\gamma}^2$
CD	$\sigma^2 + n\sigma_{\gamma\delta}^2$
Error	σ^2

Note that intelligence levels, treatment, and the interaction of intelligence levels and treatment are properly tested against the nested classroom effect; sex and the interactions AC, BC, and ABC are tested against the CD interaction; and nested classrooms and CD interaction are tested against within-cell error. The design was based on groups of varying sizes; therefore the usual computational formulas were inapplicable. Although an unweighted-means analysis would be computationally most convenient, the availability of computer services made the exact least-squares solution practical.

Although measures of both achievement and interest were gathered in the study, only the results from the interest test are presented at this time (the achievement test was administered on a pretest and posttest basis; thus, analysis of covariance was utilized for this measure). The social studies interest test involved pairing social studies with other school subjects and asking the pupil to select the member of the pair which he preferred. A pupil's score was simply the number of times he chose social studies. The analysis-of-variance summary for the interest test was:

source	df	sum of squares	mean square	F	p
A Treatment	1	7.39	7.39	.93	$>.05$
B Levels (intelligence)	2	6.88	3.44	.44	$>.05$
C Sex	1	.83	.83	.37	$>.05$
D Classrooms (nested)	12	95.28	7.94	3.52	$<.01$
AB	2	68.76	34.38	4.33	$<.05$
AC	1	1.54	1.54	.70	$>.05$
BC	2	20.34	10.17	4.62	$<.05$
ABC	2	11.22	5.61	2.55	$>.05$
CD	12	26.41	2.20	.97	$>.05$
Error	365	824.90	2.26		
Total	400	1,063.55			

Among the main effects, only that due to classrooms within treatments and intelligence levels reaches significance at a conventional level. This result merely indicates that different teachers using the same treatment and working with pupils of comparable intelligence finished the experiment with groups of pupils varying significantly in interest in the subject matter being taught (which is, of course, not an unexpected outcome). The two significant interactions (i.e., between treatments and intelligence levels, and between sex and intelligence levels) are of major interest, and they are plotted in Figure 6-1. Note that above-average female pupils have considerably higher interest scores than

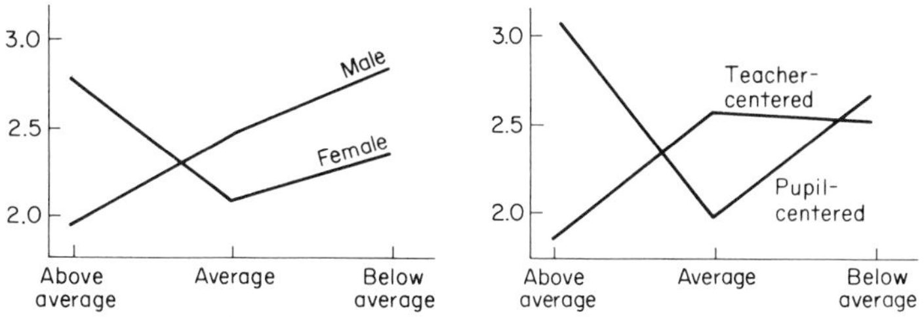

figure 6-1 *Interaction graphs from instructional methods experiment.*

above-average males, but that males at both the average and below-average intelligence levels show more interest than females. With respect to the treatment, above-average students in pupil-centered classrooms showed the highest interest and were considerably higher than above-average students in teacher-centered classrooms. However, the opposite pattern holds for average pupils, and for below-average pupils the interest scores are very similar.

example 2 This example has been selected to illustrate the possibility of an incorrect analysis resulting from a researcher's failure to take nesting into account in the analysis of an experimental design. Williams and Levy (1964) report an investigation of response mode, preliminary training, and time of testing as it affects achievement from programmed instructional materials. The design was basically a 2^3 factorial with the dimensions described as:

A. Type of content. One group given preliminary exposure to material so that experimental exposure was a review; second group given no preliminary exposure so that experimental exposure was initial exposure

B. Response mode. One group given program in standard constructed-response mode; second group given material with response already filled in so that each frame contained a complete statement and did not require a response

C. Time of testing. One group given criterion test on day following completion of training; second group given criterion test 4 weeks after completion of training

A pretest (identical with the criterion test) was given to all subjects, and the scores used in the analysis were gains (that is, criterion test score minus pretest score). The authors treated the design as a 2^3 factorial with the following plan:

	TYPE OF CONTENT (A)			
	review		new	
RESPONSE MODE (B):	constructed	reading	constructed	reading
TIME OF TESTING (C) Immediate	G_1	G_2	G_3	G_4
TIME OF TESTING (C) Delayed	G_5	G_6	G_7	G_8

However, the randomization procedures as described by the authors indicate that nesting actually took place. A total of four classrooms of sixth-grade students were utilized. The authors' description is:

> The experimenter gave two of these classes, selected at random, approximately 20 minutes of instruction on the material presented in the program. For these subjects, then, the program, given later, was essentially a review of previously learned material. The other two classes were given no such preliminary training. Within these two groups . . . , subjects were randomly assigned to one of the two training conditions (Constructed-Response mode and straight reading mode). . . . On the day following completion of the program, half the subjects in each training condition (randomly chosen) were given the posttest. The other subjects were given the posttest four weeks after completion of the program. [Williams and Levy, 1964, p. 213].

It should be noted that classrooms were nested within the type-of-content dimension of the design. However, both response mode and time of testing were crossed with the classrooms in factorial fashion. The actual structure of the design is, therefore:

		TYPE OF CONTENT (A)						
	review				new			
CLASSROOM (D):	1		2		3		4	
RESPONSE MODE (B):†	B_1	B_2	B_1	B_2	B_1	B_2	B_1	B_2
TIME OF TESTING (C) Immediate	G_1	G_2	G_3	G_4	G_9	G_{10}	G_{11}	G_{12}
TIME OF TESTING (C) Delayed	G_5	G_6	G_7	G_8	G_{13}	G_{14}	G_{15}	G_{16}

†B_1 is constructed response mode, and B_2 is reading mode.

Since classrooms were randomly assigned to the levels of A, it is reasonable to treat this dimension as representing a random effect, while the three dimensions of the basic factorial design are clearly fixed. Since D (that is, classrooms) is nested only within the levels of A, there are interaction components corresponding to BD, CD, and BCD in addition to the terms from the factorial design (that is, $A, B, C, AB, AC,$ BC, and ABC). The appropriate structural model for a score is

$$Y_{ijkms} = \mu + \alpha_m + \beta_k + \gamma_j + \delta_{i(m)} + \alpha\beta_{mk} + \alpha\gamma_{mj} + \beta\gamma_{kj} + \alpha\beta\gamma_{mkj} + \beta\delta_{ki(m)} + \gamma\delta_{ji(m)} + \beta\gamma\delta_{kji(m)} + \epsilon_{ijkms}$$

If we assume A at p levels, B at q levels, C at r levels, and D at s levels, the special operators for the Cornfield and Tukey algorithm are $D_p = 0$, $D_q = 0$, $D_r = 0$, and $D_s = 1$. The expectations, then, are (the student should confirm these by carrying out the algorithm):

source	E(MS)
A	$\sigma^2 + qrn\sigma_\delta^2 + qrsn\sigma_\alpha^2$
B	$\sigma^2 + rn\sigma_{\beta\delta}^2 + prsn\sigma_\beta^2$
C	$\sigma^2 + qn\sigma_{\gamma\delta}^2 + pqsn\sigma_\gamma^2$
$D(A)$	$\sigma^2 + qrn\sigma_\delta^2$
AB	$\sigma^2 + rn\sigma_{\beta\delta}^2 + rsn\sigma_{\alpha\beta}^2$
AC	$\sigma^2 + qn\sigma_{\gamma\delta}^2 + qsn\sigma_{\alpha\gamma}^2$
BC	$\sigma^2 + n\sigma_{\beta\gamma\delta}^2 + psn\sigma_{\beta\gamma}^2$
ABC	$\sigma^2 + n\sigma_{\beta\gamma\delta}^2 + sn\sigma_{\alpha\beta\gamma}^2$
$BD(A)$	$\sigma^2 + rn\sigma_{\beta\delta}^2$
$CD(A)$	$\sigma^2 + qn\sigma_{\gamma\delta}^2$
$BCD(A)$	$\sigma^2 + n\sigma_{\beta\gamma\delta}^2$
Error	σ^2

If we focus just on the basic 2^3 factorial design, the appropriate F ratios are MS_A/MS_D, MS_B/MS_{BD}, MS_C/MS_{CD}, MS_{AB}/MS_{BD}, MS_{AC}/MS_{CD}, MS_{BC}/MS_{BCD}, and MS_{ABC}/MS_{BCD}. Each of the effects involving the nested classroom effect is appropriately tested against within-cell experimental error. The authors of the report used MS_{error} as the denominator for all tests involving the effects from the basic 2^3 factorial design, and this is clearly inappropriate. Since data were not presented for the separate classrooms, it is impossible to reconstruct the correct analysis.

chapter seven

repeated measures designs

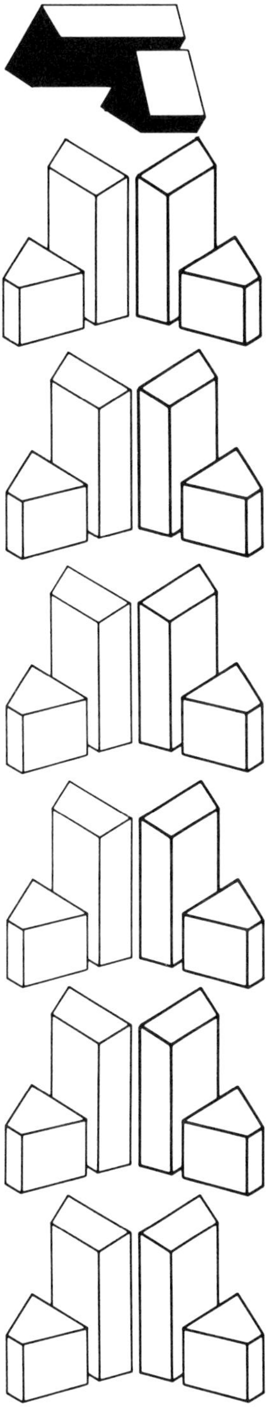

introduction

All the designs considered in earlier chapters have assumed that different treatment levels or combinations of treatment levels have been applied to independent, random groups of experimental subjects. In Chapter 5, the concept of matching groups of subjects which receive different treatment was introduced as one way in which the precision of an experiment can be increased. Although matching tends to increase the homogeneity of groups of subjects, the ideal match for a given experimental subject would be that same subject; that is, if we wished to achieve the highest degree of comparability among subjects receiving different treatment levels, the use of the *same group* of experimental subjects under all treatment levels would assure us of this goal. Experimental designs in which this stratagem is employed are referred to as *repeated measures* designs. Repeated measures may be introduced into any basic design: CRD, factorial, nested, etc. Furthermore, in designs which are basically factorial in nature, the repetition of measurements may be over the entire design or confined to certain of the treatment dimensions. Consider the two design schemes in Table 7-1. In the first part, the basic design is a CRD; just one group of experimental subjects is utilized in the experiment. This group is exposed, in turn, to each of the four levels of the treatment variable. Thus, for a total of n subjects, $4n$ observations are collected, four observations per subject. In the second part of the table, the basic design is a 3×3 factorial design. The repeated measures are carried out only for the A dimension; that is, group 1 receives, in turn, each of the three levels of A. Similarly, groups 2 and 3 each receive all three levels of A; however, with respect to the levels of B, different groups of subjects are utilized.

In practice, repeated measures designs are most often utilized for one of three purposes: First, as alluded to above, the "matching" of a subject with himself provides comparisons involving highly homogeneous material. Hence, we should expect repeated measures designs to be considerably more powerful than designs utilizing completely

table 7-1 *Examples of repeated measures designs*

TREATMENT (A)						TREATMENT (A)		
1	2	3	4			1	2	3
G_1	G_1	G_1	G_1		1	G_1	G_1	G_1
				TREATMENT (B)	2	G_2	G_2	G_2
					3	G_3	G_3	G_3

random groups of experimental subjects and, at least, somewhat more powerful than randomized blocks experiments. Second, the use of repeated measures reduces the number of experimental subjects required to conduct an experiment. Since each subject is utilized more than once, it is often possible to effect substantial savings in time and cost of experimentation through the use of repeated measures designs. And, third, in experiments in which one "treatment" dimension is actually the passage of time, the repeated measures designs are dictated. That is, the experiment involves measuring the same group of subjects repeatedly over time in order to assess effects which develop with the passage of time (e.g., experiments on forgetting and experiments involving physical growth and development).

Although repeated measures designs have these advantages, they are balanced by at least two major drawbacks. First, it is impossible to apply simultaneously all treatment combinations if the same group of subjects must appear in two or more treatment combinations. In long-term research studies, the passage of time may significantly alter general environmental conditions and render meaningless the application of more than one treatment combination to a single group of subjects. In general, the simultaneous application of treatments is highly desirable since it rules out a variety of possible contaminating conditions. The second drawback to the use of repeated measurement with the same group of experimental subjects revolves around the possibility of carry-over effects (also known as multiple treatment interference) from one treatment level to another. Consider the first design in Table 7-1. Assume that the groups receive the treatment levels in their natural order, that is, treatment level 1, then 2, and so forth. If being exposed to level 1 of the treatment in some way influences performance under level 2 (or 3 or 4), we say that the effect of level 1 *carries over* to the other levels. Similarly, of course, the effect of level 2 can carry over to levels 3 and 4, and so forth. The effect of carry-over, it should be noted, can be either accelerative or decelerative; that is, going through the experience of level 1 may either enhance or retard the effect of level 2. To illustrate these two possibilities, consider, first, an experiment which involves concept learning. Whatever the method of concept learning utilized in level 1 of the treatment, it is reasonably sure that the subject's performance on similar concept-learning tasks under level 2 will be enhanced. In studies involving learning, in general, we can assume that carry-over will occur. Now, consider an experiment which involves physical performance on the part of subjects. If the performance is at all strenuous, we should expect the subjects to become more or less fatigued as the experiment progresses; in this context, fatigue represents a carry-over effect which would retard performance at later stages, and under different levels, of the experiment.

From these considerations, it is apparent that the use of repeated measures designs in educational settings must be approached with a good deal of caution since most variables studied in education would be susceptible to carry-over effects if treatment levels were applied in sequence. As we shall see later in this chapter, it is possible to construct designs in which carry-over effects do operate, but in which they are balanced across all treatment levels so that meaningful treatment comparisons are still possible.

In this chapter, we treat a variety of designs which incorporate repeated measures on the same group of experimental subjects. The first design considered is a basic one-dimensional design (or CRD) in which each treatment level is applied to each of n subjects. We then turn to exemplary higher-dimensional designs in which repetition of measurement occurs on one or more of the treatment dimensions. Also, the use of repeated measures in Latin-square designs is explored, and the problems arising from unequal sample sizes are treated.

repeated measures in a one-dimensional design

In this design, one group of experimental subjects provides all the data for the analysis of variance. The scheme, which was presented in the first part of Table 7-1, involves successive administration of the treatment levels to each subject. In general, the order of presentation of the treatment levels should be independently randomized for each experimental subject. For example, assume that 10 subjects constitute the experimental group and that the treatment exists at four levels. For each subject, we randomly select the order of administration of the four treatment levels. This can be accomplished by means of an ordinary table of random numbers or by placing the digits 1 through 4 on separate chips, shuffling these chips in a box, and then blindly drawing the chips from the box. The order in which the digits occur in the drawing determines the order of administration of the treatment levels. Some such randomization process should be utilized whenever feasible in repeated measures designs. There are, however, circumstances in which the order of presentation is dictated by experimental arrangements (e.g., when time is one dimension of the design).

If we assume that the order of presentation of the treatment levels has been independently randomized for each experimental subject (if feasible), the data from the response variable can be summarized as in Table 7-2. In writing this data scheme, we have assumed four levels of the treatment dimension and a total of n subjects. For mnemonic reasons S will be used to stand for the subject dimension in all repeated measures designs. Note that each subject yields four scores; for subject

table 7-2 *Data notation for a one-dimensional design with repeated measures*

		TREATMENT (A) 1	2	3	4
SUBJECTS (S)	1	Y_{11}	Y_{12}	Y_{13}	Y_{14}
	2	Y_{21}	Y_{22}	Y_{23}	Y_{24}
	3	Y_{31}	Y_{32}	Y_{33}	Y_{34}
	...	...	...	...	...
	n	Y_{n1}	Y_{n2}	Y_{n3}	Y_{n4}

1, these are Y_{11}, Y_{12}, Y_{13}, and Y_{14}. The structural model for a score in a one-dimensional repeated measures design is

$$Y_{ij} = \mu + \alpha_j + \pi_i + \alpha\pi_{ji} + \epsilon_{ij}$$

where μ = usual grand-mean effect
α_j = effect of being at jth level of treatment dimension
π_i = unique effect attributable to ith individual
$\alpha\pi_{ji}$ = interaction effect due to treatments and subjects
ϵ_{ij} = experimental error, which is assumed to be normally distributed about 0 and to be homogeneous for all treatments

Note that the data do not allow an independent estimation of experimental error since there is no within-cell variability of scores (that is, per combination of treatment level and subject there is just one observation).

Although the levels of the A treatment dimension can represent either fixed values or random components, it is necessary that the subjects be randomly selected in order for any valid analysis to exist. If the levels of A are selected randomly from a possible universe of such levels, then the design represents a random-effects model; if, as is more common, A represents fixed levels of a treatment, then the design is a mixed model with A fixed and S random. The fact that S is always a random effect means that the expectation of the mean square corresponding to the treatment will contain AS interaction among its components.

A major problem which may arise in the analysis and interpretation of repeated measures designs concerns the possibility of correlated experimental errors for the same group of subjects under different treatment levels. In designs utilizing independent, random groups of subjects, it is safe to assume that the magnitudes of experimental errors will be completely random and unrelated to treatment effects. In repeated measures designs, however, it is not unlikely that the errors associated with a given individual will be similar (and, hence, correlated) across the various treatment levels. If this occurs, the partitioning of the total sum of squares into additive components cannot be accomplished with the usual ease. Indeed, without some restrictive assumptions (which will be specified shortly), the problem defies any straightforward solution. To clarify the situation with respect to possibly correlated experimental-error terms, let us begin with sample data which are strictly additive. The data below, for a repeated measures design with three treatment levels and five subjects, have been contrived to be completely

		TREATMENT (A)			
		1	2	3	π_i
SUBJECTS (S)	1	5	7	3	-3
	2	6	8	4	-2
	3	8	10	6	0
	4	11	13	9	$+3$
	5	10	12	8	$+2$
	α_j	0	$+2$	-2	

additive; the values of the treatment and subject effects are shown. Note that, by assuming a grand-mean value of 8, the values in the cells can be perfectly reproduced by adding the appropriate components to the grand mean. Thus, the entry in cell (32) is 10; the treatment effect for level 2 is $+2$, and the subject effect for subject 3 is 0; thus, $10 = \mu + \alpha_2 + \pi_3 = 8 + (+2) + (0) = 10$. The data are, thus, free of both experimental error ϵ_{ij} and treatment by subject interaction $\alpha\pi_{ji}$. Now, focus on just levels 1 and 2 of the treatment; since $\alpha_1 = 0$ and $\alpha_2 = +2$, the corresponding values in the two columns of scores differ in every case by exactly two units. Thus, $Y_{i2} = Y_{i1} + 2$, and the scores in columns 1 and 2 are perfectly correlated. Similarly, the scores in columns 1 and 3 are perfectly correlated since they differ by -2 units, and the scores in columns 2 and 3 are perfectly correlated since they differ by 4 units. However, the scores as presented contain neither experimental error nor treatment by subject interaction. If interaction components were added to each score, it is obvious that the correlations among columns of scores could be changed; similarly, if experimental error were added to each score, the column intercorrelations could be changed. In partitioning the total sum of squares for a one-dimensional repeated measures design, it is necessary to assume that these column intercorrelations are constant over the possible pairs of columns. Thus, if the magnitude of experimental error were related to the treatment effects or if the interaction components were not randomly distributed, these correlations could be quite dissimilar.

Ordinarily the condition of constant correlation among the columns in a one-dimensional repeated measures design is stated in terms of covariances rather than in terms of correlations. The correlation between two variables can be written as

$$r_{ab} = \frac{\sum_{i=1}^{n} (Y_{ia} - \bar{Y}_a)(Y_{ib} - \bar{Y}_b)}{(n-1)S_a S_b}$$

where $\bar{Y}_a$ and $\bar{Y}_b$ = mean scores for the two variables
n = number of subjects
S_a and S_b = standard deviations for the two sets of scores

The covariance of a and b is defined as

$$S_{ab} = \frac{\sum_{i=1}^{n} (Y_{ia} - \bar{Y}_a)(Y_{ib} - \bar{Y}_b)}{n-1}$$

therefore $S_{ab} = r_{ab}S_aS_b$. In an analysis-of-variance setting, the usual assumption of homogeneity of variance implies that all variances, and all standard deviations, are equal. Thus, specifying equality of correlations is equivalent to specifying equality of covariances if this is taken in conjunction with the assumption of homogeneous variances. We have, then, a homogeneity-of-covariance requirement in addition to the usual homogeneity-of-variance requirement.

The conditions of homogeneity of variance and homogeneity of covariance can be conveniently summarized by the use of matrix notation.[1] By considering the columns of a data table for a one-dimensional repeated measures design, it is possible to compute $p(p-1)/2$ intercorrelations between pairs of columns (across subjects), assuming that the treatment exists at p levels. Also, there are p different variances, one per column. These quantities are shown summarized in a $p \times p$ matrix in Table 7-3. Note that the elements below the diagonal of the matrix are equivalent to symmetrically placed elements above the diagonal since $S_{ab} = S_{ba}$ (that is, the order of entering variables in computing a covariance is irrelevant). The left part of Table 7-3 shows the matrix of variances and covariances utilizing ordinary notation, and the right part has the covariances in an equivalent form utilizing correla-

[1]For students unfamiliar with matrix notation and elementary matrix operations, a brief introduction to these topics can be found in Appendix A.

table 7-3 *Variance-covariance matrix for a repeated measures design*

$$\begin{bmatrix} S_1^2 & S_{12} & S_{13} & \cdots & S_{1p} \\ S_{21} & S_2^2 & S_{23} & \cdots & S_{2p} \\ S_{31} & S_{32} & S_3^2 & \cdots & S_{3p} \\ \cdots & \cdots & \cdots & \cdots & \cdots \\ S_{p1} & S_{p2} & S_{p3} & \cdots & S_p^2 \end{bmatrix} \qquad \begin{bmatrix} S_1^2 & r_{12}S_1S_2 & \cdots & r_{1p}S_1S_p \\ r_{21}S_2S_1 & S_2^2 & \cdots & r_{2p}S_2S_p \\ \cdots & \cdots & \cdots & \cdots \\ r_{p1}S_pS_1 & r_{p2}S_pS_2 & \cdots & S_p^2 \end{bmatrix}$$

tions and standard deviations. Under the assumptions of homogeneity of variance and homogeneity of covariance, the diagonal elements S_1^2 through S_p^2, are each independent estimates of a common population variance σ^2; and the correlations, likewise, each estimate a common population correlation ρ. Thus, the condition of homogeneity of variance and covariance is equivalent to stating that the observed matrix of variances and covariances is sampled from a population in which the true matrix is

$$\begin{bmatrix} \sigma^2 & \rho\sigma^2 & \rho\sigma^2 & \cdots & \rho\sigma^2 \\ \rho\sigma^2 & \sigma^2 & \rho\sigma^2 & \cdots & \rho\sigma^2 \\ \rho\sigma^2 & \rho\sigma^2 & \sigma^2 & \cdots & \rho\sigma^2 \\ \cdots & \cdots & \cdots & \cdots & \cdots \\ \rho\sigma^2 & \rho\sigma^2 & \rho\sigma^2 & \cdots & \sigma^2 \end{bmatrix}$$

Note that all diagonal elements are equal to σ^2, and all off-diagonal elements are equal to $\rho\sigma^2$. Although it is known that the analysis of variance is relatively robust with respect to violations of the homogeneity-of-variance assumption, Box (1953) has demonstrated that violation of the equi-covariance assumption tends to lead to an excess of false rejections of the null hypothesis (i.e., the test has positive bias). Box has also presented an adjustment to the testing procedure which overcomes this positive bias, and this approach will be described in connection with the analysis of repeated measures designs.

As with the homogeneity-of-variance assumption, it is possible to test the assumption of homogeneity of covariance. In fact, it is possible to test the hypothesis that the population variance-covariance matrix has the form specified above and, hence, to test simultaneously the homogeneity-of-variance and homogeneity-of-covariance assumptions. This test is due to Box (1950) and is a more general form of Bartlett's test for homogeneity of variance. In applying Box's test to a one-dimensional repeated measures design, the first step is the computation of variances and covariances to form the entries for a variance-covariance matrix of the type shown in Table 7-3. For reference, we shall denote this matrix as S. The second step involves finding estimates for σ^2 and $\rho\sigma^2$ under the assumptions of equality of variances and equality of covariances. These estimates are simply the mean of the sample variances for the p treatment levels and the mean of the sample covariances. That is, define $\bar{S}_0 = (S_1^2 + S_2^2 + \cdots + S_p^2)/p$ and $\bar{S}_{00} = (S_{12} + S_{13} + \cdots + S_{p-1,p})/\frac{1}{2}p(p-1)$. A matrix is then formed utilizing these values:

$$S_0 = \begin{bmatrix} \bar{S}_0 & \bar{S}_{00} & \bar{S}_{00} & \cdots & \bar{S}_{00} \\ \bar{S}_{00} & \bar{S}_0 & \bar{S}_{00} & \cdots & \bar{S}_{00} \\ \bar{S}_{00} & \bar{S}_{00} & \bar{S}_0 & \cdots & \bar{S}_{00} \\ \cdots & \cdots & \cdots & \cdots & \cdots \\ \bar{S}_{00} & \bar{S}_{00} & \bar{S}_{00} & \cdots & \bar{S}_0 \end{bmatrix}$$

Assuming n subjects and p levels of the treatment, Box's test utilizes the statistics

$$M = -(n-1) \ln \frac{|S|}{|S_0|}$$

where ln is the natural logarithm, or logarithm to base e;

$$C = \frac{p(p+1)^2(2p-3)}{6(n-1)(p-1)(p^2+p-4)}$$

$$B = (1-C)M$$

The statistic B is distributed as chi-square with degrees of freedom equal to $(p^2 + p - 4)/2$. Note that in the computation of M the determinants of S and of S_0 must be computed. For relatively small values of p, this is not a forbidding task; however, for large p the task is best left to digital computers.

In the event that the sample variance-covariance matrix cannot be concluded to be a sample from a population having a homogeneous variance-covariance matrix, two avenues are open to the researcher. He can utilize the conservative procedure which Box recommends and which is described in connection with the analysis of the one-dimensional repeated measures design, or he can turn to the use of a testing procedure which does not depend upon the equi-variance and equi-covariance assumptions. A test of this type is provided by Hotelling's T^2 statistic. This procedure is not described in this volume, but the interested reader can find a discussion of the Hotelling statistic in Morrison (1967).

We turn now to the analysis of one-dimensional repeated measures designs. The data from such a design constitute, from a formal point of view, a two-dimensional analysis of variance with one observation per cell. Thus, sums of squares for treatment, subjects, and interaction can be computed by utilizing formulas which are already familiar. In the notation introduced in Table 7-2, the column sums (corresponding to treatment levels) are $Y_{.1}$ through $Y_{.p}$. Similarly, the row sums (corresponding to subjects) are $Y_{1.}$ through $Y_{n.}$, and the grand sum of the scores is $Y_{..}$. Thus, the sums of squares can be computed from

$$\text{SS}_{\text{treat}} = \frac{\sum_{j=1}^{p} (Y_{.j})^2}{n} - \frac{(Y_{..})^2}{np}$$

$$\text{SS}_{\text{subj}} = \frac{\sum_{i=1}^{n} (Y_{i.})^2}{p} - \frac{(Y_{..})^2}{np}$$

$$\text{SS}_{\text{total}} = \sum_{i=1}^{n} \sum_{j=1}^{p} Y_{ij}^2 - \frac{(Y_{..})^2}{np}$$

$$\text{SS}_{\text{int}} = \text{SS}_{\text{total}} - \text{SS}_{\text{treat}} - \text{SS}_{\text{subj}}$$

The degrees of freedom are $p - 1$, $n - 1$, and $(p - 1)(n - 1)$ for treatment, subjects, and interaction, respectively. The results are conveniently summarized in an analysis-of-variance summary table (Table 7-4).

Under the assumptions of homogeneity of variance and homogeneity of covariance, and if subjects have been randomly selected and the treatment levels are fixed, the mean square for interaction provides the appropriate denominator in an F ratio for testing treatment effects. There is, however, no appropriate term to use in testing the among-subjects variability. In general, a test of this source is of no particular Interest to the researcher; however, a conservative test is possible through the use of MS_{int}. The expectations which justify the above testing procedures are

$$E(\text{MS}_{\text{treat}}) = \sigma^2 + \sigma_{\alpha\pi}^2 + n\sigma_\alpha^2$$

$$E(\text{MS}_{\text{subj}}) = \sigma^2 + p\sigma_\pi^2$$

$$E(\text{MS}_{\text{int}}) = \sigma^2 + \sigma_{\alpha\pi}^2$$

table 7-4 *Analysis-of-variance summary table for a one-dimensional repeated measures design*

source	*df*	sum of squares	mean square
Treatment	$p - 1$	SS_{treat}	$\text{SS}_{\text{treat}}/(p - 1)$
Subjects	$n - 1$	SS_{subj}	$\text{SS}_{\text{subj}}/(n - 1)$
Interaction	$(p - 1)(n - 1)$	SS_{int}	$\text{SS}_{\text{int}}/(p - 1)(n - 1)$

As mentioned previously, in the event that homogeneity of covariance is not an appropriate model for the data (i.e., the test for homogeneity of covariance yields a significant chi-square). Box has proposed a test which is known to be conservative in the sense that the actual level of significance is larger than the nominal level associated with the test statistic. In the event of lack of homogeneous covariances, compute the usual F ratio $\text{MS}_{\text{treat}}/\text{MS}_{\text{int}}$; however, in entering the F tables, assign 1 degree of freedom to the numerator and $n - 1$ degrees of freedom to the denominator. The critical values of F associated with 1 and $n-1$ degrees of freedom are utilized for the ratio rather than the critical values for $p - 1$ and $(p - 1)(n - 1)$ degrees of freedom. Note that for $p = 2$, the degrees of freedom are the same; however, for $p > 2$, both numerator and denominator degrees of freedom will be less than in the usual test and, hence, larger critical values will result. The fact that Box's procedure always leads to a test which is conservative (or, at least, to a test with no positive bias) suggests a stratagem which can, in many circumstances, avoid the calculation of the homogeneity-of-covariance test. Carry out the analysis in the usual way and set up the ratio $\text{MS}_{\text{treat}}/\text{MS}_{\text{int}}$. Then enter the F table first with degrees of freedom of 1 and $n - 1$. If rejection of the null hypothesis is indicated, this result can be adopted since with $p - 1$ and $(p - 1)(n - 1)$ degrees of freedom the same result would occur; thus, the data clearly imply rejection whether or not a conservative test is used. If acceptance of the null hypothesis is indicated with 1 and $n - 1$ degrees of freedom, proceed to check for significance with $p - 1$ and $(p - 1)(n - 1)$ degrees of freedom also. If acceptance is likewise indicated with these degrees of freedolm, this result can be adopted since both a conservative test and a possiby positively biased test are in agreement. Only in the case of contradictory results with the two sets of degrees of freedom need the researcher bother to perform the homogeneity-of-covariance test to determine which set of degrees of freedom is really appropriate. Note that a contradictory result can occur only when the decision of rejection is reached based on $p - 1$ and $(p - 1)(n - 1)$ degrees of freedom, while acceptance is reached based on 1 and $n - 1$ degrees of freedom.

Although the basic analysis presents nothing new, we shall consider an exemplary analysis in order to illustrate the Box test for homogeneity of covariance. The data in Table 7-5 are for 10 subjects who were each exposed to three different levels of a treatment dimension. We must assume that the 10 subjects were randomly selected from a population of available subjects; also, the subjects were independently administered the treatments in random order. In practice, the data analyst would undoubtedly utilize the stratagem based on using degrees of freedom of both 1 and $n - 1$, and $p - 1$ and $(p - 1)(n - 1)$. However, we present the computation for Box's test in order to illus-

table 7-5 *Hypothetical data for* 10 *subjects, each run under three treatment conditions*

subject	TREATMENT LEVEL 1	2	3
1	14	21	19
2	18	19	15
3	17	23	18
4	18	21	17
5	24	20	20
6	23	19	15
7	16	17	12
8	17	19	15
9	15	18	20
10	16	19	15

trate its components. Our first step is to compute the elements of the variance-covariance matrix S. From the columns of Table 7-5, the variances are found to be $S_1^2 = 10.6222$, $S_2^2 = 2.9333$, and $S_3^2 = 6.9333$. From the pairs of columns, the covariances are $S_{12} = .2444$, $S_{13} = .5778$, and $S_{23} = 2.2667$. Thus, the observed variance-covariance matrix is

$$S = \begin{bmatrix} 10.6222 & .2444 & .5778 \\ .2444 & 2.9333 & 2.2667 \\ .5778 & 2.2667 & 6.9333 \end{bmatrix}$$

If we assume homogeneity of variance and of covariance, our best estimates of corresponding population variances and covariances are found by pooling the entries in S; thus, S_0 has entries $(10.6222 + 2.9333 + 6.9333)/3 = 6.8296$ on its main diagonal and entries $(.2444 + .5778 + 2.2667)/3 = 1.0296$ in all off-diagonal positions. That is,

$$S_0 = \begin{bmatrix} 6.8296 & 1.0296 & 1.0296 \\ 1.0296 & 6.8296 & 1.0296 \\ 1.0296 & 1.0296 & 6.8296 \end{bmatrix}$$

In order to compute the statistic M, the determinants of both S and S_0 are needed. Since both of these matrices are 3×3, their determinants are relatively easy to find. Thus,

$$\begin{aligned} |S| &= (10.6222)(2.9333)(6.9333) + (.2444)(2.2667)(.5778) \\ &\quad + (.5778)(.2444)(2.2667) - (.5778)(2.9333)(.5778) \\ &\quad - (2.2667)(2.2667)(10.6222) - (6.9333)(.2444)(.2444) \\ &= 216.0285 + .3201 + .3201 - .9794 - 54.5758 - .4139 \\ &= 160.6996 \end{aligned}$$

and

$$|S_0| = 6.8296^3 + 1.0296^3 + 1.0296^3 - 3(6.8296)(1.0296)^2$$
$$= 318.5558 + 1.0915 + 1.0915 - 3(7.2401)$$
$$= 320.7388 - 21.7203 = 299.0185$$

Then,

$$M = -(10 - 1) \ln \frac{160.6996}{299.0185} = -9 \ln (.5374) = -9(-.6213) = 5.5917$$

$$C = \frac{3(3+1)^2(2 \cdot 3 - 3)}{6(10-1)(3-1)(3^2+3-4)} = \frac{144}{864} \quad \text{and} \quad 1 - C = .8333$$

Thus, $B = (.8333)(5.5917) = 4.6596$. B is distributed as chi-square with $(p^2 + p - 4)/2 = 4$ degrees of freedom. If we check the relevant chi-square distribution, the obtained value of B is not significant at the .05 level, and we conclude that the data satisfy the condition of homogeneity of variance and covariance. Therefore, when the analysis of variance is complete, the conservative testing procedure is not indicated.

Turning now to the analysis, the sums of squares are

$$SS_{total} = 14^2 + 18^2 + \cdots + 15^2 - \frac{(14 + 18 + \cdots + 15)^2}{30}$$
$$= 9{,}950 - \frac{540^2}{30} = 9{,}950 - 9{,}720.00 = 230.00$$

$$SS_{treat} = \frac{178^2 + 196^2 + 166^2}{10} - \frac{540^2}{30} = \frac{97{,}656}{10} - 9{,}720.00 = 45.60$$

$$SS_{subj} = \frac{54^2 + 52^2 + \cdots + 50^2}{3} - \frac{540^2}{30} = 9{,}800.00 - 9{,}720.00 = 80.00$$

$$SS_{int} = SS_{total} - SS_{treat} - SS_{subj} = 230.00 - 45.60 - 80.00 = 104.40$$

These results, as well as mean squares and the F ratio, are displayed in Table 7-6. By utilizing the usual testing procedure, the computed F value of 3.93 is compared with the tabular F with 2 and 18 degrees of freedom; the computed value exceeds the tabular value, and we conclude that the three treatment levels produce discriminable effects (with the conservative degrees of freedom, 1 and 9, the computed value of F is not significant at the .05 level; thus, for this example it was necessary to carry out the homogeneity of variance-covariance test).

table 7-6 *Analysis-of-variance summary table for a hypothetical one-dimensional repeated measures design*

source	df	sum of squares	mean square	F
Treatment	2	45.60	22.80	3.93
Subjects	9	80.00	8.89	
Interaction	18	104.40	5.80	
Total	29	230.00		

repeated measures in higher-dimensional designs

As illustrated earlier, in designs having two or more treatment dimensions, it is possible for the repetition of measurement to occur over all the treatment combinations or to be confined to just certain of the dimensions of the design. Persons are, in effect, nested within those dimensions over which repetition of measurement does not occur. That is, if we consider the experimental subjects as one dimension of the design, they may be nested within one or more of the major dimensions. This conceptualization of the repeated measures dimension of a design is highly useful since all the principles learned in connection with nested designs can be readily applied to repeated measures designs. Also, the Cornfield and Tukey algorithm (in its augmented form) can be used to find expected values of mean squares from repeated measures designs. To illustrate the analogy with nested designs and the use of the Cornfield and Tukey algorithm, consider the 3×3 factorial design in Table 7-7; note that there is repetition of measurement over the A dimension of the design; that is, subjects are crossed with respect to the A dimension. The first n subjects are nested within level 1 of the B treatment dimension but are crossed with the A dimension in the sense that each of these subjects receives each level of A. Similarly, the second n subjects are nested within level 2 of B, and the third n subjects within level 3 of B. In writing the structural model for a score, this nesting should be denoted by the same parenthetical usage introduced earlier. Thus, the effect due to subjects would be written as $\pi_{i(j)}$, assuming that the score Y_{ijk} denotes the ith subject nested within the jth level of B and exposed to the kth level of A. The complete model is

$$Y_{ijk} = \mu + \alpha_k + \beta_j + \pi_{i(j)} + \alpha\beta_{kj} + \alpha\pi_{ki(j)} + \epsilon_{ijk}$$

Note that the interaction effect of experimental subjects and the A dimension occurs since there is crossing of these dimensions.

Table 7-8 shows the results of the Cornfield and Tukey algorithm as applied to this example.

table 7-7 3×3 *factorial design with repeated measures on the A dimension*

			TREATMENT A		
		subjects	1	2	3
		1			
		2			
	1	3	G_1	G_1	G_1
		.			
		.			
		.			
		n			
		1			
		2			
TREATMENT B	2	3	G_2	G_2	G_2
		.			
		.			
		.			
		n			
		1			
		2			
	3	3	G_3	G_3	G_3
		.			
		.			
		.			
		n			

In the most common case, both A and B would represent fixed factors, while subjects would be randomly chosen; thus, $D_p = D_q = 0$ and $D_n = 1$. Under these conditions, the expectations become

$$E(\text{MS}_A) = \sigma^2 + \sigma_{\alpha\pi}{}^2 + nq\sigma_\alpha{}^2$$

$$E(\text{MS}_B) = \sigma^2 + p\sigma_\pi{}^2 + pn\sigma_\beta{}^2$$

$$E(\text{MS}_S) = \sigma^2 + p\sigma_\pi{}^2$$

$$E(\text{MS}_{AB}) = \sigma^2 + \sigma_{\alpha\pi}{}^2 + n\sigma_{\alpha\beta}{}^2$$

$$E(\text{MS}_{AS}) = \sigma^2 + \sigma_{\alpha\pi}{}^2$$

Since each cell of the design contains only one score, there is no estimate of experimental error, and MS_{error} cannot be computed from the data. However, the available mean squares permit tests of the effects

table 7-8 *Expectations of mean squares for a 3 × 3 factorial design with repeated measures on the A dimension*

effect	k	j	i	E(MS)
α_k	D_p	q	n	$\sigma^2 + D_n\sigma_{\alpha\pi}^2 + nD_q\sigma_{\alpha\beta}^2 + nq\sigma_\alpha^2$
β_j	p	D_q	n	$\sigma^2 + D_pD_n\sigma_{\alpha\pi}^2 + nD_p\sigma_{\alpha\beta}^2 + pD_n\sigma_\pi^2 + pn\sigma_\beta^2$
$\pi_{i(j)}$	p	1	D_n	$\sigma^2 + D_p\sigma_{\alpha\pi}^2 + p\sigma_\pi^2$
$\alpha\beta_{kj}$	D_p	D_q	n	$\sigma^2 + D_n\sigma_{\alpha\pi}^2 + n\sigma_{\alpha\beta}^2$
$\alpha\pi_{ki(j)}$	D_p	1	D_n	$\sigma^2 + \sigma_{\alpha\pi}^2$
ϵ_{ijk}	1	1	1	σ^2

from the basic 3 × 3 factorial design, and these are of primary interest to the researcher (that is, the design does not permit testing S or AS; these effects are rarely of interest since they reflect individual differences between experimental subjects). From inspection of the expectations, it is apparent that A and AB are tested against the AS interaction and B is tested against S.

As a matter of terminology, designs in which repetition of measurement is restricted to less than the total number of dimensions of the design are referred to as *mixed* designs. Also, from the point of view of the partitioning of the sums of squares (which will be shown shortly), those effects which are based on contrasts among groups of experimental subjects are called *among subjects*, and the remaining effects are called *within subjects*. For the present example, the sources of variation, degrees of freedom, and expectations would be displayed as:

source	df	E(MS)
Among subjects	$nq - 1$	
B	$q - 1$	$\sigma^2 + p\sigma_\pi^2 + pn\sigma_\beta^2$
S	$q(n - 1)$	$\sigma^2 + p\sigma_\pi^2$
Within subjects	$nq(p - 1)$	
A	$p - 1$	$\sigma^2 + \sigma_{\alpha\pi}^2 + nq\sigma_\alpha^2$
AB	$(p - 1)(q - 1)$	$\sigma^2 + \sigma_{\alpha\pi}^2 + n\sigma_{\alpha\beta}^2$
AS	$q(n - 1)(p - 1)$	$\sigma^2 + \sigma_{\alpha\pi}^2$

Note that this organization of the effects groups together those effects which are tested by means of the same denominator. Also, the among-subjects effect (with $nq - 1$ degrees of freedom) and the within-subjects effect [with $nq(p - 1)$ degrees of freedom] represent a complete partitioning of the available degrees of freedom and of the total sum of squares.

We turn, now, to the partitioning of the total sum of squares for the repeated measures design depicted in Table 7-7. The score notation and sums are shown in Table 7-9.

From inspection of Table 7-9, it is apparent that the B effect involves comparisons among groups of subjects and the A effect involves comparisons within groups of subjects. The nature of the AB interaction is less clear. It is useful to write orthogonal contrasts on the nine cells of the basic 3×3 factorial design to clarify the fact that AB is a within-subject effect. Utilizing patterned contrasts for the main effects and the usual multiplication rule for interaction, we obtain:

	CELL								
	group 1			group 2			group 3		
effect	(11)	(12)	(13)	(21)	(22)	(23)	(31)	(32)	(33)
B_1	+1	+1	+1	−1	−1	−1	0	0	0
B_2	+1	+1	+1	+1	+1	+1	−2	−2	−2
A_1	+1	−1	0	+1	−1	0	+1	−1	0
A_2	+1	+1	−2	+1	+1	−2	+1	+1	−2
AB_1	+1	−1	0	−1	+1	0	0	0	0
AB_2	+1	+1	−2	−1	−1	+2	0	0	0
AB_3	+1	−1	0	+1	−1	0	−2	+2	0
AB_4	+1	+1	−2	+1	+1	−2	−2	−2	+4

With the three groups of experimental subjects labeled, it is apparent from the contrasts that the B effect involves intergroup contrasts [e.g., contrast B_1, which is cells (11) + (12) + (13) − (21) − (22) − (23), is also a contrast of groups 1 and 2 of experimental subjects]. Also, the contrasts constituting the A effect involve within-subject contrasts [e.g., contrast A_1 involves (11) − (12), (21) − (22), and (31) − (32), which are all within-group contrasts]. Note that the contrasts constituting the AB effect are also within-subject contrasts [e.g., AB_1 involves (11) − (12) and (22) − (21), which are both within-group contrasts].

Utilizing the marginal "dot" notation for sums of scores gives the total sum of squares as

$$\text{SS}_{\text{total}} = \sum_{i=1}^{n} \sum_{j=1}^{q} \sum_{k=1}^{p} Y_{ijk}^2 - \frac{(Y_{...})^2}{npq}$$

table 7-9 *Score notation for a 3 × 3 factorial design with repeated measures on the A dimension*

		subjects	TREATMENT A 1	2	3	subject sum	treatment B sum
		1	Y_{111}	Y_{112}	Y_{113}	$Y_{11\cdot}$	
	1	2	Y_{211}	Y_{212}	Y_{213}	$Y_{21\cdot}$	
		⋯	⋯	⋯	⋯	⋯	$Y_{\cdot 1\cdot}$
		n	Y_{n11}	Y_{n12}	Y_{n13}	$Y_{n1\cdot}$	
		1	Y_{121}	Y_{122}	Y_{123}	$Y_{12\cdot}$	
TREATMENT B	2	2	Y_{221}	Y_{222}	Y_{223}	$Y_{22\cdot}$	
		⋯	⋯	⋯	⋯	⋯	$Y_{\cdot 2\cdot}$
		n	Y_{n21}	Y_{n22}	Y_{n23}	$Y_{n2\cdot}$	
		1	Y_{131}	Y_{132}	Y_{133}	$Y_{13\cdot}$	
	3	2	Y_{231}	Y_{232}	Y_{233}	$Y_{23\cdot}$	
		⋯	⋯	⋯	⋯	⋯	$Y_{\cdot 3\cdot}$
		n	Y_{n31}	Y_{n32}	Y_{n33}	$Y_{n3\cdot}$	
Treatment A sum			$Y_{\cdot\cdot 1}$	$Y_{\cdot\cdot 2}$	$Y_{\cdot\cdot 3}$	$Y_{\cdots}$	Grand sum

The among-subject effects B and S are found from the column marginal sums:

$$\text{SS}_B = \sum_{j=1}^{q} \frac{(Y_{\cdot j\cdot})^2}{np} - \frac{(Y_{\cdots})^2}{npq}$$

$$\text{SS}_S = \sum_{j=1}^{q} \left[\frac{\sum_{i=1}^{n} Y_{ij\cdot}^2}{p} - \frac{(Y_{\cdot j\cdot})^2}{np} \right]$$

These two sums of squares SS_B and SS_S represent a partitioning of the among-subjects variability:

$$\sum_{i=1}^{n} \sum_{j=1}^{q} (Y_{ij\cdot})^2 - \frac{(Y_{\cdots})^2}{npq}$$

Since there is a total of nq subjects, this among-subjects sum of squares has $nq - 1$ degrees of freedom. In the partitioning, SS_B has $q - 1$ degrees of freedom, and SS_S has the remainder, $q(n - 1)$. Lest the notation SS_S cause confusion, the student should read this term as "the sum of squares for subjects *within levels of B*" (that is, SS_S is a sum of squares due to subjects nested within levels of B).

The within-subject effects are A, AB, and AS. The last term is most easily found by subtraction after direct computation of the first two terms:

$$\text{SS}_A = \sum_{k=1}^{p} \frac{(Y_{\cdot\cdot k})^2}{nq} - \frac{(Y_{\dots})^2}{npq}$$

$$\text{SS}_{AB} = \sum_{j=1}^{q} \sum_{k=1}^{p} \frac{(Y_{\cdot jk})^2}{n} - \frac{(Y_{\dots})^2}{npq} - \text{SS}_A - \text{SS}_B$$

where $Y_{\cdot jk}$ is the sum of the n scores in cell (jk) of the basic factorial design:

$$\text{SS}_{AS} = \text{SS}_{\text{total}} - \text{SS}_A - \text{SS}_B - \text{SS}_{AB} - \text{SS}_S$$

These three sums of squares SS_A, SS_{AB}, and SS_{AS} represent a partitioning of the within-subject variability:

$$\sum_{i=1}^{n} \sum_{j=1}^{q} \left[\sum_{k=1}^{p} Y_{ijk}^2 - \frac{(Y_{ij\cdot})^2}{p} \right]$$

This sum of squares has $nq(p-1)$ degrees of freedom, of which $p-1$ is associated with A, $(p-1)(q-1)$ is associated with AB, and the remainder, $q(n-1)(p-1)$, is associated with AS. These results, and the corresponding mean squares, are shown in the form of an analysis-of-variance summary table in Table 7-10.

For higher-dimensional designs involving repeated measures, there are homogeneity-of-variance and homogeneity-of-covariance requirements, as was the case for one-dimensional designs. There is, also, a conservative testing procedure which amounts to adjusting the degrees

table 7-10 *Analysis-of-variance summary table for a 3 × 3 factorial design with repeated measures on the A dimension*

source	df	sum of squares	mean square
Among subjects	$nq-1$		
B	$q-1$	SS_B	$\text{SS}_B/(q-1)$
S	$q(n-1)$	SS_S	$\text{SS}_S/q(n-1)$
Within subjects	$nq(p-1)$		
A	$p-1$	SS_A	$\text{SS}_A/(p-1)$
AB	$(p-1)(q-1)$	SS_{AB}	$\text{SS}_{AB}/(p-1)(q-1)$
AS	$q(n-1)(p-1)$	SS_{AS}	$\text{SS}_{AS}/q(n-1)(p-1)$
Total	$pqn-1$	SS_{total}	

of freedom when these requirements are not met. For the present design, with repeated measures only for the A dimension, the adjustment is necessary only for those sources of variation which are within subjects (that is, A and AB). The conservative degrees of freedom for the ratio $\mathrm{MS}_A/\mathrm{MS}_{AS}$ are 1 and $q(n-1)$ rather than the usual $p-1$ and $q(n-1)(p-1)$; for the ratio $\mathrm{MS}_{AB}/\mathrm{MS}_{AS}$ the conservative degrees of freedom are $q-1$ and $q(n-1)$ rather than $(p-1)(q-1)$ and $q(n-1)(p-1)$. Although there are procedures for testing the necessary homogeneity conditions (to be described shortly), the statagem of checking for significance with both the usual and the conservative sets of degrees of freedom can often avoid the labor of actually carrying out the homogeneity tests. The procedure is identical with that described in connection with one-dimensional repeated measures designs.

In mixed repeated measures designs, the homogeneity-of-variance and homogeneity-of-covariance requirements imply that the same population variance-covariance matrix characterizes the subjects *at each level* of the design in which the subjects are nested. In the case of the 3×3 factorial design with repetition on the A dimension, this means that the three sample variance-covariance matrices (one per level of B) are homogeneous. We shall denote these matrices as S_1, S_2, and S_3 (or, in general, as S_1 through S_q when B exists at q levels). The hypothesized common population matrix is identical with that described in connection with one-dimensional designs; that is, each diagonal element is equal to σ^2, and each off-diagonal element is equal to $\rho\sigma^2$, where ρ is the population correlation between pairs of scores for the same group of individuals. However, the testing procedure requires an additional step since there are now two conditions to fulfill: (1) the sample variance-covariance matrices S_1 through S_q must be homogeneous, and (2) the common pattern of S_1 through S_q must conform to the equi-variance and equi-covariance condition mentioned above. In effect, condition 1 can be true while condition 2 is not met; that is, the matrices may be within sampling error of each other, but their common pattern may not be one with identical diagonal elements and identical off-diagonal elements. The test for condition 2 is the procedure presented in connection with one-dimensional repeated measures designs (except for minor changes in constant terms). The test for condition 1, which is also due to Box (1950), involves computing a pooled variance-covariance matrix from the matrices S_1, S_2, etc. The entries in the pooled matrix S_p are weighted averages of the elements in S_1, S_2, etc. (by assuming constant n per level of B, the entries in S_1, S_2, etc., can simply be averaged to yield the elements of S_p). The necessary statistics are, then,

$$M_p = n_t \ln |S_p| - \sum_{j=1}^{q} n_j \ln |S_j|$$

where $n_t = \Sigma_{j=1}^{q} n_j$ is the total number of subjects in the experiment;

$$C_p = \left[\frac{2p^2 + 3p - 1}{6(p+1)(p-1)}\right]\left[\sum_{j=1}^{q} \frac{1}{n_j} - \frac{1}{n_t}\right]$$

$$B_p = (1 - C_p)M_p$$

The statistic B_p is distributed as chi-square with degrees of freedom equal to $p(p+1)(p-1)/2$. Since the n_j are almost always equal in repeated measures designs, some slight simplifications are possible in these formulas. It should be noted that M_p involves finding the determinant of the pooled matrix S_p and the determinants of each of the separate matrices S_1 through S_q. If the value of B_p is not significant when compared with the appropriate chi-square distribution, then the second stage of the testing procedure can be undertaken. If we utilize the elements from S_p, the values of $\bar{S}_0$ and $\bar{S}_{00}$ are found by averaging all diagonal and all off-diagonal elements, respectively. Then the matrix S_0 is formed with $\bar{S}_0$ for all diagonal elements and $\bar{S}_{00}$ for all off-diagonal elements (note that this procedure is identical with that for the one-dimensional case). The statistic M is now

$$M = -(n_t - q) \ln \frac{|S_p|}{|S_0|}$$

and

$$C = \frac{p(p+1)^2(2p-3)}{6(n_t - q)(p-1)(p^2 + p - 4)}$$

$$B = (1 - C)M$$

B is distributed as chi-square with $(p^2 + p - 4)/2$ degrees of freedom. For small values of p, the matrices will all be of relatively low order, and the determinants can be computed without undue difficulty.

To illustrate the computational procedures for the homogeneity tests and for the repeated measures analysis, consider the hypothetical data in Table 7-11. The repetition of measurement is over the A dimension, and there are five subjects nested within each level of B. The structural model for a score is the same as previously presented, and the expectations of mean squares are identical with those displayed in Table 7-8. For the analysis, we assume that both A and B represent fixed dimensions, and, of course, the subjects have been randomly assigned to levels of B.

The homogeneity of variance and covariance conditions require, first, that the separate variance-covariance matrices for the two levels

table 7-11 2×3 *factorial design with repeated measures on the A dimension*

		subjects	TREATMENT A 1	2	3	subject sum	treatment B sum
TREATMENT B	1	1	12	17	9	38	
		2	16	19	5	40	
		3	11	13	7	31	193
		4	15	20	10	45	
		5	12	16	11	39	
		Cell sum	66	85	42		
	2	1	19	26	15	60	
		2	17	21	13	51	
		3	17	24	17	58	285
		4	20	28	19	67	
		5	17	21	11	49	
		Cell sum	90	120	75		
Treatment A sum			156	205	117	478	Grand sum

of B be homogeneous. From the data at level 1 of B, it is possible to compute three variances (i.e., one per cell, each cell containing the scores for the five subjects under a specific level of A) and three covariances (i.e., in terms of levels of A: 1 versus 2, 1 versus 3, and 2 versus 3). Carrying out these computations results in the matrix S_1:

$$S_1 = \begin{bmatrix} 4.70 & 5.25 & -1.85 \\ 5.25 & 7.50 & .25 \\ -1.85 & .25 & 5.80 \end{bmatrix}$$

Note that the covariance of subjects under levels 2 and 3 of A is negative, indicating a negative correlation in the sample between the scores. Turning to level 2 of B, the same computations yield

$$S_2 = \begin{bmatrix} 2.00 & 4.00 & 3.00 \\ 4.00 & 9.50 & 8.50 \\ 3.00 & 8.50 & 10.00 \end{bmatrix}$$

Since there is an equal number of subjects at each level of B, S_1 and S_2 can be pooled by simply averaging corresponding elements. Thus,

$$S_p = \begin{bmatrix} 3.350 & 4.625 & .575 \\ 4.625 & 8.500 & 4.375 \\ .575 & 4.375 & 7.900 \end{bmatrix}$$

The values of the determinants of these three matrices are

$$|S_1| = 13.7687 \qquad |S_2| = 4.0000 \qquad |S_p| = 12.3051$$

Then

$$\begin{aligned} M_p &= 10[\ln (12.3051)] - [5\cdot\ln (13.7687) + 5\cdot\ln (4.0000)] \\ &= 10(2.5100) - 5(2.6225 + 1.3863) = 25.1000 - 20.0440 = 5.0560 \end{aligned}$$

and

$$C_p = \frac{18 + 9 - 1}{6\cdot 4\cdot 2}\left(\frac{2}{5} - \frac{1}{10}\right) = .1625$$

Finally,

$$B_p = (1 - .1625)(5.0560) = 4.23$$

The statistic B_p is distributed as chi-square with 12 degrees of freedom; the computed value does not approach significance at conventional levels, and we can conclude that the variance-covariance matrices at the two levels of B are homogeneous.

Although the homogeneity of S_1 and S_2 has been established, it is still necessary to test S_p for the specific requirement of equal variances and equal covariances.

The elements of S_0 are

$$\bar{S}_0 = \frac{3.350 + 8.500 + 7.900}{3} = 6.5833$$

$$\bar{S}_{00} = \frac{4.625 + .575 + 4.375}{3} = 3.1917$$

Thus

$$S_0 = \begin{bmatrix} 6.5833 & 3.1917 & 3.1917 \\ 3.1917 & 6.5833 & 3.1917 \\ 3.1917 & 3.1917 & 6.5833 \end{bmatrix}$$

The determinant of S_0 is $|S_0| = 149.1557$; also, from the previous test, $|S_p| = 12.3051$. Then

$$M = -(10 - 2)\ \ln \frac{12.3051}{149.1557} = -8[\ln (.0825)] = -8(-2.4950) = 19.9600$$

$$C = \frac{144}{768} = .1875$$

Then

$$B = (1 - .1875)(19.9600) = .8125(19.9600) = 16.22$$

The value of B is distributed as chi-square with $(3^2 + 3 - 4)/2 = 4$ degrees of freedom; compared with the tabular values, the computed value is significant at beyond the .01 level, and we conclude that the repeated measures data of Table 7-11 deviate from the condition of homogeneity of variance and covariance. Therefore, after computation of the sums of squares, the conservative testing procedure will be utilized.

Turning to the sums of squares, we have

$$SS_{\text{total}} = 8{,}452 - \frac{478^2}{30} = 8{,}452 - 7{,}616.1333 = 835.87$$

$$SS_B = \frac{193^2}{15} + \frac{285^2}{15} - \frac{478^2}{30} = 7{,}898.2667 - 7{,}616.1333 = 282.13$$

$$\begin{aligned} SS_S &= \left(\frac{38^2 + 40^2 + 31^2 + 45^2 + 39^2}{3} - \frac{193^2}{15}\right) \\ &\qquad + \left(\frac{60^2 + 51^2 + 58^2 + 67^2 + 49^2}{3} - \frac{285^2}{15}\right) \\ &= (2{,}517.0000 - 2{,}483.2667) + (5{,}485.0000 - 5{,}415.0000) = 103.73 \end{aligned}$$

Note that the total among-subject variability is

$$\frac{38^2 + 40^2 + \cdots + 49^2}{3} - \frac{478^2}{30} = 8{,}002.0000 - 7{,}616.1333 = 385.87$$

and

$$SS_B + SS_S = 282.1334 + 103.7333 = 385.87$$

Within subjects, the sums of squares are

$$\begin{aligned} SS_A &= \left(\frac{156^2}{10} + \frac{205^2}{10} + \frac{117^2}{10}\right) - \frac{478^2}{30} \\ &= \frac{80{,}050.0000}{10} - 7{,}616.1333 = 8{,}005.0000 - 7{,}616.1333 = 388.87 \end{aligned}$$

$$\begin{aligned} SS_{AB} &= \frac{66^2 + 85^2 + 42^2 + 90^2 + 120^2 + 75^2}{5} - \frac{478^2}{30} - SS_A - SS_B \\ &= \frac{41{,}470}{5} - 7{,}616.1333 - 388.8667 - 282.1334 = 6.87 \end{aligned}$$

and

$$SS_{AS} = SS_{total} - SS_A - SS_B - SS_{AB} - SS_S = 54.27$$

Note that the total within-subject variability is the total sum of squares minus the total among-subject variability: $835.87 - 385.87 = 450.00$; also, $SS_A + SS_{AB} + SS_{AS} = 388.87 + 6.87 + 54.27 = 450.01$.

The partitioning of the total sum of squares is now complete and is displayed in Table 7-12.

The test of the B effect, with 1 and 8 degrees of freedom, is significant at the .01 level. Since B exists at only two levels, inspection of Table 7-11 reveals that level 2 of B had the greater effect. Since the condition of homogeneity of variance and covariance was not upheld for these data, the within-subject tests are carried out with conservative degrees of freedom. Thus, the test for A is evaluated with 1 and $q(n-1) = 8$ degrees of freedom rather than with 2 and 16 degrees of freedom. Nevertheless, the computed F of 57.33 is significant at beyond the .01 level (note that if the stratagem of checking this F *both* at 1 and 8 and at 2 and 16 degrees of freedom had been pursued, rejection would have been implied in both cases, and carrying out the Box test for homogeneity could have been avoided at least for the test of the A effect). For the interaction AB, the computed value of F is 1.01; and, by utilizing the conservative procedure, this is evaluated with $q - 1 = 1$ and $q(n-1) = 8$ degrees of freedom. The value is obviously nonsignificant and is also nonsignificant with the usual 2 and 16 degrees of freedom. Since the tests of both the A and the AB effects are unambiguous whether or not the conservative degrees of freedom are utilized, the computation of the statistics for the Box test of homogeneity of variance and covariance could have been completely avoided.

table 7-12 *Analysis-of-variance summary table for a* 2×3 *factorial design with repeated measures on the* A *dimension*

source	df	sum of squares	mean square	F
Among subjects	9			
B	1	282.13	282.13	21.76
S	8	103.73	12.97	
Within subjects	20			
A	2	388.87	194.43	57.33
AB	2	6.87	3.43	1.01
AS	16	54.27	3.39	
Total	29	835.87		

For dimensions existing at more than two levels, the researcher must carry out additional analysis when significant results occur. In general, any of the procedures presented in Chapter 2 can be utilized (e.g., the Newman-Keuls test or orthogonal contrasts); however, the appropriate error term must be chosen for the specific dimension along which these tests are being applied. In the present case, MS_S would enter into standard-error formulas used in testing contrasts associated with the B treatment dimension, and MS_{AS} would enter into these formulas in the case of the A and AB effects. Also, in computing standard-error terms, the appropriate denominator must reflect the dimensions of the design over which summing takes place. For example, the standard error for tests on A for the Newman-Keuls test would be $\sqrt{\text{MS}_S/np}$ since the F ratio for the B effect was MS_B/MS_S and since the sums for B involve totaling over the p levels of A and over the n "levels" of subjects. Similarly, the standard-error term for tests on A would utilize the term $\sqrt{\text{MS}_{AS}/nq}$ since the F ratio for the A effect is $\text{MS}_A/\text{MS}_{AS}$ and since the sums for A involve totaling over the q levels of B and over the n "levels" of subjects.

Since within-subject effects (for example, A and AB in the present example) involve comparisons of the same group of subjects under different treatment conditions, we should expect tests on these effects to have more sensitivity than tests of among-subject effects which involve intergroup comparisons. From Table 7-12 it can be seen that the error term for the among-subject effects, MS_S, is nearly four times as large as the error term MS_{AS} for the within-subject effects. Also, the within-subject tests have greater numbers of degrees of freedom (although, under the conservative testing procedures, the degrees of freedom are the same for among- and within-subject tests in the present example). In general, one can expect MS_{AS} to be less than MS_S; however, if subjects interact very strongly with the A treatment, this will inflate the value of MS_{AS}, and some of the advantages of the repeated measures design are lost (or, at least, abated). The differential power of the tests of the within- and among-subject effects should be borne in mind when designing a study; that is, repetition of measurement should be carried out over those dimensions of the design for which the researcher wishes to increase sensitivity.

complete repetition of measurement in factorial designs

In the previous section, we considered a two-dimensional design in which repetition of measurement was restricted to just one of the treatment dimensions. Although less common than mixed designs, a factorial experiment can be designed in which one group of experimental subjects provides all the data; that is, there is repetition of measurement under *all* the treatment conditions. For a 2×3 factorial design, the scheme would be:

		TREATMENT A		
		1	2	3
TREATMENT B	1	G_1	G_1	G_1
	2	G_1	G_1	G_1

When repetition of measurement is complete, there is no nesting of subjects within any of the treatment dimensions; therefore, subjects are crossed with all treatment dimensions, and the entire set of possible subject-by-treatment interactions is estimable from the data. In fact, the simplest conceptualization of such designs is to consider subjects as an explicit dimension of the design; then each cell of the design contains one observation. For example, the 2×3 factorial design depicted above can be written as:

		TREATMENT A					
		1		2		3	
TREATMENT B:		1	2	1	2	1	2
	1	Y_{111}	Y_{121}	Y_{112}	Y_{122}	Y_{113}	Y_{123}
	2	Y_{211}	Y_{221}	Y_{212}	Y_{222}	Y_{213}	Y_{223}
SUBJECTS(S)	3	Y_{311}	Y_{321}	Y_{312}	Y_{322}	Y_{313}	Y_{323}
	...	...	...	...	...	...	...
	n	Y_{n11}	Y_{n21}	Y_{n12}	Y_{n22}	Y_{n13}	Y_{n23}

Note that the design is formally equivalent to an $n \times 2 \times 3$ factorial design. In fact, from the point of computing sums of squares, the design may be treated as a factorial design with one replicate per cell. Thus, the sums of squares and associated degrees of freedom appropriate for a 2×3 factorial design with complete repetition of measurement would be:

sum of squares	df
A	2
B	1
AB	2
S	$n-1$
AS	$2(n-1)$
BS	$n-1$
ABS	$2(n-1)$
Total	$6n-1$

There is, of course, no estimate of pure experimental error since, with one observation per cell, there is no within-cell variability in the data table. If we assume that both A and B represent fixed dimensions while subjects are randomly chosen, the expectations for mean squares are (the student should confirm these by using the Cornfield and Tukey algorithm with the structural model $Y_{ijk} = \mu + \alpha_k + \beta_j + \alpha\beta_{kj} + \pi_i + \alpha\pi_{ki} + \beta\pi_{ji} + \alpha\beta\pi_{kji} + \epsilon_{ijk}$):

source	E(MS)
A	$\sigma^2 + q\sigma_{\alpha\pi}{}^2 + qn\sigma_\alpha{}^2$
B	$\sigma^2 + p\sigma_{\beta\pi}{}^2 + pn\sigma_\beta{}^2$
AB	$\sigma^2 + \sigma_{\alpha\beta\pi}{}^2 + n\sigma_{\alpha\beta}{}^2$
S	$\sigma^2 + pq\sigma_\pi{}^2$
AS	$\sigma^2 + q\sigma_{\alpha\pi}{}^2$
BS	$\sigma^2 + p\sigma_{\beta\pi}{}^2$
ABS	$\sigma^2 + \sigma_{\alpha\beta\pi}{}^2$

Note that the appropriate F ratios for the terms from the basic factorial design are $\text{MS}_A/\text{MS}_{AS}$, $\text{MS}_B/\text{MS}_{BS}$, and $\text{MS}_{AB}/\text{MS}_{ABS}$. There is, however, no appropriate denominator to form ratios for any of the effects involving subjects (that is, S, AS, BS, and ABS).

repeated measures in higher-dimensional factorial designs

In moving from two-dimensional factorial designs to higher-dimensional designs, the number of mixed designs increases since the repetition of measurement can be with respect to one dimension, two dimensions, etc. In this section, some exemplary designs are considered in order to illustrate the range of possibilities. In practice, it is likely that the student will encounter or construct original designs not fitting exactly any of those illustrated here. However, these should pose no particular problem if the subject dimension of the design is considered as a nested factor. Thus, the Cornfield and Tukey algorithm can always be used to determine expected values of mean squares, and the appropriate breakdown of within-subject and among-subject effects is easily determined. There will be homogeneity of variance-covariance assumptions for any repeated measures design; the Box test will always be applicable, but it will be necessary to decide how the stages of testing must be carried out. The examples will clarify these procedures so that generalization to novel designs should pose no difficulty. Also, there will always be a conservative testing procedure (that is, a procedure involving adjustment of degrees of freedom) for the within-subject effects in the case that homogeneity conditions are not met. This procedure always follows a particular pattern. For cases in which repetition of measurement is over just one dimension of the design, the degrees of freedom for that dimension are set at 1, and whenever that dimension occurs in any within-subject effect (main or interaction), its degrees of freedom are taken as equal to 1. Note that this is, in effect, the way in which the conservative degrees of freedom were found for the design shown in Table 7-9. The within-subject effects are A, AB, and AS. Their usual (i.e., nonconservative) degrees of freedom are $p-1$, $(p-1)(q-1)$, and $q(p-1)(n-1)$. Since repetition of measurement was over the A dimension in that example, we set the degrees of freedom for A equal to 1 (rather than $p-1$). Thus, whenever $p-1$ appears in the usual degrees-of-freedom expressions, it is replaced by 1. The conservative degrees of freedom for the three effects are, then, 1, $q-1$, and $q(n-1)$. If repetition of measurement is carried out over more than one dimension of a design, the same principle applies to all dimensions having repeated measures. That is, in each case, the usual degrees of freedom are replaced by the value 1 when these degrees of freedom occur for within-subject effects.

three-dimensional design with repeated measures on two dimensions

In a three-dimensional factorial design, repetition of measurement can take place on one, two, or all three dimensions of the design. The cases of repetition on one or two dimensions result in mixed designs and a division of effects into within-subject and among-subject components. When there is repetition on just one dimension, the analysis differs very little from that presented for a two-dimensional design with repetition on one factor. The among-subject effects will be (assuming repeated measures on the A dimension) B, C, BC, and S. The within-subject effects will be A, AB, AC, ABC, and AS. If we assume that all dimensions are fixed (except, of course, the subject dimension), each among-subject effect is tested against S, and each within-subject effect is tested against AS. The conservative degrees of freedom can be determined by replacing $p - 1$ by 1 whenever it occurs in the usual degrees-of-freedom formulas. The student can derive the appropriate expectations and confirm the above analysis by noting that subjects are nested within the levels of both B and C. In order to apply Box's test for homogeneity of variance and covariance, two stages must be followed:

1. Use the statistic B_p to test the homogeneity of the variance-covariance matrices within each BC cell of the design (that is, if B exists at q levels and C at r levels, there is a sample variance-covariance matrix in each of qr cells; label these S_1 through S_{qr} and then pool them to find S_p; the remainder of the procedure is identical with that previously illustrated).
2. Use the statistic B to test the hypothesis that the population variance-covariance matrix has equal diagonal elements and constant off-diagonal elements. This procedure is identical with that previously illustrated (except that, in all cases, the appropriate degrees of freedom must be accounted for; that is, q is replaced by qr wherever appropriate).

The second mixed design possible in a three-dimensional factorial design involves repetition of measurement over two dimensions of the design. Assume that repetition is over the A and B dimensions (that is, subjects are nested within the levels of C). The scheme for the design (assuming a basic 2^3 factorial design) is:

		TREATMENT A			
		1		2	
TREATMENT B:		1	2	1	2
TREATMENT C	1	G_1	G_1	G_1	G_1
	2	G_2	G_2	G_2	G_2

Note that each subject receives all possible combinations of levels of the A and B dimensions; if A and B are each at two levels, this means that each subject is exposed to four different treatment combinations in the experiment and, thus, yields four scores in the data table. On the other hand, separate groups of subjects appear for the levels of the C dimension; in effect, subjects are nested within the levels of C. The structural model for a score in such a design is

$$Y_{ijkm} = \mu + \alpha_m + \beta_k + \gamma_j + \pi_{i(j)} + \alpha\beta_{mk} + \alpha\gamma_{mj} + \beta\gamma_{kj} + \alpha\beta\gamma_{mkj} + \alpha\pi_{mi(j)} + \beta\pi_{ki(j)} + \alpha\beta\pi_{mki(j)} + \epsilon_{ijkm}$$

Since subjects are exposed to combinations of levels of both A and B, the interactions of subjects with A, with B, and with AB appear in the model (that is, the terms AS, BS, and ABS).

Expectations for mean squares associated with this mixed design can be determined by the Cornfield and Tukey algorithm. For simplicity (and because it is typical in practice), we assume that A, B, and C all represent levels of a fixed treatment dimension; of course, subjects are randomly assigned to the levels of C to form the groups. The expectations of mean squares are shown in Table 7-13.

Studying the patterns of terms for the within-subject effects reveals that more than one denominator is required in order to test all effects. The nature of the testing procedure is most clearly shown by displaying the results as:

source	df	E(MS)
Among subjects	$nr - 1$	
C	$r - 1$	$\sigma^2 + pq\sigma_\pi^2 + pqn\sigma_\gamma^2$
S	$r(n - 1)$	$\sigma^2 + pq\sigma_\pi^2$
Within subjects	$rn(pq - 1)$	
A	$p - 1$	$\sigma^2 + q\sigma_{\alpha\pi}^2 + qrn\sigma_\alpha^2$
AC	$(p - 1)(r - 1)$	$\sigma^2 + q\sigma_{\alpha\pi}^2 + qn\sigma_{\alpha\gamma}^2$
AS	$r(p - 1)(n - 1)$	$\sigma^2 + q\sigma_{\alpha\pi}^2$
B	$q - 1$	$\sigma^2 + p\sigma_{\beta\pi}^2 + prn\sigma_\beta^2$
BC	$(q - 1)(r - 1)$	$\sigma^2 + p\sigma_{\beta\pi}^2 + pn\sigma_{\beta\gamma}^2$
BS	$r(q - 1)(n - 1)$	$\sigma^2 + p\sigma_{\beta\pi}^2$
AB	$(p - 1)(q - 1)$	$\sigma^2 + \sigma_{\alpha\beta\pi}^2 + rn\sigma_{\alpha\beta}^2$
ABC	$(p - 1)(q - 1)(r - 1)$	$\sigma^2 + \sigma_{\alpha\beta\pi}^2 + n\sigma_{\alpha\beta\gamma}^2$
ABS	$r(p - 1)(q - 1)(n - 1)$	$\sigma^2 + \sigma_{\alpha\beta\pi}^2$

The only among-subjects effect is C, and this is tested against the subject effect. For the within-subject effects, the tests divide into three sets. In the first set, A and AC are each tested against AS; in the second set, B and BC are each tested against BS; and in the third set, AB and ABC are each tested against ABS.

Computation of the sums of squares for this design can most easily be accomplished in two stages. In the first stage, effects associated with the basic factorial design are found from the corresponding summary table. In the second stage, effects associated with the subjects

table 7-13 *Expectation of mean squares in a three-dimensional mixed design with repeated measures on A and B, and with A, B, and C fixed*

effect	m	k	j	i	E(MS)
α_m	0	q	r	n	$\sigma^2 + q\sigma_{\alpha\pi}^2 + qrn\sigma_\alpha^2$
β_k	p	0	r	n	$\sigma^2 + p\sigma_{\beta\pi}^2 + prn\sigma_\beta^2$
γ_j	p	q	0	n	$\sigma^2 + pq\sigma_\pi^2 + pqn\sigma_\gamma^2$
$\pi_{i(j)}$	p	q	1	1	$\sigma^2 + pq\sigma_\pi^2$
$\alpha\beta_{mk}$	0	0	r	n	$\sigma^2 + \sigma_{\alpha\beta\pi}^2 + rn\sigma_{\alpha\beta}^2$
$\alpha\gamma_{mj}$	0	q	0	n	$\sigma^2 + q\sigma_{\alpha\pi}^2 + qn\sigma_{\alpha\gamma}^2$
$\beta\gamma_{kj}$	p	0	0	n	$\sigma^2 + p\sigma_{\beta\pi}^2 + pn\sigma_{\beta\gamma}^2$
$\alpha\beta\gamma_{mkj}$	0	0	0	n	$\sigma^2 + \sigma_{\alpha\beta\pi}^2 + n\sigma_{\alpha\beta\gamma}^2$
$\alpha\pi_{mi(j)}$	0	q	1	1	$\sigma^2 + q\sigma_{\alpha\pi}^2$
$\beta\pi_{ki(j)}$	p	0	1	1	$\sigma^2 + p\sigma_{\beta\pi}^2$
$\alpha\beta\pi_{mki(j)}$	0	0	1	1	$\sigma^2 + \sigma_{\alpha\beta\pi}^2$
ϵ_{ijkm}	1	1	1	1	σ^2

and their interactions are computed. Utilizing the dot notation for sums, the sums of scores from the basic 2^3 factorial design would be:

		TREATMENT A 1		TREATMENT A 2		
TREATMENT B:		1	2	1	2	sums for C
TREATMENT C	1	$Y_{\cdot 111}$	$Y_{\cdot 121}$	$Y_{\cdot 112}$	$Y_{\cdot 122}$	$Y_{\cdot 1\cdot\cdot}$
	2	$Y_{\cdot 211}$	$Y_{\cdot 221}$	$Y_{\cdot 212}$	$Y_{\cdot 222}$	$Y_{\cdot 2\cdot\cdot}$
Sums for A		$Y_{\cdots 1}$		$Y_{\cdots 2}$		$Y_{\cdots\cdot}$ Grand sum
Sums for B		$Y_{\cdot\cdot 1\cdot}$	$Y_{\cdot\cdot 2\cdot}$			

From this summary table, the effects from the basic factorial design are easily computed:

$$\text{SS}_A = \sum_{m=1}^{p} \frac{(Y_{\cdots m})^2}{qrn} - \frac{(Y_{\cdots\cdot})^2}{pqrn}$$

$$\text{SS}_B = \sum_{k=1}^{q} \frac{(Y_{\cdot\cdot k\cdot})^2}{prn} - \frac{(Y_{\cdots\cdot})^2}{pqrn}$$

$$\text{SS}_C = \sum_{j=1}^{r} \frac{(Y_{\cdot j\cdot\cdot})^2}{pqn} - \frac{(Y_{\cdots\cdot})^2}{pqrn}$$

$$SS_{AB} = \sum_{k=1}^{q} \sum_{m=1}^{p} \frac{\left(\sum_{j=1}^{r} Y_{\cdot jkm}\right)^2}{rn} - \frac{(Y_{\cdots\cdot})^2}{pqrn} - SS_A - SS_B$$

$$SS_{AC} = \sum_{j=1}^{r} \sum_{m=1}^{p} \frac{\left(\sum_{k=1}^{q} Y_{\cdot jkm}\right)^2}{qn} - \frac{(Y_{\cdots\cdot})^2}{pqrn} - SS_A - SS_C$$

$$SS_{BC} = \sum_{j=1}^{r} \sum_{k=1}^{q} \frac{\left(\sum_{m=1}^{p} Y_{\cdot jkm}\right)^2}{pn} - \frac{(Y_{\cdots\cdot})^2}{pqrn} - SS_B - SS_C$$

$$SS_{ABC} = \sum_{j=1}^{r} \sum_{k=1}^{q} \sum_{m=1}^{p} \frac{(Y_{\cdot jkm})^2}{n} - \frac{(Y_{\cdots\cdot})^2}{pqrn} - SS_A - SS_B - SS_C - SS_{AB} - SS_{AC} - SS_{BC}$$

Note that computations involved in the two-factor interactions are equivalent to setting up two-way summary tables and computing from these.

The sum of squares associated with experimental subjects and the remaining effects involving interactions with subjects can be computed from the original data table. The notation from this table is:

		TREATMENT A				
		1		2		
TREATMENT B:		1	2	1	2	subject sums
TREATMENT C	1	Y_{1111}	Y_{1121}	Y_{1112}	Y_{1122}	$Y_{11\cdot\cdot}$
		Y_{2111}	Y_{2121}	Y_{2112}	Y_{2122}	$Y_{21\cdot\cdot}$
		⋯	⋯	⋯	⋯	
		Y_{n111}	Y_{n121}	Y_{n112}	Y_{n122}	$Y_{n1\cdot\cdot}$
	2	Y_{1211}	Y_{1221}	Y_{1212}	Y_{1222}	$Y_{12\cdot\cdot}$
		Y_{2211}	Y_{2221}	Y_{2212}	Y_{2222}	$Y_{22\cdot\cdot}$
		⋯	⋯	⋯	⋯	
		Y_{n211}	Y_{n221}	Y_{n212}	Y_{n222}	$Y_{n2\cdot\cdot}$

Because of the nesting of subjects within levels of C, each subject sum is shown with two subscripts (that is, the C level subscript and the position index for the individual within the level of C; for example, $Y_{12\cdot\cdot}$ is the sum of scores for the first subject within the second level of C). The sums of squares are

$$SS_S = \sum_{i=1}^{n} \sum_{j=1}^{r} (Y_{ij..})^2 - \frac{(Y_{....})^2}{pqrn} - SS_C$$

Note that the first part of this expression (i.e., all except the final subtraction of SS_C) involves the computation of the total among-subject variability).

$$SS_{AS} = \sum_{i=1}^{n} \sum_{j=1}^{r} \sum_{m=1}^{p} \frac{\left(\sum_{k=1}^{q} Y_{ijkm}\right)^2}{q} - \frac{(Y_{....})^2}{pqrn} - SS_A - SS_C - SS_{AC} - SS_S$$

$$SS_{BS} = \sum_{i=1}^{n} \sum_{j=1}^{r} \sum_{k=1}^{q} \frac{\left(\sum_{m=1}^{p} Y_{ijkm}\right)^2}{p} - \frac{(Y_{....})^2}{pqrn} - SS_B - SS_C - SS_{BC} - SS_S$$

Finally, SS_{ABS} can be found by subtraction after determining the total sum of squares:

$$SS_{\text{total}} = \sum_{i=1}^{n} \sum_{j=1}^{r} \sum_{k=1}^{q} \sum_{m=1}^{p} Y_{ijkm}^2 - \frac{(Y_{....})^2}{pqrn}$$

$$SS_{ABS} = SS_{\text{total}} - SS_A - SS_B - SS_C - SS_{AB} - SS_{AC} - SS_{BC} - SS_{ABC} - SS_S - SS_{AS} - SS_{BS}$$

The formulas for SS_{AS} and SS_{BS} require some explanation; the first two terms of the expression for SS_{AS} constitute the sum of squares based on the cells of the AS summary table. This table has the form:

		TREATMENT A 1	2
Level 1 of C	1		
	2		
	.		
	.		
	.		
	n		
Level 2 of C	1		
	2		
	.		
	.		
	.		
	n		

Since subjects are nested within levels of C, the subject dimension of this summary also includes the contrast of the levels of C. Furthermore, since A appears as the columns and C appears as part of the rows, the AC interaction is part of the summary. Thus, to reduce the cell sum of squares by appropriate terms to yield just that due to AS, it is necessary to subtract out sources due to A, C, AC, and S. A similar line of reasoning underlies the formula associated with SS_{BS}.

The Box tests for homogeneity of variance and covariance can be adapted to higher-dimensional designs by an appropriate specification of sample variance-covariance matrices and determination of degrees of freedom. In the present case, there would be two sample variance-covariance matrices: one for group 1 and one for group 2. Since repetition of measurement is carried out over the levels of both the A and the B dimensions, the variance-covariance matrices will each have pq rows and pq columns. Thus, in the first stage of testing, the statistic is $B_p = (1 - C_p)M_p$, where M_p is defined as usual, except for notational changes:

$$M_p = n_t \ln |S_p| - \sum_{j=1}^{r} n_j \ln |S_j| \qquad \text{and} \qquad n_t = \sum_{j=1}^{r} n_j$$

The correction term C_p must be adjusted to reflect the pattern of repeated measures:

$$C_p = \frac{2p^2q^2 + 3pq - 1}{6(pq + 1)(pq - 1)} \left(\sum_{j=1}^{r} \frac{1}{n_j} - \frac{1}{n_t} \right)$$

Note that the limits of summation have been taken to equal r since the groups of subjects exist within the levels of C. Also, p has been replaced by pq since the number of levels over which repetition occurs encompasses both the A and the B dimensions. The degrees of freedom for the chi-square test are $pq(pq + 1)(pq - 1)/2$. Again, p is replaced by pq in the formula.

For the second stage of testing, similar changes in Box's procedure are needed. In the M statistic, q is replaced by r to account for the nesting of subjects within levels of C; in the expression for C, q is replaced by r and p is replaced by pq wherever they occur. The degrees of freedom are $(p^2q^2 + pq - 4)/2$.

In practice, of course, the most judicious tactic is to complete the computations leading to the analysis-of-variance summary table without applying Box's test. Then, test each term, using *both* the usual degrees of freedom and the conservative degrees of freedom. If both sets of degrees of freedom yield the same decisions with respect to acceptance or rejection of the null hypotheses, then there is no need to apply the tests for homogeneity of variance and covariance. For the case in which repeated measures occur for both the A and the B dimensions of the design, the conservative degrees of freedom are determined from the usual formulas by replacing $p - 1$ by 1 and $q - 1$ by 1. That is, the two sets of degrees of freedom are:

source	usual df	conservative df
Among subjects	$nr - 1$	
C	$r - 1$	$r - 1$
S	$r(n - 1)$	$r(n - 1)$
Within subjects	$rn(pq - 1)$	
A	$p - 1$	1
AC	$(p - 1)(q - 1)$	1
AS	$r(p - 1)(n - 1)$	$r(n - 1)$
B	$q - 1$	1
BC	$(q - 1)(r - 1)$	$r - 1$
BS	$r(q - 1)(n - 1)$	$r(n - 1)$
AB	$(p - 1)(q - 1)$	1
ABC	$(p - 1)(q - 1)(r - 1)$	$r - 1$
ABS	$r(p - 1)(q - 1)(n - 1)$	$r(n - 1)$

table 7-14 *Hypothetical data for a* 3 × 2 × 2 *factorial design with repeated measures on the first two dimensions*

			TREATMENT A: 1		TREATMENT A: 2	
TREATMENT B:		subjects	1	2	1	2
TREATMENT C	1	1	5	7	5	7
		2	7	10	3	4
		3	4	12	0	9
	2	1	11	15	9	12
		2	9	14	7	10
		3	15	18	11	14
	3	1	20	25	18	23
		2	17	20	20	28
		3	22	30	26	28

Let us turn, now, to an example of the computations involved in finding sums of squares for a three-dimensional design with repetition of measurement on two dimensions. Sample data for $p = 2$, $q = 2$, $r = 3$, and $n = 3$ are displayed in Table 7-14. Although the Box tests could be applied immediately to these data to test the homogeneity of variance-covariance conditions, each of the three sample variance-covariance matrices is of order 4×4, and evaluating their determinants is quite laborious. Rather than attempt the tests at this time, the more efficient procedure is to complete the analysis and utilize both the usual and the conservative sets of degrees of freedom; if they yield identical outcomes, the labor of the Box tests can be avoided.

The sums of squares from the basic design can be calculated from the summary table of cell sums:

		TREATMENT A					
		1		2			
TREATMENT B:		1	2	1	2	sums for C	
	1	16	29	8	20	73	
TREATMENT C	2	35	47	27	36	145	
	3	59	75	64	79	277	
Sums for A		261		234		495	Grand sum
Sums for B		209	286				

$$SS_A = \frac{261^2 + 234^2}{18} - \frac{495^2}{36} = 6{,}826.5000 - 6{,}806.2500 = 20.25$$

$$SS_B = \frac{209^2 + 286^2}{18} - \frac{495^2}{36} = 6{,}970.9444 - 6{,}806.2500 = 164.69$$

$$SS_C = \frac{73^2 + 145^2 + 277^2}{12} - \frac{495^2}{36} = 8{,}590.2500 - 6{,}806.2500 = 1{,}784.00$$

$$SS_{AB} = \frac{110^2 + 151^2 + 99^2 + 135^2}{9} - \frac{495^2}{36} - SS_A - SS_B$$
$$= 6{,}991.8889 - 6{,}806.2500 - 20.2500 - 164.6944 = .69$$

$$SS_{AC} = \frac{45^2 + 28^2 + 82^2 + 63^2 + 134^2 + 143^2}{6} - \frac{495^2}{36} - SS_A - SS_C$$
$$= 8{,}651.1667 - 6{,}806.2500 - 20.2500 - 1{,}784.0000 = 40.67$$

$$SS_{BC} = \frac{24^2 + 49^2 + 62^2 + 83^2 + 123^2 + 154^2}{6} - \frac{495^2}{36} - SS_B - SS_C$$
$$= 8{,}759.1667 - 6{,}806.2500 - 164.6944 - 1{,}784.0000 = 4.22$$

$$SS_{ABC} = \frac{16^2 + 29^2 + 8^2 + 20^2 + 35^2 + 47^2 + 27^2 + 36^2 + 59^2 + 75^2 + 64^2 + 79^2}{3} - \frac{495^2}{36} - SS_A - SS_B - SS_C - SS_{AB} - SS_{AC} - SS_{BC}$$
$$= 8{,}821.0000 - 6{,}806.2500 - 20.2500 - 164.6944 - 1{,}784.0000 - .6945 - 40.6667 - 4.2223 = .22$$

The sums of squares associated with the subject effects must be calculated from the original table of scores (Table 7-14).

$$SS_S = \frac{24^2 + 24^2 + 25^2 + 47^2 + 40^2 + 58^2 + 86^2 + 85^2 + 106^2}{4} - \frac{495^2}{36} - SS_C$$
$$= 8{,}701.7500 - 6{,}806.2500 - 1{,}784.0000 = 111.50$$

$$SS_{AS} = \frac{12^2 + 17^2 + 16^2 + 26^2 + 23^2 + 33^2 + 45^2 + 37^2 + 52^2 + 12^2 + 7^2 + 9^2 + 21^2 + 17^2 + 25^2 + 41^2 + 48^2 + 54^2}{2} - \frac{495^2}{36} - SS_A - SS_C - SS_{AC} - SS_S$$
$$= 8{,}805.5000 - 6{,}806.2500 - 20.2500 - 1{,}784.0000 - 40.6667 - 111.5000 = 42.83$$

$$SS_{BS} = \frac{10^2 + 10^2 + 4^2 + 20^2 + 16^2 + 26^2 + 38^2 + 37^2 + 48^2 + 14^2 + 14^2 + 21^2 + 27^2 + 24^2 + 32^2 + 48^2 + 48^2 + 58^2}{2} - \frac{495^2}{36} - SS_B - SS_C - SS_{BC} - SS_S$$
$$= 8{,}899.5000 - 6{,}806.2500 - 164.6944 - 1{,}784.0000 - 4.2223 - 111.50000 = 28.83$$

The final term, SS_{ABS}, is most easily obtained by subtraction after computation of the total sum of squares.

$$SS_{\text{total}} = 5^2 + 7^2 + \cdots + 28^2 - \frac{495^2}{36} = 9{,}021.0000 - 6{,}806.2500 = 2{,}214.75$$

$$SS_{ABS} = SS_{total} - SS_A - SS_B - SS_C - SS_{AB} - SS_{AC} - SS_{BC} - SS_{ABC} - SS_{AS} - SS_{BS} - SS_S$$

$$= 2{,}214.7500 - 20.2500 - 164.6944 - 1{,}784.0000 - .6945 - 40.6667 - 4.2223 - .2221 - 42.8333 - 28.8333 - 111.5000$$

$$= 16.83$$

The analysis-of-variance summary table is shown in Table 7-15. For the present example, the conservative degrees of freedom are identical with the usual degrees of freedom since both A and B exist at two levels (i.e., the usual degrees of freedom for both A and B are 1). Among the F ratios, only those for the C effect and the B effect reach significance at conventional levels.

For designs of more than three dimensions, there will be a number of ways in which mixed designs can be set up. If the basic design has four treatment dimensions, then either one, two, or three of the dimensions can have repetition of measurement, and there will be both within-subject and among-subject components comprising the sums of squares (if all four dimensions show repetition of measurement, then all sources are within subjects). For mixed designs in general, the analytical approach is to first compute the sums of squares associated with the basic factorial design from a summary table constructed by summing over the subject dimension of the design. This part of the

table 7-15 *Analysis-of-variance summary table for a* $3 \times 2 \times 2$ *factorial design with repeated measures on A and B*

source	*df*	sum of squares	mean square	*F*
Among subjects	8			
C	2	1,784.00	892.00	48.01
S	6	111.50	18.58	
Within subjects	27			
A	1	20.25	20.25	2.84
AC	2	40.67	20.33	2.85
AS	6	42.83	7.14	
B	1	164.69	164.69	34.31
BC	2	4.22	2.11	.44
BS	6	28.83	4.80	
AB	1	.69	.69	.25
ABC	2	.22	.11	.04
ABS	6	16.83	2.80	
Total	35	2,214.75		

analysis is identical with that for any factorial design; the sums of squares can be computed by the procedures based on treatment and cell sums, as illustrated for the three-dimensional repeated measures design, or they can be determined from sets of orthogonal contrasts, as discussed in Chapter 3. In the second stage of analysis, the sums of squares associated with subjects and interactions involving subjects are computed. The analysis here is best conducted by forming summary tables involving the subjects as one dimension and a relevant treatment as the other dimension. Since the subject dimension will also comprise one or more of the treatment contrasts and certain interactions involving treatments, it is necessary to study carefully the structure of the design in order to decide on the appropriate analytical procedure. The example of the three-dimensional design will provide a useful guide during this part of the computations of sums of squares. Ordinarily, the highest-order interaction involving subjects is computed by subtraction after finding the total sum of squares.

Let us consider the composition of within-subject and among-subject sources of variation for a four-dimensional design with repetition of measurement on two of the four treatment dimensions. Assume that the repetition of measurement occurs over the A and B dimensions of the design; in effect, then, subjects are nested within the CD cells of the design. The scheme for such a design (assuming a basic 2^4 factorial) is:

		TREATMENT A							
		1				2			
TREATMENT B:		1		2		1		2	
TREATMENT C:		1	2	1	2	1	2	1	2
TREATMENT D	1	G_1	G_2	G_1	G_2	G_1	G_2	G_1	G_2
	2	G_3	G_4	G_3	G_4	G_3	G_4	G_3	G_4

Note that each of the four groups of experimental subjects occurs under each level of A and of B; however, each CD cell is comprised of a separate group of subjects. In terms of sums of squares, all effects associated with the A and B dimensions will be within subjects, while those associated with C and D will be among subjects. The sources of variation and their expectations are (assuming A at p levels, B at q levels, C at r levels, and D at t levels):

source	df	E(MS)
Among subjects	$nrt - 1$	
C	$r - 1$	$\sigma^2 + pq\sigma_\pi^2 + pqrn\sigma_\gamma^2$
D	$t - 1$	$\sigma^2 + pq\sigma_\pi^2 + pqtn\sigma_\delta^2$
CD	$(r-1)(t-1)$	$\sigma^2 + pq\sigma_\pi^2 + pqn\sigma_{\gamma\delta}^2$
S	$rt(n-1)$	$\sigma^2 + pq\sigma_\pi^2$
Within subjects	$nrt(pq - 1)$	
A	$p - 1$	$\sigma^2 + q\sigma_{\alpha\pi}^2 + qrtn\sigma_\alpha^2$
AC	$(p-1)(r-1)$	$\sigma^2 + q\sigma_{\alpha\pi}^2 + qtn\sigma_{\alpha\gamma}^2$
AD	$(p-1)(t-1)$	$\sigma^2 + q\sigma_{\alpha\pi}^2 + qrn\sigma_{\alpha\delta}^2$
ACD	$(p-1)(r-1)(t-1)$	$\sigma^2 + q\sigma_{\alpha\pi}^2 + qn\sigma_{\alpha\gamma\delta}^2$
AS	$rt(n-1)(p-1)$	$\sigma^2 + q\sigma_{\alpha\pi}^2$
B	$q - 1$	$\sigma^2 + p\sigma_{\beta\pi}^2 + prtn\sigma_\beta^2$
BC	$(q-1)(r-1)$	$\sigma^2 + p\sigma_{\beta\pi}^2 + ptn\sigma_{\beta\gamma}^2$
BD	$(q-1)(t-1)$	$\sigma^2 + p\sigma_{\beta\pi}^2 + prn\sigma_{\beta\delta}^2$
BCD	$(q-1)(r-1)(t-1)$	$\sigma^2 + p\sigma_{\beta\pi}^2 + pn\sigma_{\beta\gamma\delta}^2$
BS	$rt(n-1)(q-1)$	$\sigma^2 + p\sigma_{\beta\pi}^2$
AB	$(p-1)(q-1)$	$\sigma^2 + \sigma_{\alpha\beta\pi}^2 + rtn\sigma_{\alpha\beta}^2$
ABC	$(p-1)(q-1)(r-1)$	$\sigma^2 + \sigma_{\alpha\beta\pi}^2 + tn\sigma_{\alpha\beta\gamma}^2$
ABD	$(p-1)(q-1)(t-1)$	$\sigma^2 + \sigma_{\alpha\beta\pi}^2 + rn\sigma_{\alpha\beta\delta}^2$
$ABCD$	$(p-1)(q-1)(r-1)(t-1)$	$\sigma^2 + \sigma_{\alpha\beta\pi}^2 + n\sigma_{\alpha\beta\gamma\delta}^2$
ABS	$rt(n-1)(p-1)(q-1)$	$\sigma^2 + \sigma_{\alpha\beta\pi}^2$

It should be noted that three separate terms are necessary to test all the within-subject effects. Those sources involving A and the interactions of A with the nesting factors (that is, C and D) are all tested against AS; the sources involving B and the interactions of B with the nesting factors are all tested against BS; and the sources involving AB and the second- and third-order interactions of AB with the nesting factors are all tested against ABS. Of course, all among-subject effects are tested against the mean square for subjects. The organization of the sources of variation is a result of applying the Cornfield and Tukey algorithm to the structural model for a score and then grouping together sources which are tested against a common term (the structural model for this example is $Y_{ijkmg} = \mu + \alpha_g + \beta_m + \gamma_k + \delta_j + \pi_{i(kj)} + \alpha\beta_{gm} + \alpha\gamma_{gk} + \alpha\delta_{gj} + \beta\gamma_{mk} + \beta\delta_{mj} + \gamma\delta_{kj} + \alpha\pi_{gi(kj)} + \beta\pi_{mi(kj)} + \alpha\beta\gamma_{gmk} + \alpha\beta\delta_{gmj} + \alpha\gamma\delta_{gkj} + \beta\gamma\delta_{mkj} + \alpha\beta\pi_{gmi(kj)} + \alpha\beta\gamma\delta_{gmkj} + \epsilon_{ijkmg}$).

repeated measures in Latin-square designs

In Chapter 4 it was pointed out that Latin-square designs are a special case of incomplete factorial designs; in particular, a $k \times k$ Latin square is a $1/k$ replicate of a k^3 factorial design. It is possible to introduce repetition of measurement into Latin-square designs in a number of ways. Within a single Latin square, repetition of measurement can be carried out over the rows, columns, or letters. Also, several Latin squares can be utilized in one design and repetition of measurement introduced over an entire Latin square. In this section, we shall introduce two designs involving repetition of measurement. A more complete treatment of this topic can be found in Winer (1962).

One manner in which repeated measures can be incorporated into a Latin square is to equate one of the three dimensions, say rows, with groups of subjects (i.e., nest subjects within one of the dimensions) and carry out repeated measurement on the remaining two dimensions, say columns and Latin letters. Table 7-16 shows such an arrangement when the basic design is a 4×4 Latin square. If groups are ignored, the design is simply a 4×4 design; however, instead of independent, random groups of subjects being assigned per cell of the design, the groups are assigned to the levels of dimension B (the rows), and the same subjects are run under four combinations of treatments A and C (columns and Latin letters). It is apparent that B represents an among-subjects effects, and both A and C represent within-subject effects. As in any incomplete factorial design, there is confounding of interaction effects and main effects; that is, for example, the main effect for A is confounded with the BC interaction.

If subjects are nested within one dimension of a single Latin-square design, the analysis is quite straightforward. The effects associated with the basic design are $A, B, C,$ and the residual. In addition, a sum of squares for subjects can be found for subjects within groups and, also within groups, an estimate of experimental error is available. It must be remembered that the interpretation of such a design rests upon knowledge that the interactions among dimensions of the design are negligible.

It is interesting to compare the breakdown of degrees of freedom for the repeated measures case and the same design with independent

table 7-16 *Repeated measures in a single Latin-square design*

		TREATMENT A			
group	TREATMENT B	1	2	3	4
1	1	a	b	c	d
2	2	b	c	d	a
3	3	c	d	a	b
4	4	d	a	b	c

groups of subjects. If we assume that the basic Latin square has p rows, p columns, and p Latin letters, this breakdown is:

source	df for basic design	df for repeated measures design
Between subjects	....	$np - 1$
B	$p - 1$	$p - 1$
S	....	$p(n - 1)$
Within subjects	....	$np(p - 1)$
A	$p - 1$	$p - 1$
C	$p - 1$	$p - 1$
Residual	$(p - 1)(p - 2)$	$(p - 1)(p - 2)$
Error	$p^2(n - 1)$	$p(n - 1)(p - 1)$
Total	$np^2 - 1$	$np^2 - 1$

The only new effect is for subjects, and it accounts for $p(n - 1)$ degrees of freedom. These degrees of freedom come from within cells and reduce the error degrees of freedom from $p^2(n - 1)$ to $p(n - 1)(p - 1)$. The expected values of mean squares associated with the within-subject effects are the same as in an ordinary Latin-square design without repetition of measurement. However, the nesting of subjects within levels of B results in a component due to subjects among the terms in the expectation for B. Thus, for the repeated measures design, the expectations are:

source	E(MS)
Between subjects	
B	$\sigma^2 + p\sigma_\pi^2 + np\sigma_\beta^2$
S	$\sigma^2 + p\sigma_\pi^2$
Within subjects	
A	$\sigma^2 + np\sigma_\alpha^2$
C	$\sigma^2 + np\sigma_\gamma^2$
Residual	$\sigma^2 + n\sigma_{\text{res}}^2$
Error	σ^2

Computation of sums of squares for this repeated measures design presents no undue difficulty since the only new term is SS_S. The data can be represented as:

			TREATMENT A			
		group	1	2	3	4
			Y_{1111}	Y_{1122}	Y_{1133}	Y_{1144}
	1	1	Y_{2111}	Y_{2122}	Y_{2133}	Y_{2144}
			....	....	....	....
TREATMENT B			Y_{n111}	Y_{n122}	Y_{n133}	Y_{n144}
	.	.	....	....	....	....
			Y_{1414}	Y_{1421}	Y_{1432}	Y_{1443}
	4	4	Y_{2414}	Y_{2421}	Y_{2432}	Y_{2443}
			....	....	....	....
			Y_{n414}	Y_{n421}	Y_{n432}	Y_{n443}

Note that the last subscript for a score represents the Latin letter associated with the cell containing that score. That is, Y_{ijkm} is the ith subject receiving the jth level of B, the kth level of A, and the mth Latin letter (where level a is coded 1, b is coded 2, and so forth). The sums of squares are

$$\text{SS}_A = \sum_{k=1}^{p} \frac{(Y_{..k.})^2}{np} - \frac{(Y_{....})^2}{np^2}$$

$$\text{SS}_B = \sum_{j=1}^{p} \frac{(Y_{.j..})^2}{np} - \frac{(Y_{....})^2}{np^2}$$

$$\text{SS}_C = \sum_{m=1}^{p} \frac{(Y_{...m})^2}{np} - \frac{(Y_{....})^2}{np^2}$$

$$\text{SS}_{\text{res}} = \sum_{j=1}^{p} \sum_{k=1}^{p} \sum_{m=1}^{p} \frac{(Y_{.jkm})^2}{n} - \frac{(Y_{....})^2}{np^2} - \text{SS}_A - \text{SS}_B - \text{SS}_C$$

$$\text{SS}_S = \sum_{i=1}^{n} \sum_{j=1}^{p} \frac{(Y_{ij..})^2}{p} - \frac{(Y_{....})^2}{np^2} - \text{SS}_B$$

$$\text{SS}_{\text{error}} = \sum_{i=1}^{n} \sum_{j=1}^{p} \sum_{k=1}^{p} \sum_{m=1}^{p} Y_{ijkm}^2 - \frac{(Y_{....})^2}{np^2} - \text{SS}_A - \text{SS}_B - \text{SS}_C - \text{SS}_{\text{res}} - \text{SS}_S$$

The design just presented involves repetition of measurement within a single Latin square. It is possible to construct designs in such a way that two or more Latin squares are utilized. For example, the plan of Table 7-16 can constitute a building block when an additional dimension is added to the design. Consider the scheme presented in Table

table 7-17 *Repeated measures involving two or more Latin squares*

	TREATMENT D									
	1					2				
TREATMENT B	group	TREATMENT A 1	2	3	4	group	TREATMENT A 1	2	3	4
1	1	a	b	c	d	5	a	b	c	d
2	2	b	c	d	a	6	b	c	d	a
3	3	c	d	a	b	7	c	d	a	b
4	4	d	a	b	c	8	d	a	b	c

7-17. Each level of the D dimension is applied to a Latin-square design. If the basic Latin square is $p \times p$ and there are q levels of D, a total of pq groups of experimental subjects are required for the design. Note that the new dimension D involves only among-subject comparisons; also, the fact that the complete, basic Latin square is replicated per level of D means that interactions of D and the other dimensions of the design can be estimated and included among the sums of squares for this design. The remaining effects are the same as those shown for the previous design; that is, C is an among-subjects effect, and A and B are within-subject effects. Of course, the interpretation of the analysis depends upon the negligibility of interactions among A, B, and C, the basic Latin-square dimensions. The breakdown of degrees of freedom and the expected values are (A, B, and C at p levels; D at q levels; A, B, C, D fixed):

source	df	E(MS)
Among subjects	$npq - 1$	
B	$p - 1$	$\sigma^2 + p\sigma_\pi^2 + npq\sigma_\beta^2$
D	$q - 1$	$\sigma^2 + p\sigma_\pi^2 + np^2\sigma_\delta^2$
BD	$(p-1)(q-1)$	$\sigma^2 + p\sigma_\pi^2 + np\sigma_{\beta\delta}^2$
S	$pq(n-1)$	$\sigma^2 + p\sigma_\pi^2$
Within subjects	$npq(p-1)$	
A	$p - 1$	$\sigma^2 + npq\sigma_\alpha^2$
C	$p - 1$	$\sigma^2 + npq\sigma_\gamma^2$
AD	$(p-1)(q-1)$	$\sigma^2 + np\sigma_{\alpha\delta}^2$
CD	$(p-1)(q-1)$	$\sigma^2 + np\sigma_{\gamma\delta}^2$
Residual	$q(p-1)(p-2)$	$\sigma^2 + nq\sigma_{\text{res}}^2$
Error	$pq(p-1)(n-1)$	σ^2
Total	$p^2q(n-1)$	

Computations leading to sums of squares for B, S, A, and C are identical with those presented for the previous design, except that all sums are pooled over the levels of D. For example, the formula for SS_A is modified only in that an additional subscript is added for the D dimension (i.e., the subscript g) and the denominators are adjusted to reflect the additional summing operations:

$$SS_A = \sum_{k=1}^{p} \frac{(Y_{..k..})^2}{pqn} - \frac{(Y_{.....})^2}{nqp^2}$$

The effects involving D can be found from appropriate summary tables or from the formulas

$$SS_D = \sum_{g=1}^{q} \frac{(Y_{....g})^2}{np^2} - \frac{(Y_{.....})^2}{nqp^2}$$

$$SS_{BD} = \sum_{g=1}^{q} \sum_{j=1}^{p} \frac{(Y_{.j..g})^2}{np} - \frac{(Y_{.....})^2}{nqp^2} - SS_B - SS_D$$

$$SS_{AD} = \sum_{g=1}^{q} \sum_{k=1}^{p} \frac{(Y_{..k.g})^2}{np} - \frac{(Y_{.....})^2}{nqp^2} - SS_A - SS_D$$

$$SS_{CD} = \sum_{g=1}^{q} \sum_{m=1}^{p} \frac{(Y_{...mg})^2}{np} - \frac{(Y_{.....})^2}{nqp^2} - SS_C - SS_D$$

The residual term is found from the cell totals pooled over the levels of D:

$$SS_{\text{res}} = \sum_{j=1}^{p} \sum_{k=1}^{p} \sum_{m=1}^{p} \frac{(Y_{.jkm.})^2}{qn} - \frac{(Y_{.....})^2}{nqp^2} - SS_A - SS_B - SS_C - SS_{BD} - SS_{AD} - SS_{CD}$$

The error term can be found by subtraction after computation of the total sum of squares:

$$SS_{\text{total}} = \sum_{i=1}^{n} \sum_{j=1}^{p} \sum_{k=1}^{p} \sum_{m=1}^{p} \sum_{g=1}^{q} Y_{ijkmg}^2 - \frac{(Y_{.....})^2}{nqp^2}$$

$$SS_{\text{error}} = SS_{\text{total}} - SS_A - SS_B - SS_C - SS_{BD} - SS_{AD} - SS_{CD} - SS_{\text{res}}$$

repeated measures designs with unequal group sizes

The sizes of groups of experimental subjects in repeated measures designs can vary because of deliberate planning on the part of the researcher (e.g., the groups represent stratified random samples from populations where the strata have unequal sizes), or the researcher may begin the experiment with groups of equal size but may lose some subjects from a variety of causes (e.g., subjects' failing to complete all experimental tasks). If we assume that analysis is based solely on subjects' yielding complete data, this condition can occur in factorial and related designs but cannot happen when the basic design is a CRD (that is, in a CRD there is only the group of subjects, and if cases are eliminated which fail to provide complete data, the group sizes must be equal). Indeed, by the same reasoning, equal group sizes are guaranteed in any design where repetition of measurement is complete. However, in mixed designs the possibility of unequal group sizes must be entertained. Fortunately, the existence of disparate group sizes has little influence upon the analysis of variance since the structure of repeated measures designs forces the group sizes to be at least proportional. Consider, for example, a two-dimensional repeated measures design with repetition of measure on the A dimension:

		TREATMENT A		
		1	2	3
	1	G_1	G_1	G_1
TREATMENT B	2	G_2	G_2	G_2
	3	G_3	G_3	G_3

Assume that group 1 contains n_1 subjects, group 2 contains n_2 subjects, and group 3 contains n_3 subjects. Since the same groups provide the data across the levels of A, the group sizes must be proportional; this is evident from the table of group sizes:

		TREATMENT A		
		1	2	3
	1	n_1	n_1	n_1
TREATMENT B	2	n_2	n_2	n_2
	3	n_3	n_3	n_3

Because of the proportionality of group sizes by rows (or columns) of the table, the usual computational formulas require relatively minor modifications in order to be applicable to designs with unequal group sizes. In effect, the changes consist in replacing n by n_j (where n_j is the number of cases at the jth level of B) in the appropriate computational formulas. Thus,

$$\text{SS}_{\text{total}} = \sum_{i=1}^{n_j} \sum_{j=1}^{q} \sum_{k=1}^{p} Y_{ijk}^2 - \frac{(Y_{...})^2}{pn_t} \qquad \text{where } n_t = \sum_{j=1}^{q} n_j$$

$$\text{SS}_B = \sum_{j=1}^{q} \frac{(Y_{\cdot j \cdot})^2}{pn_j} - \frac{(Y_{...})^2}{pn_t}$$

$$\text{SS}_S = \sum_{j=1}^{q} \left[\sum_{i=1}^{n_j} \frac{(Y_{ij\cdot})^2}{p} - \frac{(Y_{\cdot j \cdot})^2}{pn_j} \right]$$

$$\text{SS}_A = \sum_{k=1}^{p} \frac{(Y_{\cdot\cdot k})^2}{n_t} - \frac{(Y_{...})^2}{pn_t}$$

$$\text{SS}_{AB} = \sum_{j=1}^{q} \sum_{k=1}^{p} \frac{(Y_{\cdot jk})^2}{n_j} - \frac{(Y_{...})^2}{pn_t} - \text{SS}_A - \text{SS}_B$$

$$\text{SS}_{AS} = \text{SS}_{\text{total}} - \text{SS}_A - \text{SS}_B - \text{SS}_{AB} - \text{SS}_S$$

The generalization of these formulas to higher-dimensional and more complex repeated measures designs should present no undue difficulty to the student.

Lindquist-type designs

Because of the labels utilized by Lindquist (1953) in his experimental design textbook, certain types of two- and three-dimensional repeated measures designs are popularly referred to as Lindquist-type designs. Lindquist identified seven mixed designs, and the names "Lindquist type I design," "Lindquist type II design," etc., are often encountered in research reports. In order to make this popular usage accessible to the reader, this section is devoted to specifying the characteristics of the seven Lindquist-type designs. Only a few of these have been treated in detail in this chapter; they are, however, the more frequently used of the designs. The original source is recommended for further study of the analyses of these designs; however, from the principles developed in this chapter, the reader should have no difficulty with the analysis of any of the Lindquist-type designs.

Lindquist type I *design* The basic design is a two-dimensional factorial with repetition of measurement on one of the two dimensions; this design was considered in this chapter (Table 7-7).

Lindquist type II *design* The basic design is a $p \times p$ Latin square with groups of subjects constituting one of the dimensions:

	TREATMENT A		
group	1	2	3
1	a	b	c
2	b	c	a
3	c	a	b

The letters a, b, c represent the three levels of the B treatment dimension. This design is similar to the first Latin-square design considered in this chapter (Table 7-16); however, the rows are not associated with a treatment dimension.

Lindquist type III *design* The basic design is a three-dimensional factorial with repetition of measurement on just one of the three dimensions. The design scheme is ($p = 2$, $q = 3, r = 3$; repetition of measurement on the C dimension):

		TREATMENT A					
		1			2		
TREATMENT B:		1	2	3	1	2	3
	1	G_1	G_2	G_3	G_4	G_5	G_6
TREATMENT C	2	G_1	G_2	G_3	G_4	G_5	G_6
	3	G_1	G_2	G_3	G_4	G_5	G_6

Lindquist type IV *design* The building blocks for this design are $p \times p$ Latin squares; the design is similar to the last Latin-square design considered in this chapter (Table 7-17); however, the rows are not associated with a treatment dimension. The design scheme is (with a, b, c the levels of the B treatment):

	TREATMENT C						
	1				2		
	TREATMENT A				TREATMENT A		
group	1	2	3	group	1	2	3
1	a	b	c	4	a	b	c
2	b	c	a	5	b	c	a
3	c	a	b	6	c	a	b

Lindquist type V *design* The basic building blocks of this design are Latin squares; groups of subjects are randomly assigned to the cells of randomly selected Latin squares. If the building-block Latin squares are $p \times p$, then p Latin squares are utilized, and each receives one level of a third treatment dimension. The scheme for such a design is ($p = 3$):

		TREATMENT C								
		1			2			3		
		TREATMENT A			TREATMENT A			TREATMENT A		
		1	2	3	1	2	3	1	2	3
	1	G_2	G_3	G_1	G_1	G_3	G_2	G_2	G_1	G_3
TREATMENT B	2	G_1	G_2	G_3	G_3	G_2	G_1	G_1	G_3	G_2
	3	G_3	G_1	G_2	G_2	G_1	G_3	G_3	G_2	G_1

Note that the Latin squares per level of C are different; that is, 3×3 Latin squares are randomly permuted from their standard form for use in this design. Also, in the cells, the groups follow a Latin-square arrangement.

Lindquist type VI *design* The basic design is a three-dimensional factorial with repetition of measurement on two of the three dimensions. This design was considered earlier in this chapter (Table 7-13).

Lindquist type VII *design* The building blocks for this type of design are $p \times p$ Latin squares. Treatments are assigned to the columns and letters, and groups constitute the rows of the basic design; in addition, the basic design is repeated over the levels of a third dimension. The scheme is:

	TREATMENT C						
	1				2		
	TREATMENT A				TREATMENT A		
group	1	2	3	group	1	2	3
1	a	b	c	1	a	b	c
2	b	c	a	2	b	c	a
3	c	a	b	3	c	a	b

The letters a, b, c represent the levels of the B treatment dimension. Note that the same groups appear under the levels of C.

In the Lindquist system, designs I and II have two treatment dimensions, and III through VII each have three treatment dimensions. Also, in design II, the two dimensions must exist at the same number of levels; in design IV, the dimensionality of treatments A and B must be the same; in design V, all three treatment dimensions must exist at the same number of levels; and in design VII, treatments A and B must exist at the same number of levels. Because of the arrangement of groups, there will, in general, be no purely among-subjects component of variation in designs II, V, and VII.

evaluating carry-over effects in repeated measures designs

The application of repeated measures designs to educational research settings is often rather limited since it is almost certain that treatment effects will carry over from one treatment setting to another. When the order of presentation of the treatment levels (or combinations of treatment levels) is independently randomized for each experimental subject, there is some cancellation of these effects, but it is not systematic, and, depending upon the specific pattern of randomization that happens to occur, the effects of carry-over can be more or less serious. If the researcher is uncertain as to whether or not his particular treatment variables are subject to carry-over effects, it is possible to construct an order-of-treatments dimension and include it in the basic repeated measures design. This approach is most promising for use in a pilot study specifically set up to study the effects of carry-over.

An order-of-treatments dimension is a pseudo-dimension representing all possible arrangements of the ordering of treatment levels (or combinations of treatment levels). For example, if repeated measures are contemplated in a one-dimensional design and if the treatment dimension exists at four levels, there is a total of 24 possible different orderings of the administrations of the treatment levels (i.e., if the levels are coded 1, 2, 3, and 4, these orders are 1234, 1243, 1324, 1342, 1432, 1423, 2134, 2143, 2314, 2341, 2413, 2431, 3124, 3142, 3214, 3241, 3412, 3421, 4123, 4132, 4213, 4231, 4312, 4321). In general, if the treatment exists at p levels, there will be $p!$ different orders. Thus, the feasibility of an order-of-treatments dimension is limited to relatively small values of p (for example, $5! = 120$, and if $p = 5$, the experiment would be difficult to actually carry out). If p is 4 or less (or $pq = 4$, as in a 2×2 factorial design), it is quite possible to arrange to run two or more subjects through the treatment levels in each of the different possible orders. The sum of squares for the order-of-treatments dimension can be utilized to test whether or not there are carry-over effects for certain of the orders in which treatment levels (or combinations of treatment levels) are presented to subjects. The scheme for the design is (for $p = 3$ treatment levels):

	TREATMENT A		
order of treatments	1	2	3
123	G_1	G_1	G_1
132	G_2	G_2	G_2
213	G_3	G_3	G_3
231	G_4	G_4	G_4
312	G_5	G_5	G_5
321	G_6	G_6	G_6

Note that subjects are nested within orders of treatments; the analysis of the design is identical with that for a two-dimensional design with repetition of measurement on the A dimension.

repeated measures in nested designs

If the basic plan for an experiment calls for the use of nesting, it is still possible to carry out repeated measurement on subjects within the groups of the design. Consider, for example, a two-dimensional partially nested design in which groups of experimental subjects are nested within the levels of the A treatment dimension and repeated measures occur over the levels of the B treatment dimension:

		TREATMENT A					
		1		2		3	
NESTED FACTOR (C):		1	2	3	4	5	6
TREATMENT B	1	G_1	G_2	G_3	G_4	G_5	G_6
	2	G_1	G_2	G_3	G_4	G_5	G_6

The nested factor C is crossed with the B dimension but not with the A dimension; if we ignore the repeated measures aspect of the design, the structural model for a score will be

$$Y_{ijkm} = \mu + \alpha_m + \beta_k + \gamma_{j(m)} + \alpha\beta_{mk} + \beta\gamma_{kj(m)} + \epsilon_{ijkm}$$

The addition of repeated measurement to the design, in effect, results in a second type of nesting, that is, nesting of experimental subjects within A and C but crossing of experimental subjects with B. Thus, the model must be expanded to include a subject effect and an interaction of subjects and B effect. If we take into account the repeated measures aspect of the design, the structural model is

$$Y_{ijkm} = \mu + \alpha_m + \beta_k + \gamma_{j(m)} + \alpha\beta_{mk} + \beta\gamma_{kj(m)} + \pi_{i(jm)} + \beta\pi_{ki(jm)} + \epsilon_{ijkm}$$

If we assume that A is at p levels, B is at q levels, there are r levels of C nested within each level of A, and there are n subjects per group, the sources, degrees of freedom, and expectations of mean squares for this design are (A and B fixed; C random):

source	df	E(MS)
Among subjects	$prn - 1$	
A	$p - 1$	$\sigma^2 + q\sigma_\pi^2 + qn\sigma_\gamma^2 + qrn\sigma_\alpha^2$
$C(A)$	$p(r - 1)$	$\sigma^2 + q\sigma_\pi^2 + qn\sigma_\gamma^2$
S	$pr(n - 1)$	$\sigma^2 + q\sigma_\pi^2$
Within subjects	$prn(q - 1)$	
B	$q - 1$	$\sigma^2 + \sigma_{\beta\pi}^2 + n\sigma_{\beta\gamma}^2 + prn\sigma_\beta^2$
AB	$(p - 1)(q - 1)$	$\sigma^2 + \sigma_{\beta\pi}^2 + n\sigma_{\beta\gamma}^2 + rn\sigma_{\alpha\beta}^2$
$BC(A)$	$p(r - 1)(q - 1)$	$\sigma^2 + \sigma_{\beta\pi}^2 + n\sigma_{\beta\gamma}^2$
BS	$pr(n - 1)(q - 1)$	$\sigma^2 + \sigma_{\beta\pi}^2$
Total	$pqrn - 1$	

Since C is a random effect, the testing procedures are somewhat more involved than for other repeated measures designs. For the among-subjects effects, A is tested against $C(A)$, and $C(A)$ is tested against the subject effect [if C were fixed, both A and $C(A)$ would be tested against the subject effect]. Turning to within-subject sources, B and AB are each tested against $BC(A)$, and $BC(A)$ is tested against the BS interaction (if C were fixed, all three effects would be tested against the BS term).

The analysis of this example should pose no problem to the student since the terms for the basic partially nested design can be computed from the usual formulas, ignoring the existence of repeated measures.

The among-subjects component S is found as in any other repeated measures design, and BS can be found by subtraction.

If relevant, repeated measures can be introduced into any two-dimensional or higher-order partially nested design. In order to determine the appropriate model for the data, the subjects are considered as representing an additional nested factor, and the Cornfield and Tukey algorithm can be applied. The computations of sums of squares follow the lines for partially nested designs in general and should present no undue difficulty to the student who has mastered these procedures.

exemplary applications of repeated measures designs

example 1 Tagatz (1967) reports a study of factors chosen in terms of their probable effects upon efficiency of concept-attainment tasks. The criterion variable analyzed by means of a repeated measures design was time-to-criterion scores (defined as "time taken to attain successfully the concept"). The experimental design was four-dimensional, with two of the dimensions representing classification variables (grade level of subjects and sex of subjects), one major treatment variable (instructions provided which were designed to predispose the subjects to a commonality strategy or a conservative strategy of concept attainment), and one dimension over which repeated measures occurred (concept-attainment problems). Since two of the dimensions were classificatory, the basic design would be categorized as randomized blocks. Subjects were randomly selected from a single elementary school; 10 males and 10 females were taken from the fifth grade, and a like number from the sixth grade. Although it is not explicitly stated by the author, the description of experimental procedures suggests that subjects were randomly and equally divided between the two concept-attainment strategies. The stimulus materials consisted of figures presenting information at five dimensions; each dimension existed at one of two levels on a given figure. Instruction was given to each subject in the relevant concept-attainment strategy; i.e., either in a commonality strategy which involved focusing only on cards which did exemplify the concept to be attained, or in a conservative strategy which involved looking both at cards exemplifying and at cards not exemplifying the concept to be attained. Each subject was run through four different problems involving attainment of concepts, and "problems" represented the repeated measures dimension of the design. In terms of groups of experimental subjects, the scheme for the design was:

	PROBLEM (D)	SEX (B):	GRADE LEVEL (A) fifth: male	fifth: female	sixth: male	sixth: female
	1		G_1	G_3	G_5	G_7
Commonality	2		G_1	G_3	G_5	G_7
	3		G_1	G_3	G_5	G_7
	4		G_1	G_3	G_5	G_7
STRATEGY (C)						
	1		G_2	G_4	G_6	G_8
Conservative	2		G_2	G_4	G_6	G_8
	3		G_2	G_4	G_6	G_8
	4		G_2	G_4	G_6	G_8

Since there is no repetition of measurement on the A (grade level), B (sex), and C (strategy) dimensions, these sources and their interactions comprise the among-subjects effects. The D dimension (problems) has repetition of measurement, and it, along with interactions involving D, comprises the within-subject components.

The analysis-of-variance summary for time-to-criterion scores was:

source	df	sum of squares	mean square	F
Among subjects	39			
(A) Grade	1	9.79	9.79	4.20
(B) Sex	1	11.66	11.66	5.03
(C) Strategy	1	36.66	36.66	15.80
AB	1	.51	.51	.22
AC	1	6.26	6.26	2.69
BC	1	12.11	12.11	5.22
ABC	1	.61	.61	.26
(S) Subjects	32	74.24	2.32	
Within subjects	120			
(D) Problems	3	94.98	31.66	20.69
AD	3	15.09	5.03	3.29
BD	3	2.97	.99	.65
CD	3	43.08	14.36	9.39
ABD	3	.81	.27	.17
ACD	3	24.18	8.06	5.27
BCD	3	4.29	1.43	.93
$ABCD$	3	2.07	.69	.45
SD	96	146.88	1.53	
Total	159	486.29		

Source: Adapted from G. E. Tagatz. Effects of strategy, sex, and age on conceptual behavior of elementary school children. *J. Educ. Psychol.*, 1967, **58**, 106. Copyright 1967 by the American Psychological Association. Reproduced by permission.

The author did not carry out tests of homogeneity of variance and covariance; rather, he utilized the conservative testing procedure which involves setting the degrees of freedom for the repeated measures dimension at 1. On this basis, the results for A, B, BC, and AD were significant at the .05 level, and the results for C, D, CD, and ACD were significant at the .01 level. All four main effects were significant, and the order of outcomes was as follows: fifth-graders were more efficient (i.e., took less time to criterion) than sixth-graders; females were more efficient than males; the commonality strategy was more efficient than the conservative strategy; and the first problem took significantly more time to solve than the fourth problem, but other differences were not significant (using the Duncan multiple-range test on the means of the four problems). In addition to the main effects, three first-order interactions and one second-order interaction reached significance. The sex-by-strategy interaction (BC) was interpreted as revealing males and females equally efficient under the commonality strategy, but males less efficient than females under the conservative strategy. For the grade-by-problems interaction (AD), the major difference involved greater efficiency by fifth-graders on the first problem, but comparable performance on problems 2 through 4. The strategy-by-problem interaction (CD) indicated that on the first problem, the commonality strategy was more efficient but that on problems 2 through 4, the two strategies did not differ significantly in efficiency. The second-order interaction, grade-by-strategy-by-problems (ACD) revealed that sixth-graders receiving training in the conservative strategy were particularly inefficient on the first problem; no other pattern appeared in this interaction. All interpretations of significant effects were based upon the use of the shortest significant range statistic from the Duncan multiple-range test (except for main effects existing at two levels, for which the interpretation is immediate from inspection of means).

example 2 Faust and Anderson (1967) report an experiment in the area of programmed instruction. The intent of their study was to determine whether or not the addition of "incidental material" (that is, content irrelevant to the correct response) to program frames would increase the likelihood of recall of the relevant response. The authors predicted that incidental material would have a beneficial effect upon recall; their rationale was, basically, that in the absence of additional material, the subject could simply copy the correct response to a frame without attending to the discrimination relevant to responding in a recall setting. On the other hand, the presence of incidental material required the subject to attend to both the stimulus and response elements in order to choose the correct response. The programs involved simple Russian vocabulary terms; one version of the program presented frames having only the Russian and English word pair. A second version of the program embedded the relevant Russian/English word pair among four other irrelevant Russian/English word pairs. A written response was required in both program versions. Typical frames from the two program versions were:

COPYING FRAME (NO CONTEXT)	COPYING FRAME (CONTEXT)
A table is a stohl.	A rag is a tryapka. A bridge is a mohst. A table is a stohl. A college is a vooz. An onion is a look.
A table is a ______.	A table is a ______.

Adapted from G. W. Faust and R. C. Anderson. Effects of incidental material in a programmed Russian vocabulary lesson. *J. Educ. Psychol.*, 1967, **58**, 5. Copyright 1967 by the American Psychological Association. Reproduced by permission.

In addition to the program types (i.e., no-context version with no incidental material and context version with incidental material), two sets of vocabulary terms were utilized (i.e., no-context and context program versions were prepared, using different Russian/English word pairs); within each list, half the words were designated as easy and half as difficult, and both an immediate posttest and a delayed-retention test of recall were administered. Although different groups of subjects were used for the different sets of vocabulary terms and for the different types of programs, there were repeated measures with respect to the word difficulty and time dimensions. The design was, then:

			PROGRAM TYPE (A)			
			context		no context	
	WORD DIFFICULTY (C)	WORD LIST (B):	1	2	1	2
TESTS (D) Posttest	Easy		G_1	G_2	G_3	G_4
	Difficult		G_1	G_2	G_3	G_4
TESTS (D) Retention test	Easy		G_1	G_2	G_3	G_4
	Difficult		G_1	G_2	G_3	G_4

Note that there are repeated measures over D (that is, the time dimension, posttest versus retention test) and C (that is, the word-difficulty dimension). However, subjects are nested within cells of A (that is, program type) and B (that is, word list). Thus, the structural model for a score is

$$Y_{ijkmg} = \mu + \alpha_g + \beta_m + \alpha\beta_{gm} + \pi_{i(gm)} + \gamma_k + \delta_j + \alpha\gamma_{gk} + \alpha\delta_{gj} + \beta\gamma_{mk} + \beta\delta_{mj} + \gamma\delta_{kj} + \alpha\beta\gamma_{gmk} + \alpha\beta\delta_{gmj} + \alpha\gamma\delta_{gkj} + \beta\gamma\delta_{mkj} + \alpha\beta\gamma\delta_{gmkj} + \gamma\pi_{ki(gm)} + \delta\pi_{ji(gm)} + \gamma\delta\pi_{kji(gm)} + \epsilon_{ijkmg}$$

With the exception of experimental subjects, all dimensions are fixed. The subjects were an intact group of 48 graduate students enrolled in an educational psychology course (with those having any knowledge of Russian excluded from the population). The four groups of subjects required for the design were randomly formed from the available pool of subjects.

The analysis of variance based on posttest and retention-test scores was:

source	df	sum of squares	mean square	F
Among subjects	47			
(A) Program type	1	13.02	13.02	2.36
(B) Word list	1	.08	.08	.00
AB	1	15.19	15.19	2.74
(S) Subjects	44	244.20	5.55	
Within subjects	144			
(C) Word difficulty	1	426.00	426.00	161.36
AC	1	16.35	16.35	6.19
BC	1	38.17	38.17	14.16
ABC	1	.69	.69	.26
SC	44	116.16	2.64	
(D) Time	1	35.02	35.02	42.19
AD	1	.01	.01	.00
BD	1	.17	.17	.20
ABD	1	3.02	3.02	3.64
SD	44	36.52	.83	
CD	1	1.35	1.35	2.33
ACD	1	.18	.18	.31
BCD	1	1.13	1.13	1.94
$ABCD$	1	1.62	1.62	2.79
SCD	44	25.52	.58	

Adapted from G. W. Faust and R. C. Anderson. Effects of incidental material in a programmed Russian vocabulary lesson. *J. Educ. Psychol.*, 1967, **58**, 7.

The different sources of variation among scores have been grouped in terms of the appropriate error term (i.e., either subjects or an interaction involving subjects). The student can easily confirm the implied tests by applying the Cornfield and Tukey algorithm to the structural model for a score.

Among the F ratios, only those for word difficulty (C), BC, and time (D) reached significance at the .01 level (note that since all treatment dimensions exist at two levels, the question of homogeneity of covariance becomes moot; the usual and the conservative degrees of freedom would be the same). Also, the AC effect was significant at the .05 level. The results for C and D are of little interest since it would be supposed that difficulty of vocabulary terms would have an effect upon learning and that fewer words would be recalled on the retention test when compared with the immediate posttest. The BC interaction implies that the word lists are not uniform with respect to difficulty of words. The only effect of primary interest to the researchers involved the AC interaction. This effect resulted from higher scores by subjects in the context groups on words which were easy. Note, however, that the overall main effect for program type (A) failed to reach significance.

chapter eight

the analysis of covariance

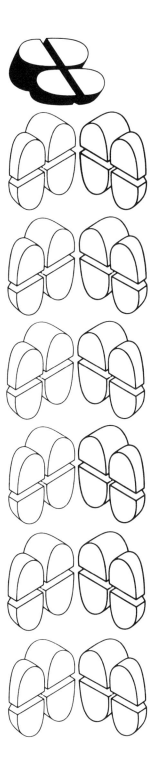

introduction

Unlike the procedures considered in earlier chapters, the analysis of covariance does not represent a specific class of experimental designs. Rather, it is a general technique for increasing the precision of any experimental design; the increase in precision is accomplished by adjusting criterion measures in terms of one or more outside variables (known as *covariables*). The adjustment of criterion scores (to the extent that it is successful) will effect a reduction in the size of experimental error and produce within-cell measures of greater homogeneity. The success of the adjustment depends upon the degree of linear correlation between the covariables and the criterion measure. Although a more precise development is presented below, the basic approach incorporated in the analysis of covariance revolves around removing from the criterion scores that part which is predictable from the covariables. The techniques utilized to make this type of adjustment are based upon ordinary linear regression (zero-order or multiple regression as appropriate to the number of covariables).

As discussed in Chapter 5, the randomized blocks designs are also capable of increasing power by increasing within-cell homogeneity of criterion measures. It is important for the researcher to be aware of the similarities and differences between analysis of covariance and randomized blocks designs so that he can use them both appropriately. For this reason, a section comparing these two procedures is presented after a discussion of some of the theory underlying the analysis of covariance. The remainder of this chapter is devoted to the treatment of specific applications of the analysis of covariance to different classes of experimental designs. We begin with the simplest case; the basic design is a completely randomized design, and just one covariable is taken into account. We turn, then, to the case of two or more covariables in a CRD. The remaining sections are devoted to such basic designs as factorial designs and nested designs.

some theory underlying the analysis of covariance

In ordinary analysis of variance, within-cell variability of subjects is conceptualized as due to random influences and is summarized by the error term in the structural model for a score. This view is reasonable since subjects are randomly assigned to cells of the design. Consider, for example, a completely randomized design; the structural model is $Y_{ij} = \mu + \alpha_j + \epsilon_{ij}$. Within a cell (treatment group) of the design, all variation among subjects is attributed to experimental error; that is, μ and α_j are constant for subjects at the same treatment level. However, it is obvious that this conceptualization is only a rough approximation to the true state of the factors operating in the experimental setting. Within a single treatment group, subjects will vary in ways which will have an effect upon their criterion scores. Depending upon the nature of the response measure, a variety of aptitude and personality variables *in addition to the treatment* would be relevant to explaining high or low criterion scores. For example, if the criterion measure were a multiple-choice examination of knowledge and comprehension of elementary concepts of atomic theory, performance on this criterion (regardless of the treatment) would be affected by previous instruction in atomic theory and related topics, general level of aptitude in scientific reasoning, general intelligence level, testwiseness (in the sense of being familiar with taking multiple-choice-type examinations), motivational factors, etc. If no cognizance is taken of these factors in the design of the experiment, and subjects are simply randomly assigned to treatment groups, then influences from these sources are randomly distributed over the treatment groups and are summarized in the error term. Of course, by ignoring individual differences among experimental subjects, a price is paid by the experimenter; namely, all these factors contribute to enlarging the magnitude of experimental error.

The analysis of covariance is a technique whereby one or more outside variables which affect criterion performance can be taken into account and, in effect, removed from the within-cell variability of the experimental subjects. In developing the rationale for the analysis of covariance, we shall utilize the CRD and take into account just one outside measure (i.e., covariable). If X is the covariable, then the linear correlation between the covariable and the criterion variable is r_{XY}. From a prediction point of view, $Y' = a + bX$, where a and b are regression constants and Y' is the predicted value of Y for a given value of X. Since $a = \overline{Y} - b\overline{X}$, the regression equation can be rewritten as $Y' = \overline{Y} + b(X - \overline{X})$. Thus, for a given individual, that part of his criterion score which is *unpredictable* on the basis of the covariable is $Y^* = Y - Y' = (Y - \overline{Y}) - b(X - \overline{X})$. Note that in regression terminology, Y^* is the error of prediction. On the basis of the covariable alone, we are able to account for all of an individual's criterion score except the amount $Y^* = Y - Y'$; although Y^* is, in terms of a regression model, an error of prediction, from the point of view of the analysis-of-variance design it is partially due to the effect of the experimental treatment. Therefore, the structural model for a score in the analysis of covariance is written in terms of Y^* rather than in terms of Y. The model is

$$Y^*_{ij} = Y_{ij} - Y'_{ij} = \mu^* + \alpha^*_j + \epsilon^*_{ij}$$

The notation utilized for the components of the model (i.e., the asterisks on all terms) is adopted in order to serve as a reminder that these terms are different from their counterparts which occur in the absence of a covariable.

If treatment groups are ignored, the adjustment of criterion scores can be carried out in terms of the total regression of Y on X. If b_t is defined as the slope constant in the prediction equation

$$b_t = \frac{\sum_{i=1}^{n} \sum_{j=1}^{p} (X_{ij} - \bar{X}_{..})(Y_{ij} - \bar{Y}_{..})}{\sum_{i=1}^{n} \sum_{j=1}^{p} (X_{ij} - \bar{X}_{..})^2}$$

then $Y'_{ij} = \bar{Y}_{..} + b_t(X_{ij} - \bar{X}_{..})$, and the error of prediction is

$$Y^*_{ij} = Y_{ij} - Y'_{ij} = (Y_{ij} - \bar{Y}_{..}) - b_t(X_{ij} - \bar{X}_{..})$$

The sum of squares associated with this error of prediction can be found by starting with the usual total sum of squares from the CRD:

$$\begin{aligned} \text{SS}_{\text{total}} &= \sum_{i=1}^{n} \sum_{j=1}^{p} (Y_{ij} - \bar{Y}_{..})^2 = \sum_{i=1}^{n} \sum_{j=1}^{p} [Y_{ij} - Y'_{ij} + b_t(X_{ij} - \bar{X}_{..})]^2 \\ &= \sum_{i=1}^{n} \sum_{j=1}^{p} (Y_{ij} - Y'_{ij})^2 + b_t\left[\sum_{i=1}^{n} \sum_{j=1}^{p} (X_{ij} - \bar{X}_{..})^2\right] \\ &= \sum_{i=1}^{n} \sum_{j=1}^{p} (Y_{ij} - Y'_{ij})^2 + \frac{\left[\sum_{i=1}^{n} \sum_{j=1}^{p} (X_{ij} - \bar{X}_{..})(Y_{ij} - \bar{Y}_{..})\right]^2}{\sum_{i=1}^{n} \sum_{j=1}^{p} (X_{ij} - \bar{X}_{..})^2} \end{aligned}$$

Thus

$$\sum_{i=1}^{n} \sum_{j=1}^{p} (Y_{ij} - Y'_{ij})^2 = \sum_{i=1}^{n} \sum_{j=1}^{p} (Y_{ij} - \bar{Y}_{..})^2 - \frac{\left[\sum_{i=1}^{n} \sum_{j=1}^{p} (X_{ij} - \bar{X}_{..})(Y_{ij} - \bar{Y}_{..})\right]^2}{\sum_{i=1}^{n} \sum_{j=1}^{p} (X_{ij} - \bar{X}_{..})^2}$$

This final expression defines an *adjusted* total sum of squares and will be denoted $\text{SS}^*_{\text{total}}$. From a regression point of view, $\text{SS}^*_{\text{total}}$ is simply the sum of squares associated with errors of prediction and is referred to as unexplained variability. Within analysis of covariance, we call $\text{SS}^*_{\text{total}}$ an adjusted total sum of squares since the predictable effect due to the covariable has been removed. Returning to the original definition of b_t, note that

$$SS^*_{\text{total}} = SS_{\text{total}} - b_t^2\left[\sum_{i=1}^{n}\sum_{j=1}^{p}(X_{ij} - \overline{X}_{..})^2\right]$$

Once SS^*_{total} has been defined, the technique of analysis of covariance consists in partitioning SS^*_{total} into components attributable to treatment and to experimental error. Recall that in an ordinary CRD, the analysis of variance consists in partitioning SS_{total} into similar components.

The partitioning of SS^*_{total} depends upon an assumption of equal within-cell regression of Y on X. That is, focusing on the treatment groups, we require that *within each group* the criterion and covariable measures are related through the same regression constant b_w. For example, if $p = 3$, it would be possible to build a prediction equation for Y in terms of X within each of the three treatment groups. Our assumption is that these three equations have the same slope (i.e., if b_1 is the slope in group 1, b_2 in group 2, and b_3 in group 3, then b_1, b_2, and b_3 must be independent estimates of the same population slope). This condition is referred to as the assumption of *homogeneity of regression*. If the regression coefficients are homogeneous, then they can be pooled to form a single estimate b_w. Note the similarity of this assumption to the homogeneity-of-variance condition required in all analysis-of-variance designs. When variances are homogeneous, this allows us to pool the variances across the treatment groups to determine a single estimate of experimental error (i.e., the mean square for error). Since the assumption of homogeneity of regression is a new principle, a schematic representation will aid in clarifying it. Figure 8-1 shows criterion and covariable scores as points on a scatter diagram for three groups of experimental subjects. Note that there is a distinct treatment effect since the clusters of points for the three groups are well separated along the Y dimension of the figure (in fact, for the sake of clarity no overlap of groups is shown in the figure; this is a highly unlikely outcome in a real data situation). The within-groups regression lines are parallel, indicating equal slopes. Explicitly, these regression lines for the three groups are

$$Y''_{i1} = \overline{Y}_{\cdot 1} + b_1(X_{i1} - \overline{X}_{\cdot 1})$$

$$Y''_{i2} = \overline{Y}_{\cdot 2} + b_2(X_{i2} - \overline{X}_{\cdot 2})$$

$$Y''_{i3} = \overline{Y}_{\cdot 3} + b_3(X_{i3} - \overline{X}_{\cdot 3})$$

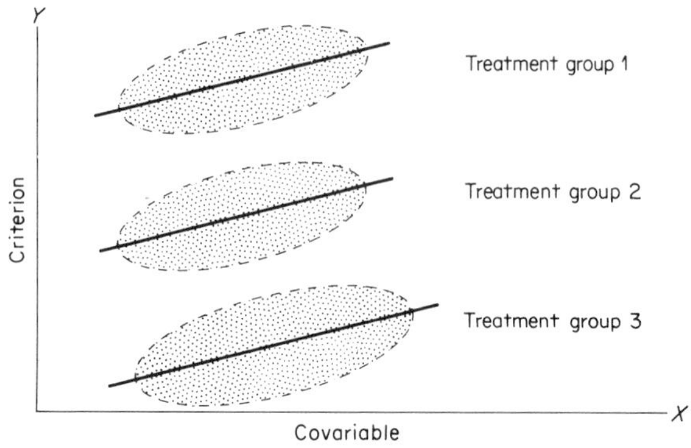

figure 8-1 *Three treatment groups displaying homogeneity of regression.*

Thus, in Figure 8-1, $b_1 = b_2 = b_3 = b_w$, although, as is apparent, the Y intercepts of the regression lines do differ. Also, the regression line resulting from the total set of scores, ignoring treatment groups, would have the same slope as the separate regression lines; that is, $b_t = b_w$. This fact is not, however, essential to the condition of homogeneity of regression. If the clusters of points are spread out along the X dimension as well as along the Y dimension, they could share common within-groups slopes, but the regression line based on the total set of scores would have a different slope (see Figure 8-2).

Returning to the partitioning of SS^*_{total}, the first step is to adjust scores within groups by utilizing the common within-group regression coefficient b_w. Then an adjusted sum of squares for error, SS^*_{error}, can be computed from these adjusted criterion scores. The adjustment can be

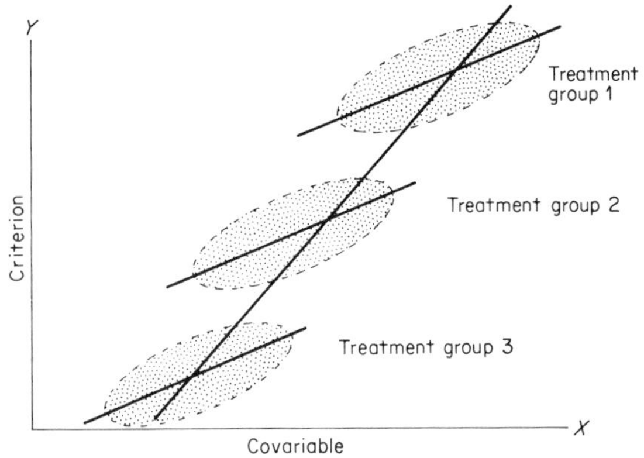

figure 8-2 *Scatter diagrams for three groups which are separated on the X dimension.*

carried out in a single step by reducing the usual SS_{error} by the amount predictable from the covariable. That is,

$$SS^*_{\text{error}} = SS_{\text{error}} - \frac{\left[\sum_{i=1}^{n}\sum_{j=1}^{p}(X_{ij} - \bar{X}_{.j})(Y_{ij} - \bar{Y}_{.j})\right]^2}{\sum_{i=1}^{n}\sum_{j=1}^{p}(X_{ij} - \bar{X}_{.j})^2}$$

Note that the expression for the sum of squares predictable from X is formally similar to that used to adjust SS_{total}. In the present case, however, the cross products $(X_{ij} - \bar{X}_{.j})(Y_{ij} - \bar{Y}_{.j})$ are taken around the group means $\bar{X}_{.j}$ and $\bar{Y}_{.j}$, rather than around the grand means $\bar{X}_{..}$ and $\bar{Y}_{..}$; also, the denominator of the expression is a within-groups sum of squares for X rather than a total sum of squares for X.

The adjustment of the sum of squares for treatment does not follow the pattern established for SS^*_{total} and SS^*_{error}. Although it would be possible to build a regression equation involving just the treatment means and to derive a regression constant for use in this equation, this approach cannot be used in determining the adjusted treatment sum of squares, because the regression constant would be largely a function of sampling error involved in the X scores. This can be seen from Figure 8-2. Note that the three treatment groups are distinct with respect to their locations along the X dimension. The correlation of the covariable means and criterion means (that is, the $\bar{X}_{.j}$ and $\bar{Y}_{.j}$ for the three groups) is a function of the degree to which the three groups *lack* comparability with respect to X. Since subjects are randomly assigned to treatment groups, these differences are due to chance. Referring back to Figure 8-1, note that in this case the three groups are comparable with respect to location on the covariable dimension but there is, nevertheless, within groups and over the entire set of subjects, a relatively high degree of correlation between covariable and criterion variable. However, the correlation based on group means would be 0. Thus, the strategy followed in the analysis of covariance is not to adjust treatment sums of squares directly in terms of the regression of Y means upon X means. Rather, the adjusted treatment sum of squares is found by subtraction; that is, $SS^*_{\text{treat}} = SS^*_{\text{total}} - SS^*_{\text{error}}$.

The interpretation of the adjusted sums of squares is aided by returning to the original expression for SS^*_{total} and showing the partitioning in terms of quantities based on differences. Thus,

$$SS^*_{\text{total}} = \sum_{i=1}^{n}\sum_{j=1}^{p}(Y_{ij} - Y'_{ij})^2$$

where $Y'_{ij} = \bar{Y}_{..} + b_t(X_{ij} - \bar{X}_{..})$ is the predicted value of Y using the regression constant based on the total set of scores. If $Y''_{ij} =$

$\overline{Y}_{.j} + b_w(X_{ij} - \overline{X}_{.j})$ is the predicted value of Y in the jth treatment group using the regression constant b_w (that is, the pooled within-groups regression constant), then it can be shown that

$$\text{SS}^*_{\text{total}} = \sum_{i=1}^{n} \sum_{j=1}^{p} (Y_{ij} - Y'_{ij} - Y''_{ij} + Y''_{ij})^2 = \sum_{i=1}^{n} \sum_{j=1}^{p} (Y_{ij} - Y''_{ij})^2$$
$$+ \sum_{i=1}^{n} \sum_{j=1}^{p} (Y''_{ij} - Y'_{ij})^2 = \text{SS}^*_{\text{error}} + \text{SS}^*_{\text{treat}}$$

The first term represents within-group variation of scores about a regression line based upon the specific group mean $(\overline{Y}_{.j})$ and utilizing the within-group regression constant b_w. The second term is based on variation due to differences between Y'_{ij} and Y''_{ij}, that is, differences between predicted values of Y based upon prediction equations utilizing b_t and b_w. A diagram will help clarify the fact that such differences are interpretable as treatment effects. In Figure 8-3 we have shown scatter diagrams for a design with two levels of treatment. The line labeled Y' is the prediction line based upon the total regression constant b_t; the lines Y''_1 and Y''_2 are the within-groups regression lines based upon b_w. Explicitly, these are

$$Y''_{i1} = \overline{Y}_{.1} + b_w(X_{i1} - \overline{X}_{.1}) \qquad \text{and} \qquad Y''_{i2} = \overline{Y}_{.2} + b_w(X_{i2} - \overline{X}_{.2})$$

For the particular individual shown (i.e., the score point Y_{ij}), the components entering into the adjusted sums of squares are noted. The usual SS_{total} (unadjusted) is based upon deviations from the grand mean, $Y_{ij} - \overline{Y}_{..}$, and is represented by the sum $C_1 + C_2 + C_3$. The adjustment of SS_{total} involves removing C_1 and yields $\text{SS}^*_{\text{total}}$, which is based on deviations represented by the sum $C_2 + C_3$. Then, the partitioning of $\text{SS}^*_{\text{total}}$ into components attributable to error and to treatment involves

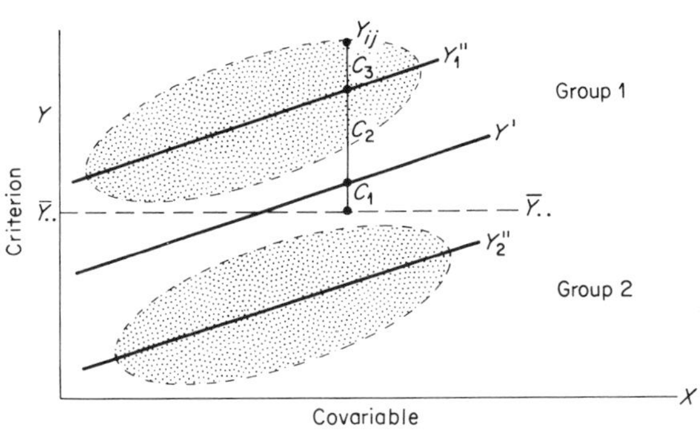

figure 8-3 *Scatter diagram for two groups which are separated on the Y dimension.*

sums of squares based on deviations corresponding to C_3 (for error) and C_2 (for treatment). Note that $C_3 = Y_{i1} - Y''_{i1}$ and $C_2 = Y''_{i1} - Y'_{i1}$. It is apparent from the figure that the magnitude of C_2 is a function of the degree to which the scores in groups 1 and 2 are separated along the Y dimension. That is, the greater the effect of the treatment in separating the two groups, the larger, on the average, will be the discrepancies between Y'_{ij} and Y''_{ij} values. Thus, $Y''_{ij} - Y'_{ij}$ is a reflection of treatment effects. On the other hand, C_3 reflects errors of prediction within the treatment groups.

The partitioning of SS^*_{total} into components attributable to treatment and to experimental error provides a basis for deriving mean squares and testing the treatment effect. Prior to turning to this, however, we must consider the appropriate degrees of freedom for each term. If we assume that the basic CRD has p treatment groups and a total of n_t subjects, the degrees of freedom for the unadjusted total sum of squares (SS_{total}) are $n_t - 1$ (or $np - 1$ for groups of the same size). In deriving the adjusted total sum of squares (SS^*_{total}) it was necessary to estimate the regression coefficient connecting the covariable and the criterion variable (this estimate being b_t). This estimation is based on 1 degree of freedom since just one parameter is involved; thus, SS^*_{total} is based on 1 less degree of freedom than is SS_{total}. That is, the degrees of freedom for SS^*_{total} are $n_t - 2$ (or $np - 2$ for groups of equal size). Since the degrees of freedom for treatment will still be $p - 1$, this 1 degree of freedom loss must be reflected in the degrees of freedom for the error term. Thus, SS^*_{error} is based on $n_t - p - 1$ degrees of freedom [or $p(n - 1) - 1$ for groups of equal size]. In order to show the full accounting of degrees of freedom, it is useful to label the *predictable* variability, which is subtracted from SS_{total} to yield SS^*_{total} as SS_{regr}. That is, $SS^*_{total} = SS_{total} - SS_{regr}$, where

$$SS_{regr} = \frac{\left[\sum_{i=1}^{n}\sum_{j=1}^{p}(X_{ij} - \bar{X}_{..})(Y_{ij} - \bar{Y}_{..})\right]^2}{\sum_{i=1}^{n}\sum_{j=1}^{p}(X_{ij} - \bar{X}_{..})^2}$$

Then the sources and their associated degrees of freedom are:

source	*df*
SS_{regr}	1
SS^*_{treat}	$p - 1$
SS^*_{error}	$n_t - p - 1$
SS_{total}	$n_t - 1$

The assumption of homogeneity of regression is rather critical to the analysis of covariance since it is known that departures from this condition can seriously affect the actual risk of a Type I error. Thus, it is necessary to test for homogeneity prior to carrying out adjustments of the sums of squares. An appropriate test can be based upon a partitioning of the adjusted sum of squares for error. In effect, SS^*_{error} is made up of two sources of variation: The first source is variability of scores around the within-groups regression lines with slopes $b_1, b_2, \ldots, b_p$. Note that it is these regression coefficients which are pooled to form b_w. The second source is the *additional* variability which is introduced when b_w is substituted for b_1, b_2, etc.; it should be noted that the prediction of Y scores will, in general, be no worse when b_1, b_2, etc., are used than when b_w is used; and, unless the values of b_1, b_2, etc., are identical, the variability accounted for by utilizing b_w will be *less than* that accounted for by using b_1, b_2, etc. The test for homogeneity of regression involves partitioning SS^*_{error} into the two components

$$S_1 = \sum_{i=1}^{n} \sum_{j=1}^{p} (Y''_{ij} - Y'''_{ij})^2 \qquad \text{and} \qquad S_2 = \sum_{i=1}^{n} \sum_{j=1}^{p} (Y_{ij} - Y''_{ij})^2$$

where $Y''_{ij} = \overline{Y}_{.j} + b_j(X_{ij} - \overline{X}_{.j})$, as previously defined (that is, Y''_{ij} is the value of Y predicted from X by using the regression equation specific to the jth treatment group), and $Y'''_{ij} = \overline{Y}_{.j} + b_w(X_{ij} - \overline{X}_{.j})$ (that is, Y'''_{ij} is the value of Y predicted from X by using the common regression constant b_w). Note that S_1 is a direct function of the discrepancy between b_w and the b_1, b_2, etc., values. If $b_1 = b_2 = \cdots = b_p$, then S_1 will equal 0. On the other hand, the greater the variability of the b_j, the larger will be the value of S_1. S_2 is strictly dependent upon the within-group errors of prediction which result from utilizing the b_1, b_2, etc., values in predicting Y from X. Since S_1 depends upon the values of the p regression coefficients b_1, b_2, etc., this sum of squares has $p - 1$ degrees of freedom. The degrees of freedom remaining for S_2 are $n_t - 2p$ [or $p(n - 2)$ for groups of equal size]. The corresponding mean squares can be denoted M_1 and M_2; thus, the test statistic for evaluating homogeneity of regression is an F ratio:

$$F = \frac{M_1}{M_2} = \frac{S_1/(p-1)}{S_2/(n_t - 2p)}$$

The degrees of freedom for entering a table of critical values of F are, of course, $p - 1$ and $n_t - 2p$. It should be noted that all formulas presented in this section are based on definitions of the quantities involved and do not, in general, represent very practical formulas for use in computing sums of squares. Computational formulas are presented in connection with the examples of the various applications of the analysis of covariance.

In general, all assumptions involved in the ordinary analysis of variance must also be met when applying the analysis of covariance. Thus, it is assumed that subjects are randomly assigned to treatment groups in a manner relevant to the specific basic design; furthermore, conditions of homogeneity of variance and normality of distributions are necessary relative to the components of the model for the analysis of covariance (e.g., if the model is $Y_{ij}^* = \mu^* + \alpha_j^* + \epsilon_{ij}^*$, then it is assumed that the ϵ^* are distributed normally and independently with equal variances across the treatment groups). In addition to these usual assumptions, the new condition of homogeneity of regression must be met.

From the point of view of the practical applicability of the analysis of covariance, two additional requirements must be considered. First, the effectiveness of the analysis of covariance is related to the degree of linear correlation between the covariable and the criterion variable. If r_{XY} is nonsignificant (that is, if the hypothesis $H_0: \rho_{XY} = 0$, where ρ_{XY} is the population correlation between X and Y, is accepted at a conventional level of significance), then the value of r_{XY} can be attributed to sampling error. In this case there would be no point in applying the analysis of covariance; an appropriate test for the significance of r_{XY} is given by the F ratio, $\text{MS}_{\text{regr}}/\text{MS}_{\text{error}}^*$ with 1 and $n_t - p - 1$ degrees of freedom. In general, the researcher should choose his covariables in terms of their *known* relationships to the criterion measure, so that it is virtually certain that a significant relationship exists. If the researcher is uncertain concerning the availability of relevant covariables, this is probably best determined in a correlational pilot study.

The second operational requirement has to do with the possible influence of treatments upon the covariable. If the measures on the covariable are taken prior to the application of the experimental treatment (e.g., in a pretest session or from available records such as school cumulative records), this requirement will be met. However, in some experiments, the researcher may choose as a covariable an early measure of performance taken from the experimental treatment groups. In this case, it is possible that the values of the covariable will reflect some effect due to the treatments. Such an outcome is also possible if the covariable measures are taken *after* the completion of the experiment. In any case, if the treatments do affect the values of the covariable, then when analysis-of-covariance adjustments are carried out, the researcher is effectively removing *treatment effects* from the criterion scores. This is, obviously, a highly undesirable state of affairs since it is these very effects which it is of interest to detect through use of the analysis.

In some research settings, the analysis of covariance has been (inappropriately) applied to equalize treatment groups with respect to some outside variable which cannot be controlled during the execution of the experiment. Thus, measures are taken on this outside variable during or after completion of the experiment, and these scores are used as a covariable. Consider an experiment, for example, which involves individualized instruction; by the nature of this instruction, the amount of time a student spends on a given subject-matter area is self-determined. In order to "control" effort, the researcher may ascertain the amount of time actually spent by each student and utilize this as a covariable when completing his analysis. This practice is clearly erroneous since it is almost certain that the treatments would, themselves, have an effect on the effort put forth by students on a subject-matter area; thus, the covariable would reflect treatment effects. The use of a covariable to "equalize" in this fashion may result in an F ratio which is too small or too large; in either case, the result is spurious and may be highly misleading. An interesting discussion of this problem, as well as some exemplary analyses, is presented in Evans and Anastasio (1968).

Occasionally, another misuse of the analysis of covariance will be encountered; some researchers utilize covariance adjustments to "equalize" groups when random assignment of subjects to treatment groups has not been accomplished. For example, in an instructional methods experiment, data may be collected from intact classrooms of students; then, by using an aptitude measure as the covariable, an analysis of covariance may be carried out on achievement scores from individual students. However, the analysis of covariance demands that subjects arrive in the cells of the design by a random process, and the use of intact groups violates this requirement. A proper analysis would involve the use of classroom mean scores rather than scores from individual students.

comparison of the analysis of covariance and randomized blocks designs

In many research situations, the use of randomized blocks and the use of analysis-of-covariance adjustments represent alternative possibilities for the control of one or more relevant, outside variables. This is generally true if the outside variables are measured prior to the formation of treatment groups and if the outside variables represent continuous dimensions. If the outside variables are measured during or after the application of experimental treatments, there will be no opportunity to form blocks, and only the analysis of covariance is appropriate. On the other hand, if the outside variables are discrete, categorical variables, the requirements for a continuous regression function embodied in the analysis of covariance will not be met. Of particular relevance to a choice between these two design possibilities are considerations of the nature of the correlation between the criterion and the outside variable, the interest of the researcher in obtaining information concerning interactions of the outside variable and the treatment variables, and relative power of the two designs. In discussing these considerations, we shall, again, use the CRD as the basic design.

The first consideration hinges on whether the outside variable and the criterion variable show significant departures from a linear relation. The analysis-of-covariance model developed in the preceding section is based on the linear regression of the criterion upon the covariable. If the regression is significantly nonlinear, the adjustment will be more or less imprecise. It is possible to develop the covariance adjustments on the basis of a more complex regression function (e.g., polynomial regression of the form $Y' = a + b_1X + b_2X^2 + \cdots + b_kX^k$, where k is chosen in terms of the apparent structure of the data; note that linear regression is the special case which occurs when $k = 1$). The necessary analysis is, however, highly complex. On the other hand, if the outside variable is utilized to form homogeneous blocks of experimental subjects prior to the application of treatments, a nonlinear relationship between the criterion and outside variable has no effect upon the efficiency of the blocking procedure. That is, within-block homogeneity of subjects does not depend upon a linear relationship, but can be achieved for virtually any functional relationship between the criterion and outside variable. Thus, in many practical situations, the researcher may lean toward the use of randomized blocks if he suspects or knows that a nonlinear regression will occur. However, the availability of suitable programs for high-speed digital computers may render the consideration of computational labor trivial. If so, the remaining danger is that no study will be made of the relationship between the criterion and outside variable, and that covariance adjustment will be carried out blindly on the basis of a linear regression model when a more appropriate model is available. A concern related to the above discussion is the time and effort involved in forming blocks prior to treatment applications. The randomized blocks designs require random assignment of subjects to treatments *after* formation of the blocks but *prior* to application of the experimental treatments. This may present great administrative problems in certain research settings in which subjects are not readily available for testing on the outside variable prior to the experimental sessions. Analysis of covariance has the advantage that information on the outside variable is not utilized until actual analysis of the data. Thus, collection of covariable scores can be accomplished before, during, or after the experimental sessions (however, care must be taken that the covariable is not contaminated by the experimental treatments).

The second consideration relative to analysis of covariance and randomized blocks designs has to do with information concerning the interaction of the treatment dimensions and the outside variable. If the outside variable is utilized to form blocks, then the analysis of variance yields sums of squares corresponding to interactions of the treatment dimensions (and combinations of treatment dimensions) with the outside, blocking variable. The analysis of covariance does not directly yield such sums of squares since the covariable is used only to adjust criterion scores and does not enter into the dimensions of the design. The problem of interaction, however, runs deeper than indicated above. If the outside variable were utilized as a blocking dimension in a randomized blocks design and if the interaction of treatment and blocks were significant, then if this same outside variable had, instead, been utilized as a covariable in an analysis of covariance, the interaction would be reflected by *lack* of homogeneity of regression. This relationship can be seen from Figure 8-4. The left panel of Figure 8-4 shows scatter diagrams for two treatment groups; heterogeneity of regression is reflected in the differing slopes of the regression lines connecting X and Y in the two groups. The right panel, based on the same data situation, shows cell means when the outside variable X has been utilized to build four blocks. The profiles of cell means over the blocks will approximate the regression trend from the left panel. The presence of interaction between treatment and blocks is reflected in the lack of parallelism of the lines of block means for the two treatment levels. Thus, it is apparent that both the analysis of covariance and the randomized blocks design are capable of yielding information concerning interactions of treatment and blocking dimensions; the crucial consideration is that the analysis of covariance reflects this interaction in heterogeneity of regression which violates an underlying assumption for the validity of tests for the treatment effects. Therefore, should

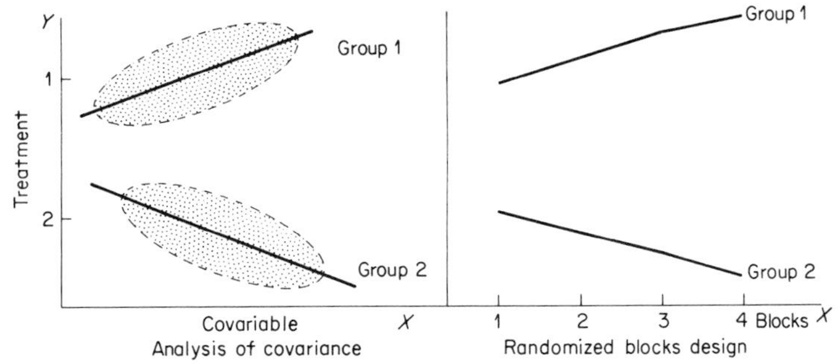

figure 8-4 *Heterogeneity of regression in analysis of covariance and treatment-by-blocks interaction in a randomized blocks design.*

interaction be present and reflected in significantly different within group regression constants, the remainder of the analysis cannot be assumed to be valid. In randomized blocks, however, significance tests can still be conducted for treatment effects even though a significant treatment-by-blocks interaction occurs.

The final consideration related to a choice between analysis of covariance and randomized blocks revolves around their relative efficiency and power. In general, the researcher is interested in making maximum use of his data in the sense of detecting significant treatment effects if, in fact, they exist. Thus, if either the analysis of covariance or the randomized blocks designs held an edge in power, this would influence the researcher's choice between them. Although the power of statistical designs depends upon a variety of factors, it appears that, overall, the analysis of covariance is more powerful when the correlation between the outside variable and the criterion is larger than .60 but that the randomized blocks design has an edge in power when this correlation is less than .40. Selected tables of "apparent imprecision ratios" for both types of designs (as well as for gain-score designs, and repeated measures designs) can be found in Ray (1960); the student may find these of interest, but the cases represented are not ones likely to typify an actual design situation since the sample sizes are very small. The power comparison also depends upon certain factors which may be difficult or impossible to assess for an actual research situation. For example, when blocks are formed, some decision must be made concerning the number of blocks to set up, and this is usually at the discretion of the experimenter. The greater the number of blocks, the greater will be the within-block homogeneity of subjects; however, there is also a loss of degrees of freedom from the error term. Thus, to a certain extent, the relative efficiency of randomized blocks and analysis of covariance will depend upon the manner in which the blocks are established for the design. Also, as mentioned above, if the relationship between the outside variable and the criterion is nonlinear, the analysis of covariance will be more seriously affected than will a randomized blocks design. From a practical point of view, the researcher will probably be more influenced by other considerations than by relative efficiency since the differences in power between the two designs are, in general, relatively small, and a clear-cut power comparison can rarely be made unless the relationship between the outside variable and the criterion has been thoroughly investigated.

A final consideration, which is trivial if computer capability is available but may be bothersome otherwise, has to do with unequal sample sizes. The analysis of covariance may be carried out on a basic CRD regardless of the sizes of the treatment groups; the computational labor for equal and unequal sample sizes is about the same. If, however, blocks are formed, it is not unlikely that missing subjects and/or unusable data will result in a design with unequal cell sizes. If this occurs, the researcher must turn to procedures appropriate to such cases, and these are computationally more burdensome than analysis with equal cell sizes. Similarly, if the basic design is factorial, it is often possible to randomly assign subjects to treatment combinations at the time of the experiment and to achieve equal or proportional cell sizes. When blocking must be done prior to the experimental sessions, it is more difficult to achieve equal cell sizes since some subjects almost invariably fail to attend experimental sessions or to yield complete data.

analysis of covariance on a CRD with one covariable

This case has been discussed rather completely in the second section of the present chapter. The purpose of this section is to present an illustrative analysis and to develop appropriate computational formulas. The general data table is shown in Table 8-1. Since cross products of X and Y scores enter into the analysis, it is necessary to have a notation analogous to that for sums of squares. The system adopted here utilizes "SP" with suitable subscripts to indicate a sum of cross products in deviation form. For example, the total sum of cross products in deviation form would be

$$\text{SP}_{\text{total}} = \sum_{i=1}^{n_j} \sum_{j=1}^{p} (Y_{ij} - \bar{Y}_{..})(X_{ij} - \bar{X}_{..})$$

Although not necessary in the analysis of covariance, the *covariance* of two variables is defined as their sum of cross products divided by appropriate degrees of freedom and is denoted S_{XY}. Thus, the total covariance of X and Y would be $S_{XY} = \text{SP}_{\text{total}}/(n_t - 1)$.

table 8-1 *Data table for analysis of covariance with a basic* CRD

	TREATMENT						
	1		2			p	
	X	Y	X	Y	...	X	Y
	X_{11}	Y_{11}	X_{12}	Y_{12}	...	X_{1p}	Y_{1p}
	X_{21}	Y_{21}	X_{22}	Y_{22}	...	X_{2p}	Y_{2p}
	X_{31}	Y_{31}	X_{32}	Y_{32}	...	X_{3p}	Y_{3p}
	.....	.....	.....	.....	...	.....	.....
	$X_{n_1 1}$	$Y_{n_1 1}$	$X_{n_2 2}$	$Y_{n_2 2}$	...	$X_{n_p p}$	$Y_{n_p p}$
Totals	$X_{.1}$	$Y_{.1}$	$X_{.2}$	$Y_{.2}$	...	$X_{.p}$	$Y_{.p}$

The total sum of squares for Y, the criterion, is computed in the usual manner; for notational purposes, a Y subscript is included:

$$\mathrm{SS}_{Y\,\mathrm{total}} = \sum_{i=1}^{n_j} \sum_{j=1}^{p} Y_{ij}^2 - \frac{(Y_{..})^2}{n_t}$$

where n_j is the number of subjects in the jth treatment group, and $n_t = \sum_{j=1}^{p} n_j$ is the total number of subjects in the experiment (for equal group sizes, n_j may be replaced by n, and n_t by np). The remainder of the analysis depends upon quantities involving both the X and the Y scores. The separate terms required from each treatment group, and overall, are:

	TREATMENT LEVEL 1	2	...	p	total
Sum of Y scores	$Y_{.1}$	$Y_{.2}$	...	$Y_{.p}$	$Y_{..}$
Sum of X scores	$X_{.1}$	$X_{.2}$	...	$X_{.p}$	$X_{..}$
Sum of squared Y scores	$\sum_{i=1}^{n_1} Y_{i1}^2$	$\sum_{i=1}^{n_2} Y_{i2}^2$	...	$\sum_{i=1}^{n_p} Y_{ip}^2$	$\sum_{i=1}^{n_j} \sum_{j=1}^{p} Y_{ij}^2$
Sum of squared X scores	$\sum_{i=1}^{n_1} X_{i1}^2$	$\sum_{i=1}^{n_2} X_{i2}^2$	...	$\sum_{i=1}^{n_p} X_{ip}^2$	$\sum_{i=1}^{n_j} \sum_{j=1}^{p} X_{ij}^2$
Sum of cross products	$\sum_{i=1}^{n_1} X_{i1} Y_{i1}$	$\sum_{i=1}^{n_2} X_{i2} Y_{i2}$	...	$\sum_{i=1}^{n_p} X_{ip} Y_{ip}$	$\sum_{i=1}^{n_j} \sum_{j=1}^{p} X_{ij} Y_{ij}$

The adjusted total sum of squares is, then,

$$\mathrm{SS}^*_{Y\,\mathrm{total}} = \mathrm{SS}_{Y\,\mathrm{total}} - \frac{(\mathrm{SP}_{\mathrm{total}})^2}{\mathrm{SS}_{X\,\mathrm{total}}}$$

where

$$\mathrm{SP}_{\mathrm{total}} = \sum_{i=1}^{n_j} \sum_{j=1}^{p} X_{ij} Y_{ij} - \frac{(X_{..})(Y_{..})}{n_t}$$

$$\mathrm{SS}_{X\,\mathrm{total}} = \sum_{i=1}^{n_j} \sum_{j=1}^{p} X_{ij}^2 - \frac{(X_{..})^2}{n_t}$$

Similar formulas required to adjust the error sum of squares are

$$\text{SS}_{Y\text{error}} = \sum_{i=1}^{n_j} \sum_{j=1}^{p} Y_{ij}^2 - \sum_{j=1}^{p} \frac{(Y_{.j})^2}{n_j}$$

$$\text{SS}^*_{Y\text{error}} = \text{SS}_{Y\text{error}} - \frac{(\text{SP}_{\text{error}})^2}{\text{SS}_{X\text{error}}}$$

where

$$\text{SP}_{\text{error}} = \sum_{j=1}^{n_j} \sum_{j=1}^{p} X_{ij}Y_{ij} - \sum_{j=1}^{p} \frac{(X_{.j})(Y_{.j})}{n_j}$$

$$\text{SS}_{X\text{error}} = \sum_{i=1}^{n_j} \sum_{j=1}^{p} X_{ij}^2 - \sum_{j=1}^{p.} \frac{(X_{.j})^2}{n_j}$$

The adjusted treatment sum of squares can be found by subtraction:

$$\text{SS}^*_{\text{treat}} = \text{SS}^*_{\text{total}} - \text{SS}^*_{\text{error}}$$

Prior to summarizing these results in an analysis-of-variance summary table, the test for homogeneity of regression must be presented. Computational formulas for the terms S_1 and S_2 are

$$S_2 = \text{SS}_{Y\text{error}} - \sum_{j=1}^{p} \frac{(SP_j)^2}{\text{SS}_{X_j}}$$

where

$$\text{SP}_j = \sum_{i=1}^{n_j} X_{ij}Y_{ij} - \frac{(X_{.j})(Y_{.j})}{n_j}$$

is the sum of cross products in deviation form for the jth treatment group, and

$$\text{SS}_{X_j} = \sum_{i=1}^{n_j} X_{ij}^2 - \frac{(X_{.j})^2}{n_j}$$

is the sum of squares for X within the jth treatment group. Then $S_1 = \text{SS}^*_{Y\text{error}} - S_2$. The test of the hypothesis $H_0: \beta_1 = \beta_2 = \cdots = \beta_p$ where β_j is the true, population regression coefficient for the jth treatment group, is given by the ratio $F = M_1/M_2$, where $M_1 = S_1/(p-1)$, and $M_2 = S_2/(n_t - 2p)$. If the hypothesis of equal regression constants is accepted, this indicates that the pooling of within-group regression information is a tenable procedure, and the analysis of covariance may be completed.

The summary table for the analysis of covariance takes the general form

source	df	sum of squares (adj)	mean square
Treatment	$p-1$	$SS^*_{Y\text{treat}}$	MS^*_{treat}
Error	$n_t - p - 1$	$SS^*_{Y\text{error}}$	MS^*_{error}
Total	$n_t - 2$	$SS^*_{Y\text{total}}$	

If the F ratio from the analysis of covariance implies rejection of the hypothesis $H_0: \alpha_1^* = \alpha_2^* = \cdots = \alpha_p^*$, then the researcher will wish to carry out additional tests on the treatment means to locate specific contrasts contributing to the significant overall test. Such tests must be carried out on adjusted means (i.e., treatment means adjusted for the covariable); these are given by the expression

$$\overline{Y}'_{.j} = \overline{Y}_{.j} + b_w(\overline{X}_{.j} - \overline{X}_{..})$$

where $\overline{Y}_{.j}$ = unadjusted mean of criterion scores in jth treatment groups
b_w = within-groups estimate of regression constant
$\overline{X}_{.j}$ = mean of covariable in jth treatment group
$\overline{X}_{..}$ = grand mean of covariable scores

Also, because there is sampling error in the estimation of b_w, the appropriate standard-error term for such tests as multiple t tests, Duncan's test, and Dunnett's test cannot be derived directly from MS^*_{error}; that is, a difference of the sort $\overline{Y}'_{.j} - \overline{Y}'_{.k}$ is a function of both treatment effects and sampling error in b_w [note that $\overline{Y}'_{.j} - \overline{Y}'_{.k} = \overline{Y}_{.j} - \overline{Y}_{.k} - b_w(\overline{X}_{.j} - \overline{X}_{.k})$]. For tests involving a specific pairwise contrast (for example, $\overline{Y}'_{.j} - \overline{Y}'_{.k}$), the appropriate standard-error term is the *square root* of

$$MS^*_{\text{error}}\left[\frac{1}{n_j} + \frac{1}{n_k} + \frac{(\overline{X}_{.j} - \overline{X}_{.k})^2}{SS_{X\text{error}}}\right]$$

Note that the mean difference $\overline{X}_{.j} - \overline{X}_{.k}$ for the covariable is involved in the expression leading to the appropriate standard-error term. For testing all pairwise contrasts, an estimate of the average standard error is given by the *square root* of

$$\frac{2\text{MS}^*_{\text{error}}}{\tilde{n}}\left[1+\frac{\text{SS}_{X\,\text{treat}}}{(p-1)\text{SS}_{X\,\text{error}}}\right]$$

where $\text{SS}_{X\,\text{treat}}$ is a treatment sum of squares for the covariable that is, $\text{SS}_{X\,\text{treat}} = \Sigma_{j=1}^{p}[(X_{.j})^2/n_j] - (X_{..})^2/n_t$, and $\tilde{n}$ is the harmonic mean of the sample sizes [that is, $\tilde{n} = p/(\Sigma_{j=1}^{p} 1/n_j)$].

If the researcher wishes to actually build the prediction equation underlying the adjustment of the total sum of squares and the within-groups sum of squares, the regression constants are given by

$$b_t = \frac{\text{SP}_{\text{total}}}{\text{SS}_{X\,\text{total}}} \quad \text{and} \quad b_w = \frac{\text{SP}_{\text{error}}}{\text{SS}_{X\,\text{error}}}$$

The equation involving b_t is $Y'_{ij} = \overline{Y}_{..} + b_t(X_{ij} - \overline{X}_{..})$. Also, for the jth treatment group, the within-groups equation is $Y''_{ij} = \overline{Y}_{.j} + b_w(X_{ij} - \overline{X}_{.j})$. Correlation coefficients corresponding to these two cases are

$$r_{\text{total}} = \frac{\text{SP}_{\text{total}}}{\sqrt{(\text{SS}_{X\,\text{total}})(\text{SS}_{Y\,\text{total}})}} \quad \text{and} \quad r_{\text{within}} = \frac{\text{SP}_{\text{error}}}{\sqrt{(\text{SS}_{X\,\text{error}})(\text{SS}_{Y\,\text{error}})}}$$

To illustrate the computational approach to an analysis-of-covariance performed on a CRD, consider the data in Table 8-2. There is a total of three treatment groups with five subjects per group. We assume that subjects have been randomly assigned to the treatment groups and that the measurements on the covariable have been procured in a manner to obviate their contamination by the treatments. The various sums required in the analysis are

	TREATMENT 1	2	3	total
Sum of Y scores	92	136	151	379
Sum of X scores	35	34	29	98
Sum of squared Y scores	1,766	3,774	4,603	10,143
Sum of squared X scores	279	290	195	764
Sum of cross products	685	986	905	2,576

The unadjusted total sum of squares for Y is

$$\text{SS}_{Y\,\text{total}} = 10{,}143 - \frac{379^2}{15} = 10{,}143 - 9{,}576.0667 = 566.9333$$

table 8-2 *Criterion and covariable data for three treatment groups*

TREATMENT 1		2		3	
X	Y	X	Y	X	Y
5	17	2	22	8	35
8	16	7	26	5	28
9	20	10	33	9	32
3	14	11	30	3	27
10	25	4	25	4	29

The terms required to adjust this sum of squares are

$$SP_{\text{total}} = 2{,}576 - \frac{(379)(98)}{15} = 2{,}576 - 2{,}476.1333 = 99.8667$$

$$SS_{X\text{total}} = 764 - \frac{98^2}{15} = 764 - 640.2667 = 123.7333$$

Then

$$SS^*_{Y\text{total}} = 566.9333 - \frac{99.8667^2}{123.7333} = 566.9333 - 80.6063 = 486.33$$

The unadjusted error sum of squares for Y is

$$SS_{Y\text{error}} = 10{,}143 - \frac{92^2}{5} - \frac{136^2}{5} - \frac{151^2}{5} = 10{,}143 - 9{,}952.20 = 190.80$$

The terms required to adjust this sum of squares are

$$SP_{\text{error}} = 2{,}576 - \frac{(92)(35)}{5} - \frac{(136)(34)}{5} - \frac{(151)(29)}{5} = 2{,}576 - 2{,}444.60$$
$$= 131.40$$

$$SS_{X\text{error}} = 764 - \frac{35^2}{5} - \frac{34^2}{5} - \frac{29^2}{5} = 764 - 644.40 = 119.60$$

Then

$$SS^*_{Y\text{error}} = 190.80 - \frac{131.40^2}{119.60} = 190.80 - 144.36 = 46.44$$

The adjusted treatment sum of squares can now be found by subtraction:

$$SS^*_{Y\text{treat}} = SS^*_{Y\text{total}} - SS^*_{Y\text{error}} = 486.33 - 46.44 = 439.89$$

Prior to setting up the analysis-of-variance summary table and computing the F ratio, the test for homogeneity of regression must be completed. The required sums of squares and cross products are

$$SP_1 = 685 - \frac{(92)(35)}{5} = 41.00 \qquad SP_2 = 986 - \frac{(136)(34)}{5} = 61.20$$

$$SP_3 = 905 - \frac{(151)(29)}{5} = 29.20 \qquad SS_{X_1} = 279 - \frac{35^2}{5} = 34.00$$

$$SS_{X_2} = 290 - \frac{34^2}{5} = 58.80 \qquad SS_{X_3} = 195 - \frac{29^2}{5} = 26.80$$

Then

$$S_2 = 190.80 - \frac{41.00^2}{34.00} - \frac{61.20^2}{58.80} - \frac{29.20^2}{26.80} = 190.80 - 49.44 - 63.70 - 31.82$$
$$= 45.84$$

$$S_1 = SS^*_{Y\text{error}} - S_2 = 46.44 - 45.84 = .60$$

Finally,

$$M_1 = \frac{S_1}{p-1} = \frac{.60}{2} = .30 \qquad \text{and} \qquad M_2 = \frac{S_2}{n_t - 2p} = \frac{45.84}{9} = 5.09$$

Then $F = M_1/M_2 = .30/5.09 = .06$, which is, obviously, a nonsignificant value of F. Thus, we can conclude that the within-group regression coefficients differ only by sampling error and that pooling them to form our estimate of b_w is a valid procedure (note that $b_w = SP_{\text{error}}/SS_{X\text{error}} = 131.40/119.60 = 1.10$; and $b_1 = 41.00/34.00 = 1.21$, $b_2 = 61.20/58.80 = 1.04$, and $b_3 = 29.20/26.80 = 1.09$).

Returning to the sums of squares, the analysis-of-variance summary table is:

source	df	sum of squares (adj)	mean square
Treatment	2	439.89	219.95
Error	11	46.44	4.22
Total	13	486.33	

The F ratio utilized in testing the null hypothesis $H_0: \alpha_1^* = \alpha_2^* = \alpha_3^*$ is $F = 219.95/4.22 = 52.12$, which is highly significant with 2 and 11 degrees of freedom (that is, $p < .01$). It is interesting to compare this result with that which would have been obtained had the covariable been ignored and the analysis run as an ordinary CRD. The total and error sums of squares, as previously computed, are $SS_{Y\,\text{total}} = 566.93$ and $SS_{Y\,\text{error}} = 109.80$; thus, $SS_{Y\,\text{treat}} = 566.93 - 190.80 = 376.13$. The relevant mean squares are $MS_{\text{treat}} = 376.13/2 = 188.07$ and $MS_{\text{error}} = 109.80/12 = 9.15$; then $F = 188.07/9.15 = 20.55$. Although highly significant, this value of F is much smaller than that resulting from the analysis of covariance.

Since $p = 3$, and a significant F resulted from the analysis, additional analyses must be completed to determine which contrasts are contributing to the overall significance. For illustration, assume that level 1 of the treatment was a control condition, and levels 2 and 3 were two variations on an experimental treatment. Then, the following a priori contrasts would have been set up:

$$2\bar{Y}'_{.1} - (\bar{Y}'_{.2} + \bar{Y}'_{.3}) \qquad \text{and} \qquad \bar{Y}'_{.2} - \bar{Y}'_{.3}$$

The adjusted means of the treatment groups are

$$\begin{aligned}\bar{Y}'_{.1} &= \bar{Y}_{.1} + b_w(\bar{X}_{.1} - \bar{X}_{..}) = 18.4 + (1.10)(7.00 - 6.53)\\ &= 18.40 + 52 = 18.92\end{aligned}$$

$$\begin{aligned}\bar{Y}'_{.2} &= \bar{Y}_{.2} + b_w(\bar{X}_{.2} - \bar{X}_{..}) = 27.2 + (1.10)(6.80 - 6.53)\\ &= 27.20 + .30 = 27.50\end{aligned}$$

$$\begin{aligned}\bar{Y}'_{.3} &= \bar{Y}_{.3} + b_w(\bar{X}_{.3} - \bar{X}_{..}) = 30.2 + (1.10)(5.80 - 6.53)\\ &= 30.20 + (-.80) = 29.40\end{aligned}$$

As mentioned above, the error mean square MS^*_{error} must be adjusted for sampling error in the estimation of b_w. The appropriate error term to utilize in orthogonal contrasts is given by

$$MS^*_{\text{error}}\left[1 + \frac{SS_{X\,\text{treat}}}{(p-1)SS_{X\,\text{error}}}\right]$$

Since $SS_{X\,\text{treat}} = SS_{X\,\text{total}} - SS_{X\,\text{error}} = 123.73 - 119.60 = 4.13$, we have

$$4.22\left(1 + \frac{4.13}{239.20}\right) = 4.22(1 + .02) = 4.30$$

Thus, the effective mean square for error to utilize in testing the orthogonal contrasts is 4.30 rather than 4.22. To complete the analysis of the contrasts, we have

$$C_1 = 2(18.92) - (27.50 + 29.40) = -19.06$$

and

$$F = \frac{n(C_1)^2}{(6)(4.30)} = \frac{1{,}816.42}{25.80} = 70.40$$

$$C_2 = 27.50 - 29.40 = -1.90$$

and

$$F = \frac{n(C_2)^2}{(2)(4.30)} = \frac{18.05}{8.60} = 2.10$$

Each of these tests has 1 and 11 degrees of freedom; although the first contrast is highly significant, the second fails to reach significance at a conventional level. As is apparent from inspection of the adjusted treatment means, the main contribution to overall significance was the low performance in the control group; the two experimental treatment groups are comparable in performance. It should be noted that, if sums of squares are computed for the two contrasts C_1 and C_2, their sum is *not* equal to $SS_{Y\,\text{treat}}$; this occurs because the adjusted treatment sum of squares is found by subtraction and not by use of the adjusted treatment means.

analysis of covariance on a CRD with two or more covariables

If adjusting the criterion scores with respect to one outside variable is beneficial in the sense of increasing the power of the analysis, it seems reasonable that including two, three, or more covariables should be even more beneficial. This is true up to a point; however, two considerations are relevant. First, 1 degree of freedom is lost from the error sum of squares for each covariable since a regression constant must be estimated for each covariable. This loss may be of concern in small experiments where the reduction in degrees of freedom may, itself, substantially *lower* the power of the test. In large experiments, however, this loss will not seriously increase the size of the critical value of F required for rejection of the null hypothesis. Second, the adjustment of sums of squares is carried out by the techniques of multiple regression, and there is a well-known phenomenon of "diminishing returns" when large numbers of predictor variables are utilized in multiple regression.

That is, because of intercorrelations among the predictor variables, little substantive increase in predictive efficiency is gained beyond the first few predictors. In most research fields in education, it seems unlikely that more than three to five covariables would be worthwhile in the sense of adding significantly to the multiple regression. This is, however, an empirical question and should be studied by the researcher when contemplating the use of large numbers of covariables. Since few researchers would attempt the analysis with multiple covariables without the aid of an electronic digital computer, the problem of increasing computational labor occasioned by including additional covariables is probably moot.

To facilitate the development of formulas, matrix notation is utilized for the case of multiple covariables. For the student who is unfamiliar with the elements of matrix algebra, reference should be made to Appendix A prior to attempting to follow the material presented in this section. Also, it is possible to develop relatively straightforward computational formulas (without matrix notation) for the case of just two covariables; these formulas are not presented in this volume, but the interested reader can find them in Winer (1962, pp. 618–621).

When multiple covariables are included in the design, the data table will take the general form of Table 8-3. In order to facilitate the denotation of the various covariables, these have been labeled $X1$, $X2$, through Xk. The criterion is labeled Y, as previously. Unadjusted sums of squares for Y can be defined in the usual manner; these are $SS_{Y_{\text{total}}}$ and $SS_{Y_{\text{error}}}$. The adjustment of these sums of squares is now in terms of k covariables instead of a single covariable, but the adjustment is based on the same model utilized in the case of a single covariable:

$$Y_{ij}^* = Y_{ij} - Y'_{ij} = \mu^* + \alpha_j^* + \epsilon_{ij}^*$$

However, the predicted criterion score is

$$Y'_{ij} = \overline{Y}_{..} + b_{t1}(X1_{ij} - \overline{X1}_{..}) + b_{t2}(X2_{ij} - \overline{X2}_{..}) + \cdots + b_{tk}(Xk_{ij} - \overline{Xk}_{..})$$

table 8-3 *Data table for analysis of covariance on a* CRD *with multiple covariables*

TREATMENT A										
1						p				
$X1$	$X2$	$\cdots$	Xk	Y	$\cdots$	$X1$	$X2$	$\cdots$	Xk	Y
$X1_{11}$	$X2_{11}$	$\cdots$	Xk_{11}	Y_{11}	$\cdots$	$X1_{1p}$	$X2_{1p}$	$\cdots$	Xk_{1p}	Y_{1p}
$X1_{21}$	$X2_{21}$	$\cdots$	Xk_{21}	Y_{21}	$\cdots$	$X1_{2p}$	$X2_{2p}$	$\cdots$	Xk_{2p}	Y_{2p}
$\cdots$	$\cdots$	$\cdots$	$\cdots$	$\cdots$	$\cdots$	$\cdots$	$\cdots$	$\cdots$	$\cdots$	$\cdots$
$X1_{n_11}$	$X2_{n_11}$	$\cdots$	Xk_{n_11}	Y_{n_11}	$\cdots$	$X1_{n_pp}$	$X2_{n_pp}$	$\cdots$	Xk_{n_pp}	Y_{n_pp}

where b_{t1} through b_{tk} are *partial multiple regression coefficients*. The adjustment of $SS_{Y\,total}$ for the k covariables $X1$ through Xk involves solving for estimates of b_{t1} through b_{tk} and for the predictable sum of squares based on the k covariables. This solution follows the lines generally utilized in multiple regression problems. Based on totals (ignoring treatment groups), a matrix of sums of squares and sums of cross products in deviation form is defined for the set of k covariables. If $SS_{Xg\,total}$ is the sum of squares for covariable g and if $SP_{XgXh\,total}$ is the sum of cross products in deviation form for covariables g and h, this matrix is

$$S_{total} = \begin{bmatrix} SS_{X1total} & SP_{X1X2total} & \cdots & SP_{X1Xktotal} \\ SP_{X2X1total} & SS_{X2total} & \cdots & SP_{X2Xktotal} \\ \cdots & \cdots & \cdots & \cdots \\ SP_{XkX1total} & SP_{XkX2total} & \cdots & SS_{Xktotal} \end{bmatrix}$$

Note that the diagonal elements are sums of squares, while the off-diagonal elements are sums of cross products in deviation form. Also, the matrix is symmetric since $SP_{XgXh\,total} = SP_{XhXg\,total}$. A second matrix (actually, a row vector) is defined as comprising the sums of cross products in deviation form between covariables and the criterion variable; utilizing similar notation, this matrix is

$$V_{total} = [SP_{X1Y\,total} \quad SP_{X2Y\,total} \quad \cdots \quad SP_{XkY\,total}]$$

The computational operations involved in finding the elements in either S_{total} or V_{total} are identical with those required to find $SS_{X\,total}$ and SP_{total} in the analysis with one covariable; however, these operations are repeated for each covariable, and the cross products involve both covariable-with-covariable and covariable-with-criterion operations. Once S_{total} and V_{total} have been found, the partial multiple regression coefficients b_{t1} through b_{tk} are determined by

$$B_{total} = V_{total}S_{total}^{-1}$$

Since V_{total} is a $1 \times k$ row vector and S_{total}^{-1} is a $k \times k$ square matrix, the product matrix B_{total} will be a $1 \times k$ row vector; the elements of B_{total} will be $B_{total} = [b_{t1} \quad b_{t2} \quad \cdots \quad b_{tk}]$. Note that if k is large (say more than 3), the inversion of S_{total} will be tedious by hand calculation.

The amount of variability attributable to the k covariables is given by $B_{total}V'_{total}$. Since B_{total} is a $1 \times k$ row vector and the transpose of V_{total} is a $k \times 1$ column vector, their product is a scalar matrix (i.e., a 1×1 matrix). The adjustment of the total sum of squares is, then,

$$SS^*_{Y\,total} = SS_{Y\,total} - B_{total}V'_{total} = SS_{Y\,total} - V_{total}S_{total}^{-1}V'_{total}$$

A parallel set of operations is required to adjust the error sum of squares. Since this procedure is dependent upon homogeneity-of-regression assumptions, the researcher must check these prior to carrying out the pooling required for the error terms. If we denote the regression constants estimated from within-group information as b_{e1} through b_{ek}, we require that the separate estimates from the p treatment groups be within sampling error of one another. The hypothesis is, therefore, complex since there are k different partial regression constants involved. The quantities involved in the computations of these estimates are all within-group sums of squares for the covariables and sums of cross products in deviation form involving pairs of covariables or for covariables with the criterion. A test of these homogeneity requirements can be constructed along the same lines pursued in the case of just one covariable. That is, the adjusted error sum of squares for the criterion is partitioned into two components; one of these components reflects the errors of predictions which result from the use of separate within-cell multiple regression equations, and the second component reflects the additional prediction error which results when pooled within-cell regression coefficients are utilized in a regression equation to predict the criterion.

Since the homogeneity-of-regression test utilizes the adjusted error sum of squares for Y, we begin by carrying out the necessary adjustment. By analogy with the matrices defined when adjusting the total sum of squares, we can set up V_{error}, which has elements involving cross products in deviation form between covariables and the criterion (that is, $V_{\text{error}} = [\text{SP}_{X1Y_{\text{error}}} \quad \cdots \quad \text{SP}_{XkY_{\text{error}}}]$); and we can also define a matrix of cross products in deviation form for pairs of covariables. This latter matrix is

$$S_{\text{error}} = \begin{bmatrix} \text{SS}_{X1\text{error}} & \text{SP}_{X1X2\text{error}} & \cdots & \text{SP}_{X1Xk\text{error}} \\ \text{SP}_{X2X1\text{error}} & \text{SS}_{X2\text{error}} & \cdots & \text{SP}_{X2Xk\text{error}} \\ \cdots & \cdots & \cdots & \cdots \\ \text{SP}_{XkX1\text{error}} & \text{SP}_{XkX2\text{error}} & \cdots & \text{SS}_{Xk\text{error}} \end{bmatrix}$$

Both matrices associated with error are obtained by pooling over the p groups constituting the CRD. For V_{error}, the entering vectors from the p treatment groups are $V_{\text{error}(1)}$ through $V_{\text{error}(p)}$. A typical element in the jth vector is $\text{SP}_{XgY_{\text{error}(j)}}$. The pooling operation is equivalent to summing the separate treatment group vectors; that is,

$$V_{\text{error}} = \sum_{j=1}^{p} V_{\text{error}(j)}$$

Similarly, S_{error} is a sum of p matrices each of which contains sums of squares and sums of cross products in deviation form for just one level of the treatment. The entering matrices are $S_{\text{error}(1)}$ through $S_{\text{error}(p)}$; a typical sum-of-squares element in $S_{\text{error}(j)}$ is $\text{SS}_{X1\text{error}(j)}$, and a typical cross-products element is $\text{SP}_{XgXh\text{error}(j)}$. Note that V_{error} and each of the $V_{\text{error}(j)}$ are $1 \times k$ row vectors, and that S_{error} and each of the $S_{\text{error}(j)}$ are $k \times k$ matrices.

The adjustment of the error sum of squares $\text{SS}_{Y\text{error}}$ involves only the matrices S_{error} and V_{error}, and this adjustment follows the same lines as for the total sum of squares. That is,

$$\text{SS}^*_{Y\text{error}} = \text{SS}_{Y\text{error}} - V_{\text{error}} S^{-1}_{\text{error}} V'_{\text{error}}$$

As was the case for one covariable, the adjusted sum of squares for treatment is found by subtraction: $\text{SS}^*_{Y\text{treat}} = \text{SS}^*_{Y\text{total}} - \text{SS}^*_{Y\text{error}}$. Since k different regression constants must be estimated, the total (and error) sums of squares must be reduced by k degrees of freedom. Thus, the adjusted total sum of squares has $n_t - k - 1$ degrees of freedom, and the adjusted error sum of squares has $n_t - p - k$ degrees of freedom. Of course, the degrees of freedom for treatment remain at $p - 1$. The computation of mean squares and the setting up of an appropriate F ratio follow the usual pattern.

Returning to the homogeneity-of-regression requirements, the F ratio resulting from the adjusted mean squares is valid only if the regression constants from the separate treatment groups are within sampling error of one another. As indicated earlier, an appropriate test requires partitioning of the adjusted error sum of squares. Utilizing the notation developed above, this partitioning is

$$S_2 = \text{SS}_{Y\text{error}} - \sum_{j=1}^{p} [V_{\text{error}(j)} S^{-1}_{\text{error}(j)} V'_{\text{error}(j)}]$$

$$S_1 = \text{SS}^*_{Y\text{error}} - S_2$$

Note that S_2 is equivalent to an adjusted error sum of squares, where the adjustment is carried out by utilizing the separate within-cell regression constants. S_1 is, then, the difference between this adjusted value and that obtained by using the pooled regression constants (that is, $\text{SS}^*_{Y\text{error}}$ was obtained by using pooled estimates of the regression constants). The appropriate degrees of freedom for S_1 and S_2 are, respectively, $k(p - 1)$ and $n_t - pk - p$, where n_t is the total number of cases in the CRD (that is, $n_t = \Sigma_{j=1}^{p} n_j$). The test statistic is $F = M_1/M_2$, where $M_1 = S_1/k(p - 1)$ and $M_2 = S_2/(n_t - pk - p)$.

It should be noted that the homogeneity-of-regression test presented for multiple covariables is a generalization of the procedures developed for the case of just one covariable. Thus, for $k = 1$ the vector

$V_{\text{error}(j)}$ contains a single element $SP_{XY_{\text{error}(j)}}$; or, in the notation utilized for the one-covariable case, $SP_{XY_{\text{error}(j)}} = SP_j$. Similarly, $S_{\text{error}(j)}$ is the single element $SS_{X_{\text{error}(j)}}$, which is equivalent to SS_{X_j} in the earlier notation. Given these equivalences, the student can easily verify that the terms S_2 and S_1 as defined above are identical with those presented for the single-covariable case.

analysis of covariance on a factorial design

Analysis of covariance can be applied to any of the basic designs considered in earlier chapters of this textbook. The labor of computation becomes very great for all except the simplest designs, and, for this reason, we restrict our attention in this section to a basic two-dimensional factorial design as a model for the utilization of analysis of covariance with factorial designs. The principles underlying the application of the analysis of covariance to factorial designs of all dimensionalities will be apparent from study of the case of two dimensions. Also, we limit ourselves to the case in which there is just one covariable; by analogy with the multiple-covariable procedure for CRDs, the student can infer how the extension to multiple covariables with a basic factorial design would be accomplished.

The structural model for a score in a two-dimensional factorial design is

$$Y_{ijk} = \mu + \alpha_k + \beta_j + \alpha\beta_{kj} + \epsilon_{ijk}$$

where the two treatment dimensions are labeled A and B. If X is an outside variable being utilized as a covariable, the predicted value of Y is $Y'_{ijk} = \overline{Y}_{...} + b_t(X_{ijk} - \overline{X}_{...})$, where, as before, $\overline{Y}_{...}$ and $\overline{X}_{...}$ are grand means taken over the entire set of scores. The model, rewritten to take into account the covariable, is

$$Y^*_{ijk} = Y_{ijk} - Y'_{ijk} = \mu^* + \alpha^*_k + \beta^*_j + \alpha\beta^*_{kj} + \epsilon^*_{ijk}$$

When the basic design was a CRD, our goal was to adjust just two sums of squares, $SS_{Y_{\text{error}}}$ and $SS_{Y_{\text{treat}}}$; however, in a two-dimensional factorial design, we must adjust four sums of squares, $SS_{Y_{\text{error}}}$, SS_{YA}, SS_{YB}, and SS_{YAB}. For this reason, we can no longer use the simple stratagem of adjusting the total sum of squares and then finding the adjusted treatment sums of squares by subtraction. However, the overall logic of the adjustment procedure is the same as that employed in the case of a CRD; that is, we shall remove adjusted error variability from a sum of squares which, initially, reflects both error variability and variability due to the treatment (or interaction) effect of interest.

The data table for a 2 × 2 factorial design with n observations per cell and a single covariable is illustrated in Table 8-4.

Within each cell of the design, we have n pairs of scores; thus, a within-cell regression equation could be built for each of the four cells. The regression constants for these equations would be b_{11} through b_{22}. An adjusted error sum of squares for Y is computed by pooling information from the four cells to form a single within-cell estimate b_w. As in the case of the CRD, this pooling depends upon the condition of homogeneity of regression; this can be tested by a procedure similar to that presented earlier. Indeed, the adjustment of the error sum of squares and the test for homogeneity of regression are carried out as if the design were a CRD with four levels rather than a 2 × 2 factorial design (i.e., each cell constitutes one "level" of the CRD). Our notation for sums of squares and sums of cross products in deviation form needs only minor modification to fit the case of two-dimensional factorial designs. Within each cell, the component of the sum of squares for error can be denoted $\text{SS}_{Y\text{error}(ij)}$, where i and j are relevant row and column subscripts. Then

$$\sum_{i=1}^{q}\sum_{j=1}^{p} \text{SS}_{Y\text{error}(ij)} = \text{SS}_{Y\text{error}}$$

(assuming A exists at p levels and B at q levels; for the exemplary case, $p = q = 2$). Similarly, for cross products, a component term is $\text{SP}_{\text{error}(ij)}$, and

$$\sum_{=1}^{q}\sum_{j=1}^{p} \text{SP}_{\text{error}(ij)} = \text{SP}_{\text{error}}$$

table 8-4 *Data table for a* 2 × 2 *factorial design with one covariable*

		TREATMENT A			
		1		2	
		X	Y	X	Y
TREATMENT B	1	X_{111}	Y_{111}	X_{121}	Y_{121}
		X_{112}	Y_{112}	X_{122}	Y_{122}
		. . . . . .	. . . . . .	. . . . . .	. . . . . .
		X_{11n}	Y_{11n}	X_{12n}	Y_{12n}
	2	X_{211}	Y_{211}	X_{221}	Y_{221}
		X_{212}	Y_{212}	X_{222}	Y_{222}
		. . . . . .	. . . . . .	. . . . . .	. . . . . .
		X_{21n}	Y_{21n}	X_{22n}	Y_{22n}

The adjusted error sum of squares for Y is, then,

$$\text{SS}^*_{Y\text{error}} = \text{SS}_{Y\text{error}} - \frac{(\text{SP}_{\text{error}})^2}{\text{SS}_{X\text{error}}}$$

where $\text{SS}_{X_{\text{error}}}$ is the error sum of squares for the covariable (as defined in the case of the CRD). However, the adjusted error sum of squares can be divided into two components which are utilized in forming statistics to test the assumption of homogeneous regression constants. The partitioning is

$$S_2 = \text{SS}_{Y\text{error}} - \sum_{i=1}^{q} \sum_{j=1}^{p} \frac{(\text{SP}_{\text{error}(ij)})^2}{\text{SS}_{X\text{error}(ij)}}$$

$$S_1 = \text{SS}^*_{Y\text{error}} - S_2$$

Since, in general, pq regression constants must be estimated (i.e., one per cell), the degrees of freedom for S_1 are $pq - 1$; the remainder for S_2 is $pq(n - 2)$. Then, $M_1 = S_1/(pq - 1)$ and $M_2 = S_2/[pq(n - 2)]$; $F = M_1/M_2$ with $pq - 1$ and $pq(n - 2)$ degrees of freedom. If the hypothesis $H_0: \beta_{11} = \beta_{12} = \cdots = \beta_{pq}$ is not rejected, the condition of homogeneity of regression can be considered to have been met by the data; thus, the pooling procedure employed for adjusting the error sum of squares for Y is valid.

The adjustments of the sums of squares due to A, B, and AB are accomplished in an indirect fashion. We begin, in each case, by pooling the unadjusted sums of squares due to the particular effect *and* due to experimental error. We then remove from this pooled sum of squares the *adjusted* error sum of squares. The remainder is a component attributable to the effect in question (note that this is, in effect, the procedure used with a CRD since $\text{SS}^*_{Y\text{total}} = \text{SS}^*_{Y\text{error}} + \text{SS}^*_{Y\text{treat}}$; $\text{SS}^*_{Y\text{total}}$ is derived by adjusting a pooled sum of squares which is made up of treatment and error since $\text{SS}_{Y\text{total}} = \text{SS}_{Y\text{error}} + \text{SS}_{Y\text{treat}}$). Consider, first, the sum of squares for the A effect. By using ordinary analysis-of-variance procedures, the unadjusted sum of squares SS_{YA} can be found from the column totals (or by any other suitable computational approach). Similarly, a sum of squares for the covariable SS_{XA} can be computed. Next, the sum of cross products in deviation form of X and Y is defined in terms of column totals; that is,

$$\text{SP}_A = \frac{\sum_{j=1}^{p} X_{.j.}Y_{.j.}}{pn} - \frac{(X_{...})(Y_{...})}{npq}$$

The pooling of A and error yields $SS_{Y_A} + SS_{Y_{\text{error}}}$. This sum is adjusted to become

$$(SS_{Y_A} + SS_{Y_{\text{error}}}) - \frac{(SP_A + SP_{\text{error}})^2}{SS_{X_A} + SS_{X_{\text{error}}}}$$

If the adjusted error sum of squares $SS^*_{Y_{\text{error}}}$ is subtracted from this last term, the result is $SS^*_{Y_A}$. In somewhat simplified form, we have

$$SS^*_{Y_A} = SS_{Y_A} - \frac{(SP_A + SP_{\text{error}})^2}{SS_{X_A} + SS_{X_{\text{error}}}} + \frac{(SP_{\text{error}})^2}{SS_{X_{\text{error}}}}$$

This expression suggests a way of setting up mechanical formulas for the adjustment of the remaining two sources of variation, B and AB. Since the logic is the same in all cases, we follow the pattern of $SS^*_{Y_A}$ to establish the adjusted sums of squares:

$$SS^*_{Y_B} = SS_{Y_B} - \frac{(SP_B + SP_{\text{error}})^2}{SS_{X_B} + SS_{X_{\text{error}}}} + \frac{(SP_{\text{error}})^2}{SS_{X_{\text{error}}}}$$

$$SS^*_{Y_{AB}} = SS_{Y_{AB}} - \frac{(SP_{AB} + SP_{\text{error}})^2}{SS_{X_{AB}} + SS_{X_{\text{error}}}} + \frac{(SP_{\text{error}})^2}{SS_{X_{\text{error}}}}$$

All terms appearing in these expressions are defined by direct analogy with those in the adjustment of SS_{Y_A}. Since there is only one covariable, just 1 degree of freedom is lost from the error term, and the analysis-of-variance summary table appears as:

source	df	sum of squares (adj)	mean square
A	$p-1$	SS^*_{YA}	MS^*_A
B	$q-1$	SS^*_{YB}	MS^*_B
AB	$(p-1)(q-1)$	SS^*_{YAB}	MS^*_{AB}
Error	$pq(n-1)-1$	$SS^*_{Y_{\text{error}}}$	MS^*_{error}

In order to illustrate the computations involved in applying analysis of covariance to factorial designs, consider the data for a 2×2 design in Table 8-5. Each treatment group comprises five subjects; since unadjusted sums of squares are required for both the X and the Y variables, we begin by finding these by utilizing ordinary analysis-of-variance techniques: in addition, appropriate cross products in deviation form are computed. These are:

	sum of squares — X	sum of squares — Y	cross products
A	1.25	806.45	−31.75
B	.05	130.05	−2.55
AB	2.45	26.45	8.05
Error	142.80	420.80	214.00
Total	146.55	1,383.75	187.75

It should be noted that (unlike sums of squares), sums of cross products in deviation form can be negative quantities. In this present case, this occurs since the rank order of sums for X and for Y is reversed on both the A and the B dimensions of the design.

The adjustment of the sum of squares for error is

$$\text{SS}^*_{Y\text{error}} = \text{SS}_{Y\text{error}} - \frac{(\text{SP}_{\text{error}})^2}{\text{SS}_{X\text{error}}} = 420.80 - \frac{214.00^2}{142.80}$$
$$= 420.80 - 320.70 = 100.10$$

table 8-5 *Criterion and covariable data for a 2 × 2 factorial design*

		TREATMENT A					
		1		2		sums	
		X	Y	X	Y	X	Y
TREATMENT B	1	8	17	9	30		
		10	27	3	25		
		6	15	4	26	64	242
		9	22	5	30		
		2	14	8	36		
	2	2	18	8	40		
		7	20	3	30		
		7	22	10	45	63	293
		11	30	5	34		
		4	19	6	35		
Sums		66	204	61	331	127	535

$$\sum_{i=1}^{q}\sum_{j=1}^{p}\sum_{k=1}^{n} X_{ijk}^2 = 953 \qquad \sum_{i=1}^{q}\sum_{j=1}^{p}\sum_{k=1}^{n} Y_{ijk}^2 = 15{,}695$$

$$\sum_{i=1}^{q}\sum_{j=1}^{p}\sum_{k=1}^{n} X_{ijk}Y_{ijk} = 3{,}585$$

This can now be partitioned into components to provide a homogeneity-of-regression test:

$$S_2 = 420.80 - \frac{57.00^2}{40.00} - \frac{34.40^2}{26.80} - \frac{60.20^2}{46.80} - \frac{62.40^2}{29.20}$$
$$= 420.80 - 81.22 - 44.16 - 77.44 - 133.35 = 84.63$$

$$S_1 = \text{SS}^*_{Y\text{error}} - S_2 = 100.10 - 84.63 = 15.47$$

Since S_1 has 3 degrees of freedom and S_2 has 12 degrees of freedom, the F ratio for testing homogeneity of regression is

$$F = \frac{S_1/3}{S_2/12} = \frac{5.16}{7.05} = .73$$

which is obviously nonsignificant. Thus, we accept the hypothesis that $\beta_{11} = \beta_{12} = \beta_{21} = \beta_{22}$ and conclude that the pooling of sums of cross products is a valid procedure in this analysis of covariance.

Turning to the main and interaction effects, the adjustments are

$$\text{SS}^*_{YA} = 806.45 - \frac{(-31.75 + 214.00)^2}{1.25 + 142.80} + \frac{214.00^2}{142.80}$$
$$= 806.45 - 230.58 + 320.70 = 896.57$$

$$\text{SS}^*_{YB} = 130.05 - \frac{(-2.55 + 214.00)^2}{.05 + 142.80} + \frac{214.00^2}{142.80}$$
$$= 130.05 - 312.99 + 320.70 = 137.76$$

$$\text{SS}^*_{YAB} = 26.45 - \frac{(8.05 + 214.00)^2}{2.45 + 142.80} + \frac{214.00^2}{142.80}$$
$$= 26.45 - 339.46 + 320.70 = 7.69$$

These results and appropriate mean squares can be summarized as:

source	df	sum of squares (adj)	mean square
A	1	896.57	896.57
B	1	137.76	137.76
AB	1	7.69	7.69
Error	15	100.10	6.67

If we assume a fixed model, the appropriate F ratios are

$F = 896.57/6.67 = 134.35$, which is highly significant for the A effect

$F = 137.76/6.67 = 20.65$, which is highly significant for the B effect

$F = 7.69/6.67 = 1.15$, which is nonsignificant for the AB effect

The analysis of covariance can be applied to designs of three or more dimensions by simple extension of the procedures presented in this section. The first step is to compute unadjusted sums of squares for Y and X, and cross products in deviation form. In effect, ordinary analysis-of-variance procedures are applied to the data three times; once for X, once for Y, and once for cross products XY. The homogeneity-of-regression test involves partitioning the adjusted error sum of squares. There will be one estimated regression constant per cell of the design, and the form of the initial statistics is

$$S_2 = \mathrm{SS}_{Y\mathrm{error}} - \sum_{\mathrm{cells}} \frac{(\mathrm{SP}_{\mathrm{error}(ij\ldots)})^2}{\mathrm{SS}_{X\mathrm{error}(ij\ldots)}}$$

where Σ_{cells} indicates summation over all cells of the design, and the subscript notation $(ij\ldots)$ indicates that the number of subscripts will equal the dimensionality of the design; and

$$S_1 = \mathrm{SS}^*_{Y\mathrm{error}} - S_2$$

The degrees of freedom for S_1 will be (number of cells -1), and S_2 will have degrees of freedom equal to df'_{error} $-$ (number of cells $-$ 1), where df'_{error} is the number of degrees of freedom for the adjusted error sum of squares $\mathrm{SS}^*_{Y\mathrm{error}}$. If the homogeneity-of-regression test indicates that the regression constants are adequately represented by the pooled value b_w, the adjustment of sums of squares for main and interaction effects follows the exact model set forth for the two-dimensional case. That is, the general form of the adjustment is

$$\mathrm{SS}^*_{Y\mathrm{effect}} = \mathrm{SS}_{Y\mathrm{effect}} - \frac{(\mathrm{SP}_{\mathrm{effect}} + \mathrm{SP}_{\mathrm{error}})^2}{\mathrm{SS}_{X\mathrm{effect}} + \mathrm{SS}_{X\mathrm{error}}} + \frac{(\mathrm{SP}_{\mathrm{error}})^2}{\mathrm{SS}_{X\mathrm{error}}}$$

Sums of squares for both X and Y are computed by using ordinary analysis-of-variance formulas; the sums of cross products in deviation form are computed by substituting cross-product terms XY for squared terms.

using analysis of covariance with factorial designs having unequal cell sizes

The computational approach of the preceding section was predicated on the assumption that each cell contained the same number n of subjects. For designs with unequal numbers in the cells, a modification of the unweighted-means analysis originally presented in Chapter 3 can be utilized. For ordinary factorial designs, without covariable adjustment, the unweighted-means analysis amounted to performing an analysis of variance on the cell means and then adjusting the resulting sums of squares by multiplying by the harmonic mean of the cell sizes. The same approach can be used in the presence of a covariable; however, multiplication by the harmonic mean is included with all terms, sums of squares and sums of cross products in deviation form. For example, if the basic design were two-dimensional, the error sums of squares for X and Y and the error sum of cross products in deviation form would be computed directly from within-cell information; that is, the usual computational formulas apply to find $SS_{Y_{\text{error}}}$, $SS_{X_{\text{error}}}$, and SP_{error}. Then $SS^*_{Y_{\text{error}}}$ can be found immediately from these terms. Next, a table of cell means is set up, and computations for main and interaction effects are carried out as if there were just one observation per cell. This applies to sums of squares for X and Y and to sums of cross products in deviation form. Each of these terms is multiplied by the harmonic mean of the cell sizes, and the resulting quantities are utilized in the formulas of the preceding section to obtain adjusted sums of squares for Y.

analysis of covariance on nested designs

The approach embodied in the general adjustment formula for sums of squares resulting from factorial designs can be generalized to designs having nesting at one or more levels. For the simplest case of one nested factor (B) at q levels and one major treatment factor (A) at p levels, there are three sums of squares which require adjustment: SS_{Y_A}, $SS_{Y_{B(A)}}$, and $SS_{Y_{\text{error}}}$. In addition, there are corresponding sums of squares for X and sums of cross products in deviation form. The adjustments for error and for A are especially straightforward since they are identical with that for a two-dimensional factorial design; that is,

$$SS^*_{Y_A} = SS_{Y_A} - \frac{(SP_A + SP_{\text{error}})^2}{SS_{X_A} + SS_{X_{\text{error}}}} + \frac{(SP_{\text{error}})^2}{SS_{X_{\text{error}}}}$$

$$SS^*_{Y_{\text{error}}} = SS_{Y_{\text{error}}} - \frac{(SP_{\text{error}})^2}{SS_{X_{\text{error}}}}$$

The adjustment for the nested effect $B(A)$ must be carried out *within* the levels of A since B is nested within A. Thus, using previous definitions for components of sums of squares and sums of cross products in deviation form, we have

$$\mathrm{SS}^*_{Y_{B(A)}} = \mathrm{SS}_{Y_{B(A)}} - \sum_{j=1}^{p} \frac{(\mathrm{SP}_{B(j)} + \mathrm{SP}_{\mathrm{error}(j)})^2}{\mathrm{SS}_{X_{B(j)}} + \mathrm{SS}_{X_{\mathrm{error}(j)}}} + \sum_{j=1}^{p} \frac{(\mathrm{SP}_{\mathrm{error}(j)})^2}{\mathrm{SS}_{X_{\mathrm{error}(j)}}}$$

This result is equivalent to applying the usual adjustment formula to the sum of squares for B within the first level of A, then to the sum of squares for B within the second level of A, etc., and then pooling the results over all levels of A. In more complex nested designs, the same strategy is utilized; however, care must be taken to adjust nested factors properly, following the model above.

exemplary applications of the analysis of covariance

example 1 Instruction by computer was compared in effectiveness with two other instructional approaches by Schurdak (1967). Forty-eight undergraduate and graduate students enrolled in a course on FORTRAN programming were randomly divided among the three modes of instruction. Group I (computer group) received all instruction except some initial orientation material by means of computer-assisted instruction. Group II (programmed-text group) studied the same content as Group I, but the material was presented in programmed-textbook form. Group III (textbook group) used a standard textbook to cover the same material as Group I. Prior to the application of treatments, all subjects completed the Henmon-Nelson Tests of Mental Ability (which is an intelligence test, or test of academic aptitude) since it is reasonable to assume that general ability would be predictive of success in learning the course content. At the completion of instruction, all subjects completed a comprehensive achievement test which represented the criterion for success in the course. The group means for the Henmon-Nelson test and the group means and standard deviations for the criterion test were:

group	mean Henmon-Nelson test scores	mean criterion test scores	SD criterion test scores
I Computer	77.1	87.4	9.5
II Programmed text	79.0	76.2	13.0
III Textbook	77.4	70.6	21.2

There is a very noticeable disparity in variability of the groups on the criterion test, and, if we utilize a Bartlett statistic, the group variances are significantly different (reported value of chi-square was 9.51 with 2 degrees of freedom). Since the criterion test scores were recorded as percentage-correct scores, they were transformed by utilizing a $\sin^{-1}$ transformation. In order to compare the group means, an analysis of covariance, with Henmon-Nelson test scores as the covariable, was performed on the transformed criterion test scores. The resulting analysis-of-covariance summary table was:

source	df	sum of squares (adj)	mean square
Treatment	2	1.56	.78
Error	44	5.29	.12
Total	46	6.86	

The value of the F statistic is $F = .78/.12 = 6.50$, which, with 2 and 44 degrees of freedom, is highly significant. Also, the author indicates that a homogeneity-of-regression test was applied and that a nonsignificant value of F was obtained. For the interested student, it should be noted that the original research report presents the within-group scatter diagrams for the three treatment groups.

example 2 Potts (1967) reports a study of "interference effects" in which two covariables were utilized to adjust criterion scores. The purpose of the study was to determine whether oral learning of a foreign language interfered with written achievement in the native language (English). A total of 80 first- and second-grade students participated in the experiment; they were randomly divided between two treatment conditions. In the first treatment condition, students were given instruction in the French language using an audio-lingual teaching approach. The second group of students received no foreign language instruction, but during the time the first group was being instructed, they were given dance instruction. Prior to the experiment, all subjects were tested with the California Test of Mental Maturity (which is an intelligence test, or test of academic aptitude); two scores were obtained from this test: language mental age and nonlanguage mental age. The treatment was applied over an entire school year, and in June all students completed the California Achievement Test and the California Reading Test.

Both the language mental age and nonlanguage mental age scores from the California Test of Mental Maturity were used as covariates in separate analyses of covariance for the two criterion measures. The results for the California Reading Test were:

source	df	sum of squares (adj)	mean square
Treatment	1	86.90	86.90
Error	76	11,146.00	146.66
Total	77	11,232.90	

The F ratio of treatment and error mean squares is less than 1 and, obviously, nonsignificant. Thus, the foreign-language instruction did not "interfere" with (in either an accelerating or a depressing sense) achievement in ordinary reading of English. The author reports that both homogeneity-of-variance and homogeneity-of-regression assumptions were tested and found to be satisfied by the data. Also, the within-groups multiple regression equation is given as

$$Z'_{Yij} = .21Z_{X1_{ij}} + .44Z_{X2_{ij}}$$

where $Z'_{Y_{ij}}$ is the predicted z score of subject (ij) on the criterion, and $Z_{X1_{ij}}$ and $Z_{X2_{ij}}$ are z scores on the two covariables.

The second criterion variable represented a measure of overall achievement (i.e., scores from the California Achievement Test); again conditions of homogeneity of variance and homogeneity of regression were met by the data. The analysis-of-variance summary was:

source	df	sum of squares (adj)	mean square
Treatment	1	979.51	979.51
Error	76	71,721.66	943.71
Total	77	72,701.17	

The F ratio, which is slightly greater than 1 (that is, 1.04) is, again, nonsignificant. For this criterion, the reported regression equation is

$$Z'_{Yij} = .52Z_{X1_{ij}} + 1.34Z_{X2_{ij}}$$

chapter nine

key to additional designs

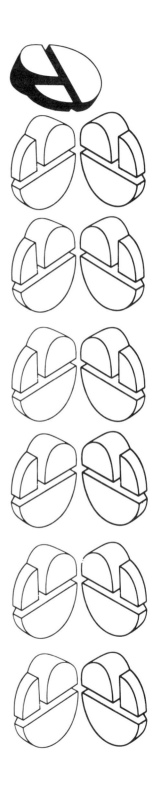

introduction

In this volume the author has attempted to give coverage to all univariate experimental designs which are in general use in educational research and related research fields. In addition, certain designs have been described even though their present level of usage in education is very low (e.g., fractional replicates of factorial designs and incomplete-block designs). Nevertheless, the student may encounter designs in educational settings which are not treated in detail in this volume, or his research interests may carry him to fields removed from education where other designs are frequently employed. The intent of this chapter is to overview briefly some design possibilities not considered in earlier chapters and to give references so that the student can pursue a more detailed treatment. Also, a brief survey of multivariate experimental designs is presented.

response-surface methodology

When all the treatment dimensions of a factorial design represent quantitative variables, it is possible to investigate the functional relationships between the treatment dimensions and the response variable. In Chapter 2 we treated the technique of orthogonal polynomials, which is designed to evaluate the fit of one or more polynomial functions to the means for the treatment groups in a CRD or the means for one dimension of a higher-order design. Response-surface methodology is essentially an extension of orthogonal-polynomial analysis to a case in which information from all dimensions of the design is utilized in building an expression for the relationship between response means and the quantitative treatment variables. Basically, the procedures used are those of multiple regression with the treatment variables representing predictors and the response measure representing the criterion. The simplest model for response-surface analysis is that of a plane or hyperplane, depending upon the number of treatment dimensions (such designs are referred to as *first-order designs*). Second- and higher-order designs are also possible, but the interpretation of response surfaces becomes difficult when the surface is of higher order than quadratic.

A problem very much related to response-surface methodology is that of determining an optimal combination of treatment levels for two or more treatment dimensions (that is, determining which combination of treatment levels causes the greatest average response). In most applications, the solution is sought sequentially in the sense that a series of interrelated experiments is designed to "zero in" on the optimal combination.

One of the most complete presentations of response-surface analysis and methods for determining optimal combinations of treatment levels is given in Chapter 8A of *Experimental Designs* by Cochran and Cox (1957). Also, Peng (1967, Chapter 8) treats these topics and sets up the relevant multiple regression procedures in matrix form. In addition, Peng presents excellent diagrams of exemplary quadratic surfaces, and these should be useful in interpreting a response-surface analysis.

Although very rarely encountered in educational research, response-surface methodology has a potentially important role in education. Much of educational research is aimed at locating the most "effective" method of instruction, counseling, etc., within specific frameworks; the use of response-surface analysis to determine optimal combinations of treatment levels seems very natural in these settings. The major limitation is, however, that in order to study polynomial regression, the treatment variables must be quantitative; at the present stage of research sophistication in education, most treatment variables are merely qualitative.

use of a single control group with a factorial design

Control groups are included in many experimental settings in order to provide base-line data relative to the effects of treatments and combinations of treatments. When the basic experimental plan calls for a factorial design, it is sometimes reasonable to include just one control group which is not exposed to any of the treatment combinations. If the basic factorial design is 2 × 3, the scheme will be:

		TREATMENT A 1	2	3	control group
TREATMENT B	1	G_{11}	G_{12}	G_{13}	G_{00}
	2	G_{21}	G_{22}	G_{23}	

Group G_{00} is randomly formed from the subject supply at the time groups G_{11} through G_{23} are formed. It is, however, not exposed to any treatment, and represents a pure control condition. The reader can construct an appropriate analysis for this case by the use of a set of orthogonal contrasts. The total experiment comprises seven treatment groups (or, in general, for a two-dimensional design, $pq + 1$ treatment groups). Thus, there are 6 degrees of freedom to consume in the analysis. Five of these are associated with the basic factorial design, and one is based on a contrast of group G_{00} with the sum of groups G_{11} through G_{23}. The relevant contrast matrix is:

effect	CELL (11)	(12)	(13)	(21)	(22)	(23)	(00)
A	+1	+1	+1	−1	−1	−1	0
B_1	+1	−1	0	+1	−1	0	0
B_2	+1	+1	−2	+1	+1	−2	0
AB_1	+1	−1	0	−1	+1	0	0
AB_2	+1	+1	−2	−1	−1	+2	0
Control	+1	+1	+1	+1	+1	+1	−6

The analysis of this general case is described and illustrated by Winer (1962, pp. 263–267).

test reliability methods based on analysis of variance

Although somewhat afield from experimental design per se, the reliability of psychometric tests, rating scales, etc., is sometimes estimated by means of models based upon the analysis of variance. For example, if a rating scale comprising p items is administered to n subjects, the resulting data (that is, p rating responses by each of the n subjects) can be summarized in an $n \times p$ table which is formally analogous to a one-dimensional repeated measures design. The columns of the table represent items on the rating scale and are analogous to "treatment" levels; the rows represent experimental subjects. The total variability of the scores can be partitioned into components which are then utilized to estimate the reliability (or, more specifically, the internal consistency) of the rating scale. The specific model and computational procedures are described in Myers (1966, pp. 294–299) and in Winer (1962, pp. 124–132). In part, the analysis is equivalent to the Hoyt reliability estimate and to Cronbach's coefficient alpha.

combining results from a series of experiments

Research rarely consists of a single isolated experiment. More commonly, the researcher or research team will plan studies and collect data relative to a specific problem over a period of time and with different subject populations. When the experimental designs are comparable and involve the same experimental treatments, it may be desirable to combine the results from two or more independent experiments. Such an approach has at least two appealing characteristics. First, estimates of experimental effects will be based on larger sample sizes and, hence, will contain less experimental error. And, second, the evidence from a series of experiments may substantiate a research finding with a much higher level of significance than the single experiment. Indeed, the combined results of two or more experiments in which a null hypothesis is accepted by the separate experiments may lead to rejection of that hypothesis. This is, of course, more likely to occur when the hypothesis is "nearly" significant in the separate experiments.

If the separate experiments have the same design, they can ordinarily be combined into one experimental plan with repetitions of the experiment as one dimension of the plan. For example, if a 2×2 factorial design is repeated on two occasions with different samples of subjects, the scheme will be:

	CELLS OF FACTORIAL DESIGN (11)	(12)	(21)	(22)
Experiment 1	G_{111}	G_{112}	G_{121}	G_{122}
Experiment 2	G_{211}	G_{212}	G_{221}	G_{222}

The analysis appropriate to this plan is the same as a three-dimensional factorial design with "experiments" as one dimension of the design. If subjects were not randomly selected for participation in the two experiments (that is, if subjects were available groups on the two occasions which were randomly divided among the four treatment combinations but not randomly selected from a larger population), then there would be no appropriate test for the main effect corresponding to "experiments." However, all other effects, main and interaction, can be tested in the usual way.

A different approach to combining results from a series of experiments which does not require identical experimental designs involves computing an overall level of significance (or probability of a Type I error) for the series. In this approach, an exact probability for the test statistic (ordinarily, an F ratio) is determined for each experiment (values may be interpolated from ordinary tables of F, or computer routines exist which compute exact probabilities for F ratios and other statistics). These probabilities are denoted p_1, p_2, through p_k for a series of k experiments. A combined probability (of a Type I error) can be determined by evaluating the following statistic as a chi-square with $2k$ degrees of freedom:

$$-2 \sum_{i=1}^{k} (\ln p_i)$$

where $\ln p_i$ is the natural logarithm for the ith probability. The topic of combining results from two or more independent experiments is discussed in more detail in Cochran and Cox (1957, Chapter 14) and Ray (1960, pp. 228–233).

multivariate experimental designs

The techniques covered in this volume are all univariate in the sense that there is just one response measure (or criterion measure) involved in the analysis. However, both the repeated measures designs and the analysis of covariance involve procedures which are fundamentally multivariate; in the case of repeated measures designs each subject yields two or more scores on the criterion, and their intercorrelations must be considered in the analysis. The analysis of covariance involves one or more predictor variables (i.e., covariables) in addition to the criterion, and the intercorrelations of covariables and the criterion are involved in the analysis. There are, in addition, many research situations in which it is natural to define the criterion as essentially multivariate in nature, and there exist multivariate extensions of the designs presented in this volume appropriate to the analysis of multiple criteria. In this section we name, briefly describe, and give exemplary applications for some of the more widely used multivariate experimental designs.

The best approach to understanding the majority of multivariate experimental designs is to consider them by analogy with the corresponding univariate design. Thus, we shall consider the multivariate analogs of the one- and two-sample t tests, of the one-way analysis of variance (or CRD), and of factorial designs and higher-dimensional designs in general.

Although of extremely limited usefulness in actual research analysis, the one-sample t test provides one of the simplest examples of experimental design and is a natural starting point. In practice, this design is ordinarily used when the experiment involves pretesting a random sample of subjects, then subjecting them to an experimental treament, and then posttesting them with the same instrument utilized in the pretest. If gain scores (that is, posttest-pretest scores) are used as the response measure, the one-sample t test is an appropriate procedure to test for treatment effects. If there is more than one pretest (or the pretest yields two or more subscores) and corresponding posttests, we have a multivariate analog to the univariate design. Thus, the multiple criterion variables are actually gains from pretest to posttest on two or more measures. If there are k premeasures and k corresponding postmeasures, there will be k gain scores for the ith individual: Y_{i1}, Y_{i2}, . . . , Y_{ik}. The hypothesis of interest is that there is no gain from pretest to posttest in the population. However, this hypothesis is complex since we are simultaneously considering k variables instead of just one variable. Utilizing matrix notation, the null hypothesis may be written H_0: $\mathbf{m} = \mathbf{0}$, where $\mathbf{m}$ is a $k \times 1$ vector of population means, and $\mathbf{0}$ is a $k \times 1$ null vector (i.e., a vector of 0s). An appropriate test for this hypothesis involves the computation of Hotelling's T^2 statistic and the evaluation of this statistic as a chi-square value. Actually, Hotelling's T^2 test is also appropriate when the null hypothesis is in a more general form: H_0: $\mathbf{m} = \mathbf{m}'$, where the elements of $\mathbf{m}'$ are any *specific* numerical values relevant to the experimental situation.

Although computational formulas will not be presented in this volume, the general logic by which T^2 and many other test statistics in multivariate analysis are derived is relevant in order to give the student a sense of the close parallel between the multivariate and univariate procedures. The logic, in brief, is: Form a linear composite of the criterion measures:

$$W_i = a_1Y_{i1} + a_2Y_{i2} + \cdots + a_kY_{ik}$$

where the a_j values are weights chosen in such a manner that if a *univariate* t test were applied to the W_i, the value of the univariate t statistic would be maximum. Then, if this maximum univariate t proves to be statistically significant through the usual t test procedures, we conclude that the multivariate null hypothesis should be rejected; otherwise, we accept the multivariate null hypothesis. In order to apply this logic, the major obstacle is determination of the values of the weights a_j which maximize the univariate statistic. When Hotelling's T^2 test is used, we are evaluating the equivalent of the maximum univariate t value (although the procedure is equivalent, the value of T^2 is *not* equal to t^2); also, from intermediate computational products of T^2, the values of the weights a_j can be determined, although they do not occur explicitly in the testing routine.

The two-sample t test is more widely used than the one-sample case. In the univariate case we are interested in the hypothesis H_0: $\mu_1 = \mu_2$, where μ_1 and μ_2 are population means corresponding to the two samples involved in the experimental design. When the criterion is multiple, the hypothesis becomes H_0: $\mathbf{m}_1 = \mathbf{m}_2$, where $\mathbf{m}_1$ and $\mathbf{m}_2$ are $k \times 1$ vectors of population means. Once again, a Hotelling's T^2

statistic is relevant to testing this hypothesis. Also, the logic underlying this statistic is similar to the one-sample case in that weights are selected which maximize the two-sample univariate t value.

In addition to Hotelling's T^2 test, there are at least two other *equivalent* procedures which the student may encounter applied to the two-sample case. The first of these, Mahalanobis' D^2 (known as generalized distance) is a simple transformation of Hotelling's T^2 statistic. The second procedure is ordinarily known as two-group (or "simple") discriminant analysis. This technique is equivalent to a multiple regression analysis with a discrete criterion variable (the criterion being treatment groups; thus, the conceptualization of the multiple regression is inverse to its usual mode). In discriminant analysis, one is usually also interested in the degree to which treatment-group membership can be predicted on the basis of the multiple criterion measures. Thus, a classification equation is constructed, and the success of prediction can be judged in terms of the percentages of correct classifications. Usually, either the T^2 or D^2 statistic is used in conjunction with discriminant analysis to provide a test of the multivariate null hypothesis concerning mean vectors.

Although there are few examples of the use of Hotelling's T^2 test (or equivalent procedures) in the research literature of education, the technique logically applies to many published studies. At present, the strategy in educational research analysis seems to be to apply univariate tests separately to each criterion measure when two or more such measures are gathered during an experiment. Many examples of this approach can be found when, for example, during a learning study, response measures are taken on overall achievement and on two or more specific achievement areas; or, in studies involving attitude measurement in which the measuring instrument yields several subscores. Wider availability of high-speed electronic digital computers (without which multivariate analysis of experiments is extraordinarily tedious) will, however, greatly accelerate the use of relevant multivariate procedures.

The one-way analysis of variance, or CRD, involves, in general, p treatment groups exposed to different experimental treatments. The multivariate extension involves two or more criterion variables and the null hypothesis H_0: $\mathbf{m}_1 = \mathbf{m}_2 = \cdots = \mathbf{m}_p$, where, as in the case of Hotelling's T^2 test, the $\mathbf{m}_j$ are $k \times 1$ vectors of population means. Similarly, as in any univariate design such as factorial designs or nested designs, we can consider the possibility of collecting two or more criterion measures and expressing relevant null hypotheses in the form of vector equations. For example, to extrapolate the hypotheses for a two-dimensional factorial design to a multivariate form, we should substitute mean vectors for the usual scalar quantities appearing in these hypotheses. Collectively, these techniques are known as the multivariate analysis of variance (MANOVA); although a variety of relevant test statistics have been developed, the most widely used is Wilk's lambda (λ), which involves forming a ratio of determinants of matrices and for which there exist transformations to approximate F statistics.

When computer capability is available, the researcher will have occasion to turn to the techniques of multivariate analysis for experimental designs. Among the textbooks treating multivariate techniques which would be appropriate to the student who has studied the content of this volume are Morrison (1967), Cooley and Lohnes (1962), and Rao (1952).

appendix a

some basic principles of matrix algebra

introduction

Matrix algebra provides a convenient tool for dealing with situations in which complex arrays of data and statistical constants must be manipulated. In this textbook, reference to matrix notation and operations was necessary in connection with repeated measures designs (i.e., Box's tests) and with analysis of covariance (i.e., the multiple covariate case). For the student who intends further pursuit of applied statistics, an elementary understanding of matrix algebra is essential. The purpose of this appendix is to provide the basis for such an understanding; emphasis is placed on the use of matrix symbols as a shorthand method for summarizing complex expressions and equations, and upon the elementary operations which may be performed on matrices.

definitions

A matrix is simply a rectangular array of algebraic quantities. The *order* of a matrix is the size of the rectangular array in terms of the number of rows and columns constituting the array. For example, a matrix of order 3×4 is

$$\begin{bmatrix} 5 & X & 2M & 8 \\ R & 7 & 3-X & 9 \\ X^2 & 5 & 18 & 2 \end{bmatrix}$$

Note the use of brackets to enclose the elements constituting the matrix. The elements, in general, can be real numbers (such as 5 or 8), algebraic variables (such as X or R), algebraic expressions (such as $2M$ or $3 - X$), or any other algebraic quantities. A matrix does not represent any particular numerical value; it is, simply, an *arrangement* of quantities. As we shall see later, operations involving matrices can be carried out, and the result may be a specific numerical value or a single algebraic expression; however, matrices per se represent a conventional organization of algebraic quantities.

Matrices are denoted by uppercase letters; thus, the 3×4 matrix presented above might be called A, and we should write

$$A = \begin{bmatrix} 5 & X & 2M & 8 \\ R & 7 & 3-X & 9 \\ X^2 & 5 & 18 & 2 \end{bmatrix}$$

In some applications, it will be convenient to indicate the order of the matrix in its name; thus, $A_{3\times4}$ will sometimes be written in place of A.

In most elementary applications of matrices, the elements will be all of a kind. That is, a single matrix will typically consist entirely of real numbers, or entirely of algebraic constants, etc., but the student will not often encounter matrices which are mixtures of different kinds of elements. For this reason, the examples in this section show elements of a kind.

Certain types of matrices are given special names because of their properties. The more important of these are as follows:

SCALAR MATRIX

A matrix of order 1×1 is called a scalar matrix; examples are $[2]$, $[a]$, $[X^2 + 2]$, and $[-305]$.

VECTOR

A vector is a matrix with either just one row or just one column. Examples of *row vectors* are $[3 \quad 4 \quad 9 \quad 2]$ and $[a \quad m \quad e]$. Some *column vectors* are

$$\begin{bmatrix} 6 \\ 4 \\ 9 \end{bmatrix} \quad \text{and} \quad \begin{bmatrix} -3 \\ 14 \\ 0 \end{bmatrix}$$

In denoting vectors, lower case roman letters are used, and these letters are boldface to distinguish them from ordinary algebraic quantities: thus,

$$\mathbf{a} = [3 \quad 4 \quad 9 \quad 2] \quad \text{and} \quad \mathbf{b} = [a \quad m \quad e]$$

SQUARE MATRIX

A matrix of order $r \times r$ (that is, with the same number of rows and columns) is called a square matrix. An example of such a matrix is

$$\begin{bmatrix} 5 & 1 & 3 & 9 & 6 \\ 2 & 4 & 6 & 8 & -2 \\ 3 & 1 & 1 & 7 & 0 \\ -5 & 7 & 3 & 1 & 6 \\ 0 & 4 & 8 & 5 & 9 \end{bmatrix}$$

Within the category of square matrices, a number of additional special types are identified. A *symmetric matrix* is a square matrix in which elements symmetrically placed about the main diagonal of the matrix are equal. The main diagonal is the set of elements running from upper left to lower right. If a_{ij} is the element in the ith row and jth column of matrix A, then in a symmetric matrix, $a_{ij} = a_{ji}$ for all i and j. For example, A is a square, symmetric matrix:

$$A = \begin{bmatrix} 3 & 4 & 7 \\ 4 & 6 & 9 \\ 7 & 9 & 0 \end{bmatrix}$$

The main diagonal of A consists of the elements $a_{11} = 3$, $a_{22} = 6$, and $a_{33} = 0$. Note the symmetry of A about this diagonal; thus, $a_{12} = 4$ and $a_{21} = 4$; $a_{13} = a_{31} = 7$; and $a_{23} = a_{32} = 9$. A *diagonal matrix* is a square matrix with elements of 0 everywhere except on the main diagonal. An example is

$$D = \begin{bmatrix} 3 & 0 & 0 \\ 0 & 5 & 0 \\ 0 & 0 & -2 \end{bmatrix}$$

Note that every diagonal matrix is, of necessity, also a square matrix. A diagonal matrix which has all its diagonal elements equal to 1 is called an *identity matrix*. When we turn to a consideration of elementary matrix manipulations, we shall see that identity matrices play a role similar to that played by "1" in ordinary algebra. Identity matrices are denoted by I; thus, the 4×4 identity matrix is

$$I_{4\times 4} = \begin{bmatrix} 1 & 0 & 0 & 0 \\ 0 & 1 & 0 & 0 \\ 0 & 0 & 1 & 0 \\ 0 & 0 & 0 & 1 \end{bmatrix}$$

elementary operations on matrices

EQUALITY OF MATRICES

Two matrices are said to be "equal" if they are of the same order and corresponding elements are equal. Thus, $A = B$ only if A and B are of the same order and $a_{ij} = b_{ij}$ for all i and j. For example,

$$A = \begin{bmatrix} 3 & 2 & 0 \\ 5 & 1 & -4 \\ 7 & 7 & 8 \end{bmatrix} \qquad B = \begin{bmatrix} 3 & 2 & 0 \\ 5 & 1 & -4 \\ 7 & 7 & 8 \end{bmatrix}$$

Since all corresponding elements are equal, $A = B$.

TRANSPOSE OF A MATRIX

The operation of transposing a matrix involves interchanging the rows and columns of the matrix. Upon transposition, a matrix of order $r \times c$ becomes a matrix of order $c \times r$. The matrix which results from the operation of transposition of the matrix A is called the transpose of matrix A and is denoted A'. For example, if

$$A = \begin{bmatrix} 1 & 7 \\ 2 & 1 \\ 3 & 5 \end{bmatrix} \qquad \text{then } A' = \begin{bmatrix} 1 & 2 & 3 \\ 7 & 1 & 5 \end{bmatrix}$$

Note that the first row of A' (that is, the elements 1, 2, and 3) is identical with the first column of A and that the second row of A' (that is, the elements 7, 1, and 5) is identical with the second column of A. In terms of elements, $a_{ij} = a'_{ji}$ for all i and j. Now consider a symmetric matrix

$$A = \begin{bmatrix} 1 & 5 & 8 \\ 5 & 2 & 6 \\ 8 & 6 & 3 \end{bmatrix}$$

Since corresponding rows and columns are already identical in a symmetric matrix, the operation of transposition results in the original matrix. Thus, for any symmetric matrix, $A = A'$. Also, the transposition of a row vector of order $1 \times c$ results in a column vector of order $c \times 1$.

ADDITION AND SUBTRACTION OF MATRICES

Two matrices are said to be *conformable to addition* (*or subtraction*) if they are of the same order. The sum of two conformable matrices is found by addition of corresponding elements. Similarly, the difference of two conformable matrices is found by subtraction of corresponding elements. Consider the two matrices A and B:

$$A = \begin{bmatrix} 4 & 5 & 9 \\ 7 & 1 & 2 \end{bmatrix} \qquad B = \begin{bmatrix} 1 & 8 & 3 \\ 3 & -2 & 0 \end{bmatrix}$$

The sum of A and B is

$$A + B = \begin{bmatrix} 4+1 & 5+8 & 9+3 \\ 7+3 & 1-2 & 2+0 \end{bmatrix} = \begin{bmatrix} 5 & 13 & 12 \\ 10 & -1 & 2 \end{bmatrix}$$

Their difference is

$$A - B = \begin{bmatrix} 4-1 & 5-8 & 9-3 \\ 7-3 & 1+2 & 2-0 \end{bmatrix} = \begin{bmatrix} 3 & -3 & 6 \\ 4 & 3 & 2 \end{bmatrix}$$

Note that, as in ordinary algebra, $A + B = B + A$; however, $A - B$ and $B - A$ are, in general, not the same (as, also, is true in ordinary algebra; for example, $5 - 3 = 2$, but $3 - 5 = -2$). Addition and/or subtraction of matrices can be carried out for a series of more than two matrices. If A, B, and C are conformable matrices, then $A + B + C$ is defined as the matrix with elements equal to the sum of corresponding elements in all three matrices. Similarly, addition and subtraction can be combined. For example, $A + B - C$ and $A - B - C$ are permissible if A, B, and C are conformable.

MULTIPLICATION OF MATRICES

Although the operations of addition and subtraction of matrices conform to our "common-sense" notions of these operations, the process of multiplication presents a quite new approach. Two matrices are said to be *conformable to multiplication* if the number of rows in the second matrix is equal to the number of columns in the first matrix. Since multiplication of matrices is not commutative (that is, in general, $AB \neq BA$), it is important to consider the order in which the matrices are multiplied together. To satisfy conformability requirements, the product AB is defined only if the order of A is $r \times c$ and the order of B is $c \times m$. The product matrix will, then, be of order $r \times m$. To be specific, the matrices $A_{4\times3}$ and $B_{3\times8}$ are conformable to multiplication in the order AB since A has three columns and B has three rows. However, they are *not* conformable in the order BA since B (which is now the first matrix in the product) has eight columns and A has only three rows. The product in the conformable order AB will be a matrix with four rows and eight columns; that is, $A_{4\times 3}B_{3\times 8} = C_{4\times 8}$.

The process of multiplication of matrices involves calculating row-by-column sums of products; that is, element c_{ij} in the product matrix is found by multiplying together, and then summing, corresponding elements in row i of the first matrix and column j of the second matrix. Formally, for the product $AB = C$, element $c_{ij} = \Sigma_{k=1}^{c} a_{ik}b_{kj}$, where c is the number of columns in A (and number of rows in B). To illustrate this procedure, consider $A_{2\times 3}$ and $B_{3\times 3}$:

$$A = \begin{bmatrix} 5 & 8 & 2 \\ 1 & 6 & 4 \end{bmatrix} \qquad B = \begin{bmatrix} 4 & 6 & 3 \\ 1 & 4 & 0 \\ 1 & 5 & 3 \end{bmatrix}$$

Note that A and B are conformable to multiplication in the order AB since A has three *columns* and B has three *rows*. Also, the product matrix C will be of order 2×3. The first element c_{11} of C is found from the first row of A and the first column of B. The sum of products is $5 \cdot 4 + 8 \cdot 1 + 2 \cdot 1 = 30$. The second element in the first row of C is found from the first row of A and the second column of B; this sum of products is $5 \cdot 6 + 8 \cdot 4 + 2 \cdot 5 = 72$. The completed product matrix C is shown below, along with the details of the computations for the elements of C:

$$AB = \begin{bmatrix} 5 & 8 & 2 \\ 1 & 6 & 4 \end{bmatrix} \begin{bmatrix} 4 & 6 & 3 \\ 1 & 4 & 0 \\ 1 & 5 & 3 \end{bmatrix} = \begin{bmatrix} 30 & 72 & 21 \\ 14 & 50 & 15 \end{bmatrix} = C$$

$$\begin{aligned} c_{11} &= 5 \cdot 4 + 8 \cdot 1 + 2 \cdot 1 = 30 \qquad & c_{21} &= 1 \cdot 4 + 6 \cdot 1 + 4 \cdot 1 = 14 \\ c_{12} &= 5 \cdot 6 + 8 \cdot 4 + 2 \cdot 5 = 72 & c_{22} &= 1 \cdot 6 + 6 \cdot 4 + 4 \cdot 5 = 50 \\ c_{13} &= 5 \cdot 3 + 8 \cdot 0 + 2 \cdot 3 = 21 & c_{23} &= 1 \cdot 3 + 6 \cdot 0 + 4 \cdot 3 = 15 \end{aligned}$$

Products of certain types of matrices are of special interest:

Product of a row vector and a column vector A $1 \times c$ row vector and a $c \times 1$ column vector are conformable to multiplication; for example,

$$AB = [4 \quad 9 \quad 0 \quad 5] \begin{bmatrix} 6 \\ 2 \\ 5 \\ 1 \end{bmatrix} = [47] = C$$

where C is a scalar matrix with the single element 47.

Product of a column vector and a row vector A $c \times 1$ column vector and a $1 \times c$ row vector are conformable to multiplication; for example,

$$BA = \begin{bmatrix} 6 \\ 2 \\ 5 \\ 1 \end{bmatrix} [4 \quad 9 \quad 0 \quad 5] = \begin{bmatrix} 24 & 54 & 0 & 30 \\ 8 & 18 & 0 & 10 \\ 20 & 45 & 0 & 25 \\ 4 & 9 & 0 & 5 \end{bmatrix} = C$$

Product of any matrix with a conformable identity matrix If A is $r \times c$, then a conformable identity matrix would be of order $c \times c$. The product of $A_{r\times c}$ and $I_{c\times c}$ is the original matrix $A_{r\times c}$; for example,

$$A_{2\times 3} = \begin{bmatrix} 5 & 8 & 2 \\ 1 & 6 & 4 \end{bmatrix} \qquad I_{3\times 3} = \begin{bmatrix} 1 & 0 & 0 \\ 0 & 1 & 0 \\ 0 & 0 & 1 \end{bmatrix}$$

Then

$$AI = \begin{bmatrix} 5 & 8 & 2 \\ 1 & 6 & 4 \end{bmatrix}$$

Similarly, a second conformable identity matrix is $I_{r\times r}$ if A is of order $r \times c$; that is, the product $I_{r\times r}A_{r\times c}$ is defined and, again, $IA = A$. In order to be able to indicate more explicitly the order of multiplication, it is useful to speak of any product AB as the product of A *postmultiplied by* B or as B *premultiplied by* A. Thus, $A_{r\times c}$ postmultiplied by $I_{c\times c}$ is A, and $A_{r\times c}$ premultiplied by $I_{r\times r}$ is also A.

Premultiplication of a matrix by a conformable diagonal matrix If A is $r \times c$, then A is conformable to multiplication with a diagonal matrix of order $r \times r$. The product is equivalent to multiplication of each element in a given *row* of A by the corresponding diagonal element in D, the diagonal matrix. For example,

$$A_{2\times 3} = \begin{bmatrix} 5 & 8 & 2 \\ 1 & 6 & 4 \end{bmatrix} \qquad D_{2\times 2} = \begin{bmatrix} 3 & 0 \\ 0 & 4 \end{bmatrix}$$

Then

$$DA = \begin{bmatrix} 15 & 24 & 6 \\ 4 & 24 & 16 \end{bmatrix}$$

Note that the first *row* of DA has elements equal to three times the elements in the first row of A, and the second *row* of DA has elements equal to four times the elements of the second row of A.

Postmultiplication of a matrix by a conformable diagonal matrix If A is $r \times c$, then A is conformable to multiplication with a diagonal matrix of order $c \times c$. The product is equivalent to multiplication of each element in a given *column* of A by the corresponding diagonal element in D, the diagonal matrix. For example,

$$A_{2\times 3} = \begin{bmatrix} 5 & 8 & 2 \\ 1 & 6 & 4 \end{bmatrix} \qquad D_{3\times 3} = \begin{bmatrix} 3 & 0 & 0 \\ 0 & 2 & 0 \\ 0 & 0 & 4 \end{bmatrix}$$

Then

$$AD = \begin{bmatrix} 15 & 16 & 8 \\ 3 & 12 & 16 \end{bmatrix}$$

Note that the first *column* of AD has elements equal to three times the elements in the first column of A, the second column has elements equal to two times the elements in the second column of A, and the third column has elements equal to four times the elements in the third column of A.

Multiplication by an algebraic quantity The elements of ordinary algebra are referred to as *scalars*. Recall that a matrix such as [2] was defined as a scalar matrix; the algebraic constant 2 is per se a scalar. Multiplication is defined in matrix algebra for cases involving matrices and scalars (but note that a *scalar matrix* is of order 1×1 and is conformable to multiplication only with another scalar matrix or a column vector). In general, multiplying a matrix by a scalar is equivalent to multiplying each element in the matrix by the scalar; for example,

$$\text{If } A = \begin{bmatrix} 5 & 8 & 2 \\ 1 & 6 & 4 \end{bmatrix} \quad \text{then} \quad 3A = \begin{bmatrix} 3\cdot 5 & 3\cdot 8 & 3\cdot 2 \\ 3\cdot 1 & 3\cdot 6 & 3\cdot 4 \end{bmatrix} = \begin{bmatrix} 15 & 24 & 6 \\ 3 & 18 & 12 \end{bmatrix}$$

INVERSION OF MATRICES

In matrix algebra, the operation of division is undefined; therefore, it is not permissible to form quotients of matrices (e.g., the expression A/B is meaningless if A and B are matrices). However, some of the results ordinarily associated with the process of division can be accomplished by utilization of the *inverse* of a matrix. Inverses exist only for square matrices; if A is a $c \times c$ matrix, then there *may* exist a matrix A^{-1} for which $AA^{-1} = A^{-1}A = I$. Note the analogy between A^{-1} and the reciprocal (or multiplicative inverse) of a number in ordinary algebra. It is important to realize that A^{-1} may not exist even if A is a square matrix.

The conditions for the existence of A^{-1} are discussed at the end of this section.

Returning to the definition, we see that the inverse A^{-1} of a matrix A is a matrix of the same order as A and which, when premultiplied or postmultiplied by A, results in an identity matrix. To illustrate such a matrix, consider

$$A = \begin{bmatrix} 2 & 4 \\ 5 & 8 \end{bmatrix} \quad \text{and} \quad A^{-1} = \begin{bmatrix} -2 & 1 \\ 1\frac{1}{4} & -\frac{1}{2} \end{bmatrix}$$

Then,

$$AA^{-1} = \begin{bmatrix} 2(-2) + 4(1\frac{1}{4}) & 2(1) + 4(-\frac{1}{2}) \\ 5(-2) + 8(1\frac{1}{4}) & 5(1) + 8(-\frac{1}{2}) \end{bmatrix} = \begin{bmatrix} 1 & 0 \\ 0 & 1 \end{bmatrix}$$

The student can easily demonstrate that the order of multiplication is irrelevant and that $A^{-1}A = I$.

For small matrices, it is not difficult to determine the value of the elements in an inverse matrix. However, inversion of matrices of large order is an extremely tedious operation by hand procedures. Fortunately, there are iterative procedures which can be utilized with electronic digital computers for the efficient inversion of matrices. We shall illustrate the direct inversion of a small matrix by purely algebraic methods but leave other solutions to outside references (Faddeeva, 1959; Horst, 1963).

Consider the 2×2 matrix shown previously:

$$A = \begin{bmatrix} 2 & 4 \\ 5 & 8 \end{bmatrix}$$

By the definition of an inverse, $AA^{-1} = I$. Assuming that the elements of A^{-1} are unknown, we can write the expression (where a^{ij} is an unknown element of A^{-1})

$$AA^{-1} = \begin{bmatrix} 2 & 4 \\ 5 & 8 \end{bmatrix} \begin{bmatrix} a^{11} & a^{12} \\ a^{21} & a^{22} \end{bmatrix} = \begin{bmatrix} 1 & 0 \\ 0 & 1 \end{bmatrix} = I$$

From the definition of matrix multiplication, we can set up four equations:

(1) $2a^{11} + 4a^{21} = 1$ (2) $2a^{12} + 4a^{22} = 0$

(3) $5a^{11} + 8a^{21} = 0$ (4) $5a^{12} + 8a^{22} = 1$

Since the four equations contain four unknowns, it is ordinarily possible to solve them uniquely. Starting with the first equation, we can solve for a^{11} in terms of $a^{21}: 2a^{11} = 1 - 4a^{21}$, or $a^{11} = \frac{1}{2} - 2a^{21}$. Next, this value for a^{11} can be substituted in the third equation to yield $5(\frac{1}{2} - 2a^{21}) + 8a^{21} = 0$, or $2.5 - 2a^{21} = 0$, or $a^{21} = 2.5/2 = 1\frac{1}{4}$. Since the value of a^{21} is now known explicitly, it can be substituted in the first equation and the value of a^{11} determined: $2a^{11} + 4(1\frac{1}{4}) = 1$, or $2a^{11} = 1 - 5 = -4$; therefore, $a^{11} = -2$. Now a similar set of operations can be performed on the second and fourth equations to yield the values of a^{12} and a^{22}. Thus, $a^{12} = -2a^{22}$ and $5(-2a^{22}) + 8a^{22} = 1$, or $-2a^{22} = 1$, and $a^{22} = -\frac{1}{2}$. Substituting this value in the second equation yields $2a^{12} + 4(-\frac{1}{2}) = 0$, or $a^{12} = 1$. Now, all four unknown elements of A^{-1} have been calculated. It should be emphasized, however, that this method is unsuitable for large matrices since the number of solution and substitution steps grows enormous.

The inverse of a matrix will exist only if the set of equations analogous to (1) through (4) above have distinct solutions for the unknowns. In some cases, the elements can be determined only proportionally, and no unique solution exists. For example, the matrix B, below, yields equations with no unique solutions:

$$B = \begin{bmatrix} 3 & 6 \\ 6 & 12 \end{bmatrix}$$

The equations based on the unknown elements of the inverse are

$$(5)\quad 3b^{11} + 6b^{21} = 1 \qquad (6)\quad 3b^{12} + 6b^{22} = 0$$

$$(7)\quad 6b^{11} + 12b^{21} = 0 \qquad (8)\quad 6b^{12} + 12b^{22} = 1$$

From Eq. (5), $3b^{11} = 1 - 6b^{21}$, or $b^{11} = \frac{1}{3} - 2b^{21}$. Substituting in Eq. (7), $6(\frac{1}{3} - 2b^{21}) + 12b^{21} = 0$, or $2 - 12b^{21} + 12b^{21} = 0$, or $2 = 0$, which is an absurd result. Since we cannot uniquely solve for the elements of B^{-1}, this inverse does not exist and B cannot be inverted.

determinants

A determinant is a specific numerical or algebraic value which can be calculated from any square array of elements. The theory of determinants and the theory of matrices are separate mathematical systems, and the operations explained above for matrices do not apply to determinants. However, for any square matrix, there is a corresponding determinant with the same elements as the matrix, and the value of this determinant can be calculated. In advanced statistical applications, situations arise which involve both matrices and determinants. For simplicity of reference, the determinant with elements identical with those of a square matrix A will be called the *determinant of* A and denoted $|A|$.

The value of a determinant is defined as the *sum* of $c!$ terms (where $n!$ is the continued product of the first n integers; for example, $5! = 5 \cdot 4 \cdot 3 \cdot 2 \cdot 1 = 120$). These $c!$ terms represent all the possible combinations of products of elements from the square array with the condition that one and only one element is taken from any single row or any single column of the square array. In addition, each term has a sign which is determined by the number of "inversions" which occur during the selection of the elements to form the term; an inversion occurs when an element from a row higher than the previously chosen element is selected. If an even number of inversions (including 0) occurs, the sign of the term is $+$; if an odd number of inversions occurs, the sign of the term is $-$. For small square arrays, the definition can be directly utilized to compute values for determinants. However, for larger determinants this is laborious, and other methods will be explained.

Consider the 2×2 matrix A:

$$A = \begin{bmatrix} 3 & 8 \\ 2 & 10 \end{bmatrix}$$

The determinant of this matrix is

$$|A| = \begin{vmatrix} 3 & 8 \\ 2 & 10 \end{vmatrix}$$

The value of this determinant will consist of the sum of $2! = 2$ terms. To form these terms, we must choose one element from each row and each column of the array. The two terms are $3 \cdot 10$ and $2 \cdot 8$; the first term has the element 3 chosen from the first row and first column and the element 10 chosen from the second row and second column. The second term has the element 2 chosen from the second row and first column and the element 8 chosen from the first row and second column. The sign of the first term $3 \cdot 10$ is positive since there is no row inversion; that is, in terms of subscripted notation, the product is $a_{11}a_{22}$. Thus, in moving from the first element in the product, a_{11}, to the second element in the product, a_{22}, the row subscripts fall in their natural order. The second term carries a negative sign; in subscripted notation, this term is $a_{21}a_{12}$, and there is one inversion of row subscripts (that is, 2 occurs as a row subscript prior to the occurrence of 1 as a row subscript). Thus, the value of $|A|$ is $a_{11}a_{22} - a_{21}a_{12} = 3 \cdot 10 - 2 \cdot 8 = 30 - 16 = 14$. In general, for any 2×2 determinant, the value will be $a_{11}a_{22} - a_{21}a_{12}$.

Consider, now, the 3 × 3 matrix A:

$$A = \begin{bmatrix} 5 & 1 & 8 \\ 3 & 4 & 2 \\ 5 & 0 & 2 \end{bmatrix}$$

The determinant of A is

$$|A| = \begin{vmatrix} 5 & 1 & 8 \\ 3 & 4 & 2 \\ 5 & 0 & 2 \end{vmatrix}$$

In terms of element notation, the $c! = 3! = 6$ terms are

$$a_{11}a_{22}a_{33} \qquad a_{11}a_{32}a_{23} \qquad a_{21}a_{32}a_{13} \qquad a_{21}a_{12}a_{33} \qquad a_{31}a_{22}a_{13} \qquad a_{31}a_{12}a_{23}$$

The signs of these terms must be determined by considering the number of inversions of row subscripts. The first term $a_{11}a_{22}a_{33}$ contains 0 inversions, and its sign is +; the second term $a_{11}a_{32}a_{23}$ contains one inversion (note that interchanging a_{32} and a_{23} results in the natural order of row subscripts: $a_{11}a_{23}a_{32}$), and its sign is −; the third term $a_{21}a_{32}a_{13}$ contains two inversions (to obtain a natural order of row subscripts, it is necessary to interchange a_{13} and a_{32} and then to interchange a_{13} and a_{21}), and its sign is +; the fourth term $a_{21}a_{12}a_{33}$ contains one inversion (to obtain a natural order of row subscripts, it is necessary only to interchange a_{21} and a_{12}), and its sign is −; the fifth term $a_{31}a_{22}a_{13}$ contains three inversions (to obtain a natural order of row subscripts, it is necessary to interchange elements a_{13} and a_{22}, then to interchange elements a_{13} and a_{31}, and then to interchange elements a_{31} and a_{22}), and its sign is −; finally, the sixth term $a_{31}a_{12}a_{23}$ contains two inversions (to obtain a natural order of row subscripts, it is necessary to interchange elements a_{31} and a_{12} and then to interchange elements a_{31} and a_{23}), and its sign is +. Considering the foregoing, the determinant of A is the sum

$$a_{11}a_{22}a_{33} + a_{21}a_{32}a_{13} + a_{31}a_{12}a_{13} - a_{11}a_{32}a_{23} - a_{21}a_{12}a_{33} - a_{31}a_{22}a_{13}$$

There is a distinctive pattern to the elements which are positive and those which are negative in this sum. In the two diagrams below, the patterns of positive elements and of negative elements are shown by the lines.

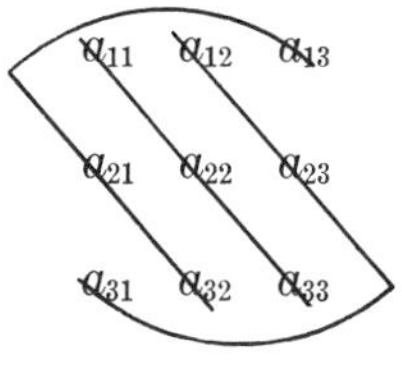

Positive terms

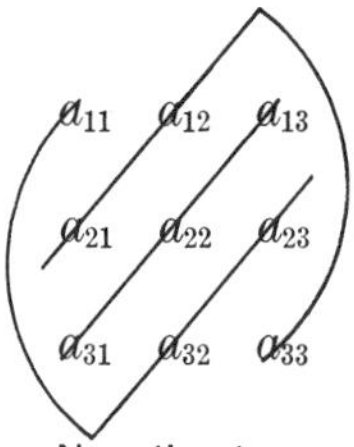

Negative terms

For the specific example of matrix A, the value of the determinant is

$$5 \cdot 4 \cdot 2 + 3 \cdot 0 \cdot 8 + 5 \cdot 1 \cdot 2 - 5 \cdot 0 \cdot 2 - 3 \cdot 1 \cdot 2 - 5 \cdot 4 \cdot 8 = 50 - 166 = -116$$

For matrices larger than 3×3, the corresponding determinantal values are difficult to compute by the direct procedure since it is time-consuming to enumerate the possible terms. For this reason, a different approach, which involves some new definitions, is recommended.

A *minor* of a determinant is a new determinant of order $(c - 1) \times (c - 1)$ (where the order of the original determinant is $c \times c$) derived by striking from the original determinant a row and column. The element at the intersection of the row and column stricken from the original determinant is the *defining element* for the minor. Thus, in a $c \times c$ determinant, each element can be made the defining element for a minor, and there is a total of c^2 minors which can be computed. Consider the 3×3 determinant $|A|$:

$$\begin{vmatrix} 5 & 1 & 8 \\ 3 & 4 & 2 \\ 5 & 0 & 2 \end{vmatrix}$$

If we choose $a_{11} = 5$ as the defining element, the corresponding minor is

$$\begin{vmatrix} 4 & 2 \\ 0 & 2 \end{vmatrix}$$

In arriving at the minor for a_{11}, the first row and first column of A were deleted. Similarly, the minor for a_{32} is the matrix

$$\begin{vmatrix} 5 & 8 \\ 3 & 2 \end{vmatrix}$$

which results when the third row and the second column of $|A|$ are deleted. If the defining element is a_{ij}, then the corresponding minor will be denoted M_{ij}.

For each minor of a determinant, there is defined the *cofactor*, which is simply the value of the minor with appropriate sign. Thus, the cofactor of the minor M_{ij} is defined as

$$A_{ij} = (-1)^{i+j} M_{ij}$$

The cofactor A_{ij} contains the term $(-1)^{i+j}$, which is a sign operator; that is, $(-1)^{i+j}$ is $+1$ if $i + j$ is an even number and is -1 if $i + j$ is an odd number. In effect, the cofactor corresponding to a minor is the value of the minor multiplied by either $+1$ or -1. For example, from the previous illustration, consider the minor M_{11}, which has the defining element $a_{11} = 5$:

$$M_{11} = \begin{vmatrix} 4 & 2 \\ 0 & 2 \end{vmatrix} = 4 \cdot 2 - 0 \cdot 2 = 8 - 0 = 8$$

The cofactor of M_{11} is $A_{11} = (-1)^{1+1}M_{11} = +1(8) = 8$. Similarly, for the minor M_{32} we have

$$M_{32} = \begin{vmatrix} 5 & 8 \\ 3 & 2 \end{vmatrix} = 5 \cdot 2 - 3 \cdot 8 = 10 - 24 = -14$$

The cofactor of M_{32} is $A_{32} = (-1)^{3+2}M_{32} = -1(-14) = 14$.

The value of a determinant can be found by a procedure known as expansion by the cofactors. If we choose any single row (or column) of the determinant, the value of the determinant is a sum of products of defining elements and cofactors. In symbols, choosing the kth row of the determinant for expansion, we have

$$|A| = \sum_{j=1}^{c} a_{kj}A_{kj}$$

or choosing the gth column, we have

$$|A| = \sum_{i=1}^{c} a_{ig}A_{ig}$$

Returning to the example of a 3×3 determinant whose determinant was determined by the direct procedure, we can choose any row or any column for expansion. For simplicity, we select the elements of the first column. Since the determinant is 3×3, the expansion will contain three terms. In detail, the computation of cofactors and the expansion process are

$$\begin{aligned} |A| &= 5(-1)^{1+1}\begin{vmatrix} 4 & 2 \\ 0 & 2 \end{vmatrix} + 3(-1)^{2+1}\begin{vmatrix} 1 & 8 \\ 0 & 2 \end{vmatrix} + 5(-1)^{3+1}\begin{vmatrix} 1 & 8 \\ 4 & 2 \end{vmatrix} \\ &= 5(+1)(4 \cdot 2 - 0 \cdot 2) + 3(-1)(1 \cdot 2 - 0 \cdot 8) + 5(+1)(1 \cdot 2 - 4 \cdot 8) \\ &= 5 \cdot 8 + 3(-2) + 5(-30) = 40 - 6 - 150 = -116 \end{aligned}$$

and this value agrees with that computed by the direct method. Any other column or any row of $|A|$ could have been chosen for expansion, and the same result would have been found.

For large determinants, the process of expansion may have to be applied at several levels in order to arrive at the value of the original determinant. Consider the 5×5 determinant $|B|$:

$$|B| = \begin{vmatrix} 3 & 1 & 4 & 2 & 8 \\ 1 & 0 & 2 & 5 & 5 \\ 3 & 1 & 6 & 2 & 6 \\ 2 & 6 & 5 & 4 & 2 \\ 9 & 1 & 3 & 3 & 2 \end{vmatrix}$$

If the first column is chosen for expansion, the cofactors are each 4×4 matrices. In order to determine the values of these cofactors, each cofactor can be expanded in terms of its first column since the resulting cofactors (of the cofactors) are of order 3×3 and they can be solved by the previous formula. In general, for determinants of any order, successive applications of the expansion by cofactors will eventually result in 3×3 determinants which can be easily solved. Of course, the total process can be rather lengthy, but it is mechanical and easy to carry out.

rank of a matrix

A square matrix is said to be of *full rank* (or of *rank* c if the matrix is $c \times c$) if the value of the determinant of the matrix is not equal to 0. If the value of the determinant is 0, then the matrix is described as *singular*. A matrix of full rank is also called *nonsingular*. If a $c \times c$ matrix is singular (that is, *not* of full rank), then the rank of the matrix is equal to the order of the highest-order nonzero cofactors of the determinant of the matrix. For example, assume that the determinant of a 5×5 matrix is found to equal 0; then, this matrix is singular. If, in the determinant of this matrix, any 4×4 cofactor (for any defining element) is not equal to 0, then the rank of the matrix is 4. However, if all 4×4 cofactors vanish (that is, are equal to 0), then the rank of the matrix is 3 if any nonzero cofactors of order 3×3 exist. In other words, the rank of a matrix is equal to the order of the largest cofactors which do not all vanish. In many statistical applications of matrices, the ranks of matrices will play an important role; it is, however, typically quite tedious to determine the ranks of matrices since large numbers of determinants must be evaluated.

matrix equations

Since equality of matrices is defined within matrix algebra and since we have available such operations as addition, subtraction, multiplication, and inversion of matrices, it is possible to build equations which express relationships among matrices. In statistical applications, of course, it is such equations which play a fundamental role. When matrix equations are set up, it is necessary to carefully consider conformability of the matrices entering into such operations as addition and multiplication. In this section, we present some examples of common statistical manipulations expressed in matrix form.

STRUCTURAL MODEL FOR A SCORE IN MATRIX FORM

A one-dimensional analysis-of-variance design has the structural model $Y_{ij} = \mu + \alpha_j + \epsilon_{ij}$. For any specific design, it is possible to write an expression in matrix notation which includes the structural models for all the Y_{ij}. Consider the CRD at just two levels with three observations per cell:

TREATMENT 1	2
Y_{11}	Y_{12}
Y_{21}	Y_{22}
Y_{31}	Y_{32}

The observations can be written as a column vector of, in general, order $pn \times 1$, where p is the number of levels of the treatment and n is the number of observations per level (if the n are not equal, the order is $n_t \times 1$, where n_t is the total number of observations). We refer to this score vector as **y**, and for the example

$$\mathbf{y} = \begin{bmatrix} Y_{11} \\ Y_{21} \\ Y_{31} \\ Y_{12} \\ Y_{22} \\ Y_{32} \end{bmatrix}$$

A second column vector, known as a parameter vector and denoted **p**, is defined as having the elements $\mu, \alpha_1, \alpha_2, \ldots, \alpha_p$. Thus, the order of this matrix for a one-dimensional design is $(p + 1) \times 1$. For the example

$$\mathbf{p} = \begin{bmatrix} \mu \\ \alpha_1 \\ \alpha_2 \end{bmatrix}$$

A third column vector of order $np \times 1$ (or $n_t \times 1$ for unequal sample sizes) and denoted **e** is defined with elements consisting of the error terms ϵ_{ij}. For the example, this is

$$\mathbf{e} = \begin{bmatrix} \epsilon_{11} \\ \epsilon_{21} \\ \epsilon_{31} \\ \epsilon_{12} \\ \epsilon_{22} \\ \epsilon_{32} \end{bmatrix}$$

Finally, a matrix of order $np \times (p + 1)$ [or, more generally, of order $n_t \times (p + 1)$] and denoted A is referred to as the *design matrix*. This matrix contains elements equal to either 0 or 1; the choice of coefficients is made so as to combine the elements of the parameter vector properly. The form of the matrix equation involving the various vectors and the design matrix is $\mathbf{y} = A\mathbf{p} + \mathbf{e}$. Since A is postmultiplied by **p**, the choice of elements for A is quite straightforward. Consider the equation in its complete form for the example:

$$\begin{bmatrix} Y_{11} \\ Y_{21} \\ Y_{31} \\ Y_{12} \\ Y_{22} \\ Y_{32} \end{bmatrix} = \begin{bmatrix} 1 & 1 & 0 \\ 1 & 1 & 0 \\ 1 & 1 & 0 \\ 1 & 0 & 1 \\ 1 & 0 & 1 \\ 1 & 0 & 1 \end{bmatrix} \begin{bmatrix} \mu \\ \alpha_1 \\ \alpha_2 \end{bmatrix} + \begin{bmatrix} \epsilon_{11} \\ \epsilon_{21} \\ \epsilon_{31} \\ \epsilon_{12} \\ \epsilon_{22} \\ \epsilon_{32} \end{bmatrix}$$

Carrying out the operations on the right results in

$$\begin{bmatrix} \mu + 1\alpha_1 + 0\alpha_2 \\ \mu + 1\alpha_1 + 0\alpha_2 \\ \mu + 1\alpha_1 + 0\alpha_2 \\ \mu + 0\alpha_1 + 1\alpha_2 \\ \mu + 0\alpha_1 + 1\alpha_2 \\ \mu + 0\alpha_1 + 1\alpha_2 \end{bmatrix} + \begin{bmatrix} \epsilon_{11} \\ \epsilon_{21} \\ \epsilon_{31} \\ \epsilon_{12} \\ \epsilon_{22} \\ \epsilon_{32} \end{bmatrix} = \begin{bmatrix} \mu + \alpha_1 + \epsilon_{11} \\ \mu + \alpha_1 + \epsilon_{21} \\ \mu + \alpha_1 + \epsilon_{31} \\ \mu + \alpha_2 + \epsilon_{12} \\ \mu + \alpha_2 + \epsilon_{22} \\ \mu + \alpha_2 + \epsilon_{32} \end{bmatrix} = \begin{bmatrix} Y_{11} \\ Y_{21} \\ Y_{31} \\ Y_{12} \\ Y_{22} \\ Y_{32} \end{bmatrix}$$

The pattern of the 0s and the 1s in the design matrix A was chosen to include the appropriate α_j, to exclude the inappropriate α_j, and to include μ in every term.

The structural models for scores in any experimental design can be summarized by a matrix equation of the form $\mathbf{y} = A\mathbf{p} + \mathbf{e}$. For, say, a factorial design, the score vector **y** contains all the observations; the parameter matrix **p** contains all different parameters (including interaction terms; for example, the elements of the parameter vector for a 2×2 factorial design would be μ, α_1, α_2, β_1, β_2, $\alpha\beta_{11}$, $\alpha\beta_{21}$, $\alpha\beta_{12}$, and $\alpha\beta_{22}$); the elements of **e**, the error vector, are the error terms corresponding to the Y_{ij}; and A, the design matrix, contains 0s and 1s in a pattern to pick up appropriate parameters, exclude inappropriate ones, and to include μ in all equations (note that the first column of A contains only 1s).

NORMAL EQUATIONS FROM MULTIPLE REGRESSION IN MATRIX FORM

The solution for "beta-weights" in multiple correlational analysis involves solving a set of normal equations, each of which is of the general form

$$\beta_1 r_{i1} + \beta_2 r_{i2} + \beta_3 r_{i3} + \cdots + \beta_p r_{ip} = r_{iY}$$

where Y = criterion variable
r_{ij} = intercorrelation of ith and jth predictor variables
r_{iY} = intercorrelation (validity coefficient) of ith predictor with criterion
β_j = unknown beta-weights

For p predictor variables, there is a total of p normal equations. By utilizing matrix notation, this system of equations can be compactly represented. The complete correlation matrix for the predictor variables is square and of order $p \times p$; a $p \times 1$ vector can be utilized to represent the beta-weights; and a $p \times 1$ vector will also represent the validity coefficients. These three matrices are denoted, respectively, R, **b**, and **v**. In their complete form, they are (for $p = 4$)

$$R = \begin{bmatrix} 1 & r_{12} & r_{13} & r_{14} \\ r_{21} & 1 & r_{23} & r_{24} \\ r_{31} & r_{32} & 1 & r_{34} \\ r_{41} & r_{42} & r_{43} & 1 \end{bmatrix} \qquad \mathbf{b} = \begin{bmatrix} \beta_1 \\ \beta_2 \\ \beta_3 \\ \beta_4 \end{bmatrix} \qquad \mathbf{v} = \begin{bmatrix} r_{1Y} \\ r_{2Y} \\ r_{3Y} \\ r_{4Y} \end{bmatrix}$$

Postmultiplying R by **b** and setting the result equal to **v** yields the entire set of normal equations; that is, $R\mathbf{b} = \mathbf{v}$. To illustrate some algebraic possibilities with matrix equations, we shall solve this equation for the unknown beta-weights. Since $R\mathbf{b} = \mathbf{v}$, we can premultiply both sides of the equation by R^{-1} (note that the conformability requirements are met in this multiplication); then $R^{-1}R\mathbf{b} = R^{-1}\mathbf{v}$. On the left, $R^{-1}R$ appears, and, by definition, $R^{-1}R = I$. Thus, the equation becomes $I\mathbf{b} = R^{-1}\mathbf{v}$. But $I\mathbf{b} = \mathbf{b}$, and the solution is $\mathbf{b} = R^{-1}\mathbf{v}$. In words, the beta-weights are found by determining the inverse of the correlation matrix and then postmultiplying this inverse by the vector of validity coefficients.

CALCULATION OF VARIANCE-COVARIANCE MATRIX

A variance-covariance matrix is a square matrix with variances for p variables on the main diagonal and covariances among the p variables as off-diagonal elements. For any set of p variables, the calculations required to generate the elements of a variance-covariance matrix can be represented in terms of elementary matrix operations. For this development, deviation formulas for variances and covariances will be used; that is,

$$S_j^2 = \sum_{i=1}^{n} \frac{(Y_{ij} - \bar{Y}_{.j})^2}{n-1} \qquad \text{and} \qquad S_{jk} = \sum_{i=1}^{n} \frac{(Y_{ij} - \bar{Y}_{.j})(Y_{ik} - \bar{Y}_{.k})}{n-1}$$

The original set of scores for n individuals on p variables can be represented by an $n \times p$ matrix of the general form

$$Y = \begin{bmatrix} Y_{11} & Y_{12} & \cdots & Y_{1p} \\ Y_{21} & Y_{22} & \cdots & Y_{2p} \\ \cdots & \cdots & \cdots & \cdots \\ Y_{n1} & Y_{n2} & \cdots & Y_{np} \end{bmatrix}$$

In order to generate deviations from means, an $n \times p$ matrix M of means is formed; each column of M is composed of identical mean values; that is, $\bar{Y}_{.1}$ occurs n times and constitutes the first column of M, and so forth. The difference $Y - M$ is

$$\begin{bmatrix} Y_{11} & Y_{12} & \cdots & Y_{1p} \\ Y_{21} & Y_{22} & \cdots & Y_{2p} \\ \cdots & \cdots & \cdots & \cdots \\ Y_{n1} & Y_{n2} & \cdots & Y_{np} \end{bmatrix} - \begin{bmatrix} \bar{Y}_{.1} & \bar{Y}_{.2} & \cdots & \bar{Y}_{.p} \\ \bar{Y}_{.1} & \bar{Y}_{.2} & \cdots & \bar{Y}_{.p} \\ \cdots & \cdots & \cdots & \cdots \\ \bar{Y}_{.1} & \bar{Y}_{.2} & \cdots & \bar{Y}_{.p} \end{bmatrix}$$

$$= \begin{bmatrix} Y_{11} - \bar{Y}_{.1} & Y_{12} - \bar{Y}_{.2} & \cdots & Y_{1p} - \bar{Y}_{.p} \\ Y_{21} - \bar{Y}_{.1} & Y_{22} - \bar{Y}_{.2} & \cdots & Y_{2p} - \bar{Y}_{.p} \\ \cdots & \cdots & \cdots & \cdots \\ Y_{n1} - \bar{Y}_{.1} & Y_{n2} - \bar{Y}_{.2} & \cdots & Y_{np} - \bar{Y}_{.p} \end{bmatrix} = Y - M$$

The numerators of the variances and covariances (that is, the sums of squares and sums of cross products) can be generated by a simple multiplication: $(Y - M)'(Y - M)$. Note that the first term is the transpose of $Y - M$. Thus, the first *row* of $(Y - M)'$ contains the elements $Y_{11} - \bar{Y}_{.1}, Y_{21} - \bar{Y}_{.1}, \ldots, Y_{n1} - \bar{Y}_{.1}$; and the first *column* of $Y - M$ contains the same elements. Since $(Y - M)'(Y - M)$ involves the sum of products of the elements in the first row of $(Y - M)'$ and the first column of $Y - M$, the result is $\Sigma_{i=1}^{n} (Y_{i1} - \bar{Y}_{.1})^2$. In like manner, the remaining sums of squares and sums of cross products are generated by the row-by-column multiplications and sums. The resulting matrix has the order $p \times p$ and is

$$\begin{bmatrix} \sum_{i=1}^{n} (Y_{i1} - \overline{Y}_{.1})^2 & \cdots & \sum_{i=1}^{n} (Y_{i1} - \overline{Y}_{.1})(Y_{ip} - \overline{Y}_{.p}) \\ \sum_{i=1}^{n} (Y_{i2} - \overline{Y}_{.2})(Y_{i1} - \overline{Y}_{.1}) & \cdots & \sum_{i=1}^{n} (Y_{i2} - \overline{Y}_{.2})(Y_{ip} - \overline{Y}_{.p}) \\ \cdots & \cdots & \cdots \\ \sum_{i=1}^{n} (Y_{ip} - \overline{Y}_{.p})(Y_{i1} - \overline{Y}_{.1}) & \cdots & \sum_{i=1}^{n} (Y_{ip} - \overline{Y}_{.p})^2 \end{bmatrix}$$

The only remaining step is to divide each element of $(Y - M)'(Y - M)$ by $n - 1$. This can be accomplished by premultiplying $(Y - M)'(Y - M)$ by a diagonal matrix in which each diagonal element is $1/(n - 1)$. This is, note, equivalent to the product $[1/(n - 1)]I$. Thus, if the variance-covariance matrix is denoted by S, the operations leading to its computation can be expressed by the equation

$$S = \frac{1}{n - 1} I(Y - M)'(Y - M)$$

latent roots and latent vectors of a square matrix

For any square matrix, an equation involving a determinant of the form $|A - \lambda I| = 0$ can be set up. If A is nonsingular and is a $p \times p$ matrix, then there will be p solutions in the sense that p values of λ can be calculated which satisfy the determinantal equation. For reference, we shall denote these solutions as $\lambda_1, \lambda_2, \ldots, \lambda_p$. Each λ is known as a *latent root* (or *eigenvalue*) of the matrix A. The product of the p latent roots is equal to the value of the determinant of A; that is, $\Pi_{i=1}^{p} \lambda_i = |A|$. Also, $\Sigma_{i=1}^{p} \lambda_i = \text{tr}(A)$, where $\text{tr}(A)$ is known as the *trace* of the matrix A and is defined as the sum of the diagonal elements of A [that is, $\text{tr}(A) = \Sigma_{i=1}^{p} a_{ii}$].

The solution of the equation $|A - \lambda I| = 0$ involves evaluation of the determinant; this will turn out to be a polynomial of the pth order. That is, it will be an expression having powers of λ from λ, λ^2, through λ^p. Once the polynomial is obtained, any standard procedure can be utilized to find its solutions, and these will be the λ's or latent roots. This process can be easily illustrated for a 2×2 matrix:

$$A = \begin{bmatrix} 3 & 5 \\ 4 & 2 \end{bmatrix}$$

The determinantal equation is

$$|A - \lambda I| = \left| \begin{bmatrix} 3 & 5 \\ 4 & 2 \end{bmatrix} - \lambda \begin{bmatrix} 1 & 0 \\ 0 & 1 \end{bmatrix} \right| = \left| \begin{bmatrix} 3 & 5 \\ 4 & 2 \end{bmatrix} - \begin{bmatrix} \lambda & 0 \\ 0 & \lambda \end{bmatrix} \right|$$

$$= \begin{vmatrix} 3 - \lambda & 5 \\ 4 & 2 - \lambda \end{vmatrix} = (3 - \lambda)(2 - \lambda) - 5 \cdot 4 = \lambda^2 - 5\lambda - 14 = 0$$

The solution of this quadratic is $(\lambda + 2)(\lambda - 7)$; thus, $\lambda_1 = -2$ and $\lambda_2 = 7$. Note that the product of these solutions for the latent roots is $(-2)(7) = -14$, and the value of the determinant of A is $3 \cdot 2 - 4 . 5 = 6 - 20 = -14$; also, the trace of A is $3 + 2 = 5$, and $\lambda_1 + \lambda_2 = 7 - 2 = 5$.

Corresponding to each latent root of a matrix is a vector (denoted **x**) known as the latent vector (or eigenvector) which is a solution to the matrix equation $(A - \lambda I)\mathbf{x} = 0$. In general, there is no unique solution for the elements of **x**, but it is possible to solve for the ratios among the elements of **x**, and then an arbitrary choice of solutions may be made. Consider, again, the 2×2 matrix with latent roots $\lambda_1 = -2$ and $\lambda_2 = 7$. The matrix equation involving the latent vector for the first latent root is $(A - (-2)I)\mathbf{x} = 0$. In detail, this is

$$\left\{\begin{bmatrix} 3 & 5 \\ 4 & 2 \end{bmatrix} - (-2)\begin{bmatrix} 1 & 0 \\ 0 & 1 \end{bmatrix}\right\}\begin{bmatrix} x_1 \\ x_2 \end{bmatrix} = \begin{bmatrix} 3 - (-2) & 5 \\ 4 & 2 - (-2) \end{bmatrix}\begin{bmatrix} x_1 \\ x_2 \end{bmatrix}$$

$$= \begin{bmatrix} 5 & 5 \\ 4 & 4 \end{bmatrix}\begin{bmatrix} x_1 \\ x_2 \end{bmatrix} = \begin{bmatrix} 5x_1 + 5x_2 \\ 4x_1 + 4x_2 \end{bmatrix} = 0$$

Thus, $5x_1 + 5x_2 = 0$ and $4x_1 + 4x_2 = 0$; hence, $x_1 = -x_2$. Note that it is impossible to solve these equations explicitly for unique values of x_1 and x_2. One possible solution (out of infinitely many) is $x_1 = 1$ and $x_2 = -1$. In general, some other condition must be imposed on the x's before a unique solution is possible. The latent vector corresponding to the second latent root $\lambda_2 = 7$ can be similarly determined. The final equation is

$$\begin{bmatrix} -4 & 5 \\ 4 & -5 \end{bmatrix}\begin{bmatrix} x_1 \\ x_2 \end{bmatrix} = 0$$

Thus, $-4x_1 + 5x_2 = 0$, $4x_1 - 5x_2 = 0$, and $x_2 = (4/5)x_1 = .8x_1$. One possible solution is $x_1 = 5/9 = .5556$ and $x_2 = 4/9 = .4444$; this solution can be found by adding the condition that $x_1 + x_2 = 1$.

appendix b

tables

use of appendix tables

TABLE B-1: RANDOM DIGITS

This table contains 10,000 digits which were generated by a pseudo-random generator on an IBM 7094 electronic computer. In randomly selecting subjects from an available population and randomly dividing subjects among two or more treatment groups, the table can be used to yield an essentially random sequence of numbers. The user should enter the table in an arbitrary fashion; since the table has 200 rows and 50 columns, one can arbitrarily choose any three-digit number (no greater than 200) and any two-digit number (no greater than 50) to define a starting point for reading random numbers. Numbers of any size (i.e., one-digit, two-digit, etc.) can be formed by combining adjacent columns of the table. Also, the user may read up, down, or sideways, whichever is his pleasure. For example, assume that we wished to draw a sample of size 10 from a population of 220 available subjects. The subjects could be given numerical labels from 000 through 219; thus, in reading from the table, all numbers larger than 219 would be ignored. Arbitrarily, say that we enter the table at row 76 and column 30 (these numbers were taken from the serial number of a dollar bill which the author happened to have in his pocket). The digit located at the intersection of row 76 and column 30 is a "9"; since we need three-digit numbers, we combine this with the next two columns and arrive at the number 973. Since this is larger than 219, we discard it and continue down the table. Note that it is not until row 88 that we encounter a number in the acceptable range; that is, 122 is between 000 and 219. If we continue the process, the first 10 three-digit numbers between 000 and 219 are 122, 92, 100, 102, 68, 131, 104, 184, 56, and 145 (note that 104 occurred twice but is recorded only once since we are sampling without replacement).

TABLE B-2: AREAS AND ORDINATES OF THE UNIT NORMAL DISTRIBUTION

This table gives information concerning the normal distribution with mean of 0 and standard deviation of 1. The figures above columns $A(Z)$ through $F(Z)$ indicate the nature of the information supplied. Thus, $A(Z)$ is the area 0 to z (note that because of symmetry, negative values of z can be entered in absolute value); $B(Z)$ is the tail area more extreme than z; $C(Z)$ is the area in the less extreme direction than z (that is, for positive z, it is the percentile rank of z; for negative z, it is the complement of the percentile rank); $D(Z)$ is the area from $+z$ to $-z$; $E(Z)$ is the sum of the areas more extreme than $+z$ and less extreme than $-z$; and $F(Z)$ is the height of the ordinate at z.

TABLE B-3: PERCENTILES OF t DISTRIBUTIONS

The rows are labeled in terms of degrees of freedom (df), and the columns represent selected percentage points of the corresponding t distributions. For "two-tailed" tests (or nondirectional tests), symmetrically placed percentiles are utilized; for example, with 10 degrees of freedom, the critical values for a "two-tailed" .05-level significance test would be -2.228 and 2.228 since these values represent the 2.5 and 97.5 percentiles of the appropriate t distribution. Note that the final row, for infinite degrees of freedom, corresponds to percentage points of the unit normal distribution.

TABLE B-4: PERCENTILES OF CHI-SQUARE DISTRIBUTIONS

The arrangement of this table is the same as that of Table B-3 for the t distributions.

TABLE B-5: PERCENTILES OF F DISTRIBUTIONS

Both right and left tail values are shown for a variety of combinations of degrees of freedom. Each different numerator degrees of freedom defines a separate panel of the table within which all the selected denominator degrees of freedom are shown. Although the relationship $F_{m,n,1-\alpha} = 1/F_{n,m,\alpha}$ is true, the present form of the tables is a convenience in making two-tailed tests when they are appropriate.

TABLE B-6: PERCENTAGE POINTS OF THE RATIO S_{max}^2/S_{min}^2

The entries in this table are critical values for the Hartley F_{max} test for homogeneity of variance. The test is always one-tailed, and the tabled values are for .05 and .01 levels of significance.

TABLE B-7: PERCENTAGE POINTS OF THE STUDENTIZED RANGE STATISTIC

Critical values at the .10, .05, and .01 levels are presented; the columns represent the number of rank positions spanned by a pair of means, and the rows represent the degrees of freedom per mean. In a typical application, if four treatment groups are involved and each contains 10 subjects, the row for 9 degrees of freedom will be utilized; the appropriate studentized ranges will be those spanning two, three, and four rank positions, which are 3.20, 3.95, and 4.41 at $\alpha = .05$.

TABLE B-8: PERCENTAGE POINTS FOR DUNCAN'S MULTIPLE RANGE TEST

This table is similar to Table B-7, except that the values shown are specifically for the Duncan test. As above, the columns represent the number of rank positions spanned by a pair of means, and the rows are degrees of freedom per mean.

TABLE B-9: PERCENTAGE POINTS OF DUNNETT'S t STATISTIC

These tables, as prepared by Dunnett, present one-tailed and two-tailed critical values at the .05 and .01 levels of significance. Note that the column heading is in terms of the number of treatment groups, *excluding* the control group, and that the degrees of freedom (i.e., rows of the table) are for the error mean square. Dunnett (1964) has extended the use of his tables to cases in which the control group contains a larger number of observations than the remaining treatment groups. The superscripts on the entries for the two-tailed critical values are utilized when n_c, the number of observations in the control group, is larger than n_t, the number of observations in each treatment group. The value of the superscript multiplied by the quantity $1 - n_t/n_c$ gives the percentage increase of the corresponding critical value which is necessary in the unequal control group size situation. If we let the superscript be s_0 and the critical value t_0, the adjustment is

$$t_0' = \left[1 + \frac{s_0}{100}\left(1 - \frac{n_t}{n_c}\right)\right]t_0$$

where t_0' is the appropriate critical value after adjustment for differences in group sizes. For example, if there are three treatment groups besides the control group and each treatment group contains five observations while the control group contains 19 observations, the error degrees of freedom are $(3)(4) + (18) = 30$. The tabular critical value with $\alpha = .05$ is 2.47, and the superscript on this value is 2.3. Thus, the adjusted critical value is

$$t_0' = \left[1 + \frac{2.3}{100}\left(1 - \frac{5}{19}\right)\right][2.47] = [1 + (.023)(.7368)][2.47]$$
$$= (1.0169)(2.47) = 2.51$$

TABLE B-10: CRITICAL VALUES OF L

This table presents .05, .01, and .001 critical values for Page's L test. The table heading presents an additional explanation.

TABLE B-11: ORTHOGONAL-POLYNOMIAL COEFFICIENTS

For four or more treatment levels, only coefficients for linear, quadratic, and cubic trends are shown since higher-order trends are generally pooled and tested together (and the corresponding sum of squares found by subtraction). For convenience, the sum of squared coefficients is presented for each case.

TABLE B-12: INVERSE SINE TRANSFORMATIONS

For proportions from .01 through .99, the table presents values of $2 \sin^{-1} \sqrt{X}$, which is a popular variance stabilizing transformation, as explained in the text. The entries are reported in radians rather than degrees; since 1 radian is equivalent to 57.2958 degrees, conversion to degrees is easily accomplished.

table B-1 *Random digits*

row number										
1	50691	91653	88574	08675	12700	32027	41034	56912	34264	77769
2	19787	66937	91769	13399	96096	43165	72096	86350	23062	99419
3	16746	77983	18061	23664	64557	78213	43857	68009	20483	00618
4	91039	16099	38824	00778	23058	76539	50584	71810	52589	32778
5	11075	62081	88977	78676	53855	56472	13090	01708	89016	45111
6	41230	92934	30342	29933	24597	72632	21727	63861	80454	47243
7	59028	24399	05075	64775	59803	45737	19025	46696	18914	03062
8	42957	25204	00753	60284	85482	34984	86637	95354	80698	87650
9	45881	59475	64445	97261	55252	50788	31295	16437	49497	22493
10	75104	45819	88471	75440	55309	63481	23616	64950	73291	10964
11	78614	07347	63528	84643	19455	95596	38158	75758	65628	10498
12	69279	59274	67459	53563	98241	18097	65297	49803	99145	25320
13	58626	91259	13832	75095	08333	53845	74223	82690	89320	89565
14	81630	00339	07996	65249	66792	05555	79169	12136	44621	95904
15	74330	13688	02044	65910	96007	82692	40473	56437	35671	95073
16	70829	66963	86390	26458	02385	41505	06239	68990	32915	89542
17	55084	58581	60759	20627	86682	76542	03648	28183	29823	68134
18	98845	17428	97397	62400	51284	92211	40593	82713	06067	46190
19	48116	91870	16346	97406	54649	42039	58407	84248	45780	60547
20	82778	31709	71564	26258	07522	03825	92087	21809	25678	39987
21	86615	67618	07446	63129	07111	70516	67289	09457	48995	08043
22	82558	99260	69136	35099	68187	85382	09569	94211	57824	98100
23	08290	70291	74090	96503	56140	27794	27765	51740	07712	29816
24	95062	76310	81603	86828	68370	46001	79205	35511	91239	52961
25	30361	66712	86801	29556	91232	98295	87322	99172	50009	27224
26	17390	96107	70391	78715	61943	33315	39778	97149	08122	86388
27	05390	33046	63920	28733	42644	38972	98161	79861	88282	28279
28	06624	21114	33869	20940	03732	39973	89948	81060	36381	06027
29	38146	77295	33742	00135	26587	54775	94846	18587	39327	71711
30	76430	23645	62335	60393	71813	52677	09917	89100	93855	75617
31	16664	30164	22546	63538	79376	26865	61996	60418	37777	84170
32	56424	64680	81038	79364	23815	44002	38480	09864	35960	10760
33	95954	15540	18554	63349	70259	03212	91950	16214	80378	56421
34	59007	56364	49965	61970	32493	55404	85950	99606	46328	17887
35	19341	87208	99853	40202	08553	78731	83463	19624	82512	13556
36	24505	87007	35748	54865	40209	49466	94574	31406	64422	87185
37	15086	92183	84632	36790	59608	00371	67456	55364	80669	75402
38	65664	02188	09164	70939	25856	24344	58859	10454	19212	59078
39	40397	76835	14062	96067	70645	23695	59140	75812	18804	55529
40	31700	24753	22919	43207	83387	27820	12494	30041	88927	22668
41	14472	19372	23759	47116	81647	44946	97716	41157	30913	30842
42	18018	57089	98428	89075	77511	15194	69634	68269	52292	63404
43	16752	54266	76103	05268	41145	36100	73916	32462	01658	68565
44	47184	33660	96555	56656	18238	56888	29315	99813	47831	81385
45	93884	63945	06606	45545	29237	21040	43552	02749	19963	23705

table B-1 (continued)

row number										
46	44112	25139	62540	20200	38793	76373	01711	24722	31244	04592
47	10991	67616	96339	56810	45766	21363	57946	50002	07989	06215
48	34402	12087	57559	65862	61811	36453	08438	84435	75085	45182
49	61419	45244	32623	86912	50091	53116	59088	27149	21605	33372
50	84037	25936	04836	37968	76851	39660	10274	70731	45720	55762
51	09972	74036	23610	90926	40167	51840	28408	33628	08432	68823
52	33572	72847	33140	69914	24682	57531	45518	32677	92151	94317
53	30984	29000	17100	19781	66887	67105	69941	69372	19946	49800
54	89424	61867	19788	84639	53944	83243	82694	96001	64194	70157
55	24763	42978	60331	80383	54125	29033	06934	41093	20723	48368
56	07490	76138	66608	17033	38218	09644	71566	58802	05890	84586
57	82030	72988	67673	11222	92849	37809	96596	61390	01108	16682
58	97742	68757	19023	27385	17544	17864	54500	51633	03194	71313
59	79030	92963	47196	11086	30843	62492	54188	06620	03386	70633
60	38997	82566	49333	95115	13313	19834	21068	02220	85744	89839
61	46568	55793	11188	20315	77746	94728	56500	61830	52174	11940
62	93128	61298	66162	64056	62650	85380	99568	00405	69673	24628
63	35870	34391	38423	12661	27185	63510	16199	78509	90668	90576
64	70701	17920	27932	50483	35058	50046	74424	33870	18591	60743
65	96275	23095	03760	22726	86330	53401	25052	58613	54639	86291
66	72519	76643	42531	10541	37698	30425	57251	11422	82956	10225
67	74151	92320	23956	31195	63386	55061	64108	80458	92690	43296
68	75905	02954	96799	43216	38647	77622	92459	31591	66892	77107
69	43316	24900	89553	85381	17191	08877	97772	34217	67821	85869
70	16420	47818	47240	87248	95579	81827	12880	06675	95361	32638
71	06967	17690	57935	44906	81676	97422	76612	20982	99730	18202
72	48151	29322	95666	91640	06518	62750	25953	99952	78277	53471
73	76842	10952	17446	94373	82479	75033	29970	53456	75480	48236
74	24374	91591	90748	32626	48879	44069	09003	30063	33513	94415
75	41897	84117	12238	73620	30566	02551	20260	22856	41530	82461
76	24171	75567	24492	51938	99770	62409	73230	39369	76231	95334
77	99175	07282	69983	18034	48565	91887	68909	71676	61937	06058
78	16273	90115	31362	10790	39524	76553	96074	22961	67515	97496
79	57555	69188	94966	56747	29428	24038	31136	75291	76045	76800
80	97741	76023	94092	27731	26921	86064	13121	83138	99525	91197
81	94337	22615	30349	42081	34739	24945	42200	06363	93259	77790
82	14038	70566	38431	19588	31832	65743	20009	84270	67370	44576
83	93639	69040	44900	62764	09029	64724	91245	29387	91749	99954
84	63826	83075	38461	63094	73666	63837	24549	89216	11686	21311
85	69178	91277	88645	83872	69880	23453	07189	60745	37212	77893
86	66238	84321	89312	16843	79880	23039	53334	44580	80370	59090
87	65142	19947	41183	95771	03005	33494	42078	65561	85287	95298
88	89932	98248	43809	30641	75089	40531	22015	01368	52197	28177
89	99005	27907	53544	27451	38789	09604	22999	15259	66875	01236
90	01271	98890	09743	33953	84817	94536	36219	99873	52202	61070

table B-1 (continued)

row number										
91	91594	30774	26558	87137	45299	49010	92118	63330	83850	24098
92	49057	07409	24436	89740	36064	55875	14382	28297	76429	26706
93	75619	01003	40763	50803	87045	52475	25088	58912	48159	85432
94	91012	64270	26773	07419	65734	53608	14437	44524	35656	64196
95	25606	81035	19169	85082	52592	80341	00523	12387	21804	73385
96	52550	85022	01735	60130	14493	58805	84679	55287	35106	35271
97	87784	37528	84134	08067	63558	95706	66662	86033	59471	70393
98	01358	58995	32987	74145	23704	11259	22323	75820	62867	00985
99	13587	34452	22682	99577	12180	42534	19294	93614	83140	24863
100	65452	03054	62082	91152	29404	84458	88331	95510	93731	33865
101	29377	92654	16050	67662	79326	71902	24281	61582	42914	98515
102	29631	52576	84125	92393	37906	96505	37932	63447	51923	10435
103	79860	68088	66125	20523	13119	00461	02775	15220	64734	06424
104	85142	62453	22014	36547	57497	92186	39415	95108	72770	93320
105	53205	76546	87850	61066	99040	56572	40891	63455	96259	89186
106	28315	11856	05562	58335	72507	62210	68945	02576	30775	35309
107	86382	99013	05265	30141	98751	37272	31315	68363	97597	14310
108	20663	86017	72482	73942	44089	50251	31965	09872	00559	69700
109	87809	24691	35104	08993	75548	95579	72987	90505	37636	31725
110	29148	36682	27676	65061	94309	25409	67453	40887	64668	98019
111	26466	20219	38160	97023	40495	15941	04154	41248	84184	29969
112	70269	61982	40959	43190	90037	25552	64531	77203	99565	79820
113	78142	22127	76929	41275	51049	04042	23557	96779	66090	80501
114	32062	93330	01442	12382	49541	84701	84220	47763	86011	58754
115	13448	16335	92331	28151	83667	73804	08059	64630	36045	77082
116	04823	78538	91558	31821	77764	72765	52284	16749	25387	10313
117	16629	13553	43373	76924	16251	72501	04106	34691	15925	14447
118	75203	83865	11140	52755	27047	16070	56702	69870	67885	58133
119	75572	04700	49960	44099	24914	93037	48409	34268	05635	71738
120	71810	45872	55319	65963	22225	72986	34204	08331	84245	86956
121	71669	52192	79748	68287	42994	57813	17786	94999	68430	45318
122	73180	00869	64564	38830	75167	76855	07107	43893	88677	67682
123	94803	70892	03417	48408	92036	80869	47654	81124	31104	82485
124	32139	28146	91218	96167	93209	08028	91222	81373	23638	30262
125	48730	57557	43790	23723	88625	59158	44723	12268	49385	27529
126	91325	46822	82401	14504	71478	98772	76948	03481	33216	28454
127	09284	14508	98088	06714	95831	91461	45503	24084	42647	43729
128	07905	01338	34750	41426	63539	49613	91258	79146	17930	38689
129	89616	70054	75299	23685	54844	71640	48093	42916	27914	71947
130	78845	25762	13358	74285	24273	00111	01496	93323	00495	20261
131	69662	35846	00185	65650	57267	52365	60174	72634	98085	77689
132	27278	94450	46967	48816	36012	90068	94652	43683	65562	79692
133	73112	28587	88277	84056	47726	53198	82093	77509	69151	28983
134	05719	48476	35810	27207	84341	70599	34539	14381	10430	24404
135	51528	44241	44746	66132	69207	75724	60641	82077	76349	30308

table B-1 (continued)

row number										
136	84276	45150	44498	00220	60147	20131	75124	50023	23830	84629
137	77983	29368	38102	81516	46693	20586	80891	15408	96524	74525
138	49261	18008	55102	70516	10736	12187	67793	87139	85802	99319
139	63428	12397	66415	19953	68031	52395	19511	63914	82942	67452
140	89693	19532	94859	43709	94408	51477	54724	02567	71917	35816
141	70200	67205	46393	67355	65519	65268	86537	01757	79293	49148
142	97248	66980	60512	80216	43360	58252	65999	09810	44903	28592
143	41725	73141	32052	01991	30819	15304	91454	93285	54536	76825
144	97908	11006	07744	47451	15372	73359	64181	99072	25430	00383
145	24084	67802	82706	95797	32224	24759	96230	31699	64040	17136
146	49360	06562	36074	78792	23019	47047	24040	80223	33983	97628
147	10612	20943	90144	18510	95095	31411	64243	70708	46531	54251
148	78725	77546	42925	15538	77131	99647	61885	44499	47506	15546
149	04540	01150	23326	32149	41303	44862	03153	95341	95418	93674
150	10663	89501	26595	26796	42837	96609	93230	93394	08651	80618
151	65272	38499	96793	80589	51174	10156	58268	91636	23190	11101
152	78459	68233	13469	36132	80947	27031	30335	80142	24574	02155
153	72164	71687	45407	69032	57797	93757	84257	21725	11227	40230
154	26512	06875	26728	33340	21533	97842	80508	16015	33723	03570
155	81480	35368	29198	48232	01674	27272	39659	85224	99372	87308
156	38464	57010	83093	77480	28151	71807	76821	93602	78613	62428
157	97234	36368	33400	61207	82780	39089	35637	74898	69285	50315
158	17480	38257	49450	05855	44004	13895	27322	21617	75595	38624
159	09048	45024	07202	07299	69244	38103	53360	18922	87391	45133
160	66386	97952	50029	16896	35331	67880	38581	37969	78294	92987
161	26272	75004	97156	00792	61530	93256	82387	69564	91483	81478
162	16102	99149	37954	34992	71208	31322	38885	99965	47961	92204
163	44839	19704	18657	63676	21959	16774	20951	38441	38846	25975
164	83695	16552	80539	18501	80695	30893	32435	19135	16418	36700
165	30932	35803	47426	71684	64534	65679	92533	85572	28784	05004
166	98707	47640	32647	97391	57836	58144	46702	71421	14014	87065
167	85075	96591	32030	95074	82342	13745	92743	33545	25568	43285
168	78586	02855	74039	16002	65090	96947	77381	18847	23698	48409
169	60018	81550	01515	75036	44559	97496	64102	10696	27637	37976
170	03436	35460	54872	28490	65799	16885	59569	17045	00135	29657
171	15340	76172	28893	23280	53730	70023	97862	78133	48827	72191
172	24672	68346	82835	83623	73040	09399	35967	82675	81857	27704
173	01510	48348	04883	87642	19580	13368	34163	28445	01158	54442
174	41606	14133	86495	97688	09838	87236	95915	42275	39665	10140
175	43785	40205	37965	11663	13004	56450	21037	31525	98405	60700
176	79270	46610	08590	41965	04179	74597	54912	51876	91441	08018
177	86177	79172	91012	28880	00425	07299	82583	33466	19740	79508
178	78045	43693	64458	00833	04869	58703	63701	16071	32311	96311
179	08574	22387	89753	36974	67655	50868	19960	90933	42360	31601
180	08925	05223	05714	29346	76288	65896	64754	97688	08436	39111

table B-1 (continued)

row number										
181	65273	94220	31038	72771	76560	54578	06444	64137	80984	27816
182	21738	04873	49608	59079	64484	74840	06952	41185	36292	87856
183	79467	45326	06895	15454	95817	38193	73860	58886	41894	46629
184	20799	61762	51383	54561	03379	18288	53431	71906	84219	08197
185	94723	25692	95932	60928	75510	51550	65656	18977	46469	38035
186	81799	48162	08627	73735	66531	77940	97552	41652	19682	10858
187	14459	55696	08731	89641	16137	84450	93624	95042	89877	52899
188	81355	37244	13083	45940	42841	96343	99502	98515	44767	26518
189	35144	79758	41117	86768	22571	14382	20253	59145	28899	05151
190	70500	54808	48070	05573	75753	59300	71793	38236	05807	16149
191	39198	07833	68262	10739	86024	23698	77127	15244	66298	70626
192	51908	75125	82165	96461	22291	44661	17703	60871	40221	38509
193	29705	53710	73159	81532	27040	95855	11956	40020	62018	43917
194	37663	02396	05029	96591	77922	25125	34612	14166	97608	40338
195	71299	98550	59540	24059	26389	57105	31966	15566	47646	91876
196	67846	72595	98348	23568	15475	36956	35093	16196	69700	29433
197	11271	54834	95839	68780	43862	20557	78933	38753	74926	53301
198	09128	92500	06940	32898	90703	40651	86875	76099	65979	21452
199	94712	16151	53308	39463	94686	68734	01192	10969	56632	58379
200	79304	54680	55149	05124	86379	25936	46239	89785	24477	88084

table B-2 *Areas and ordinates of the unit normal distribution*

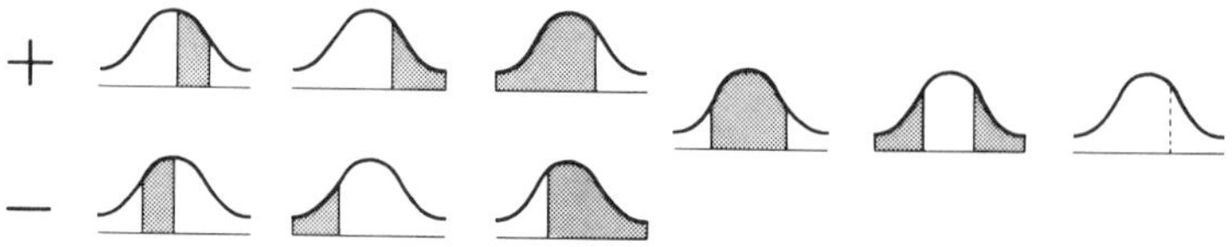

Z	A(Z)	B(Z)	C(Z)	D(Z)	E(Z)	F(Z)
0.00	.0000	.5000	.5000	.0000	1.0000	.3989
0.01	.0040	.4960	.5040	.0080	.9920	.3989
0.02	.0080	.4920	.5080	.0160	.9840	.3989
0.03	.0120	.4880	.5120	.0240	.9760	.3988
0.04	.0160	.4840	.5160	.0320	.9680	.3986
0.05	.0199	.4801	.5199	.0398	.9602	.3984
0.06	.0239	.4761	.5239	.0478	.9522	.3982
0.07	.0279	.4721	.5279	.0558	.9442	.3980
0.08	.0319	.4681	.5319	.0638	.9362	.3977
0.09	.0359	.4641	.5359	.0718	.9282	.3973
0.10	.0398	.4602	.5398	.0796	.9204	.3970
0.11	.0438	.4562	.5438	.0876	.9124	.3965
0.12	.0478	.4522	.5478	.0956	.9044	.3961
0.13	.0517	.4483	.5517	.1034	.8966	.3956
0.14	.0557	.4443	.5557	.1114	.8886	.3951
0.15	.0596	.4404	.5596	.1192	.8808	.3945
0.16	.0636	.4364	.5636	.1272	.8728	.3939
0.17	.0675	.4325	.5675	.1350	.8650	.3932
0.18	.0714	.4286	.5714	.1428	.8572	.3925
0.19	.0753	.4247	.5753	.1506	.8494	.3918
0.20	.0793	.4207	.5793	.1586	.8414	.3910
0.21	.0832	.4168	.5832	.1664	.8336	.3902
0.22	.0871	.4129	.5871	.1742	.8258	.3894
0.23	.0910	.4090	.5910	.1820	.8180	.3885
0.24	.0948	.4052	.5948	.1896	.8104	.3876
0.25	.0987	.4013	.5987	.1974	.8026	.3867
0.26	.1026	.3974	.6026	.2052	.7948	.3857
0.27	.1064	.3936	.6064	.2128	.7872	.3847
0.28	.1103	.3897	.6103	.2206	.7794	.3836
0.29	.1141	.3859	.6141	.2282	.7718	.3825
0.30	.1179	.3821	.6179	.2358	.7642	.3814
0.31	.1217	.3783	.6217	.2434	.7566	.3802
0.32	.1255	.3745	.6255	.2510	.7490	.3790
0.33	.1293	.3707	.6293	.2586	.7414	.3778
0.34	.1331	.3669	.6331	.2662	.7338	.3765
0.35	.1368	.3632	.6368	.2736	.7264	.3752
0.36	.1406	.3594	.6406	.2812	.7188	.3739
0.37	.1443	.3557	.6443	.2886	.7114	.3725
0.38	.1480	.3520	.6480	.2960	.7040	.3712
0.39	.1517	.3483	.6517	.3034	.6966	.3697

table B-2 (continued)

Z	$A(Z)$	$B(Z)$	$C(Z)$	$D(Z)$	$E(Z)$	$F(Z)$
0.40	.1554	.3446	.6554	.3108	.6892	.3683
0.41	.1591	.3409	.6591	.3182	.6818	.3668
0.42	.1628	.3372	.6628	.3256	.6744	.3653
0.43	.1664	.3336	.6664	.3328	.6672	.3637
0.44	.1700	.3300	.6700	.3400	.6600	.3621
0.45	.1736	.3264	.6736	.3472	.6528	.3605
0.46	.1772	.3228	.6772	.3544	.6456	.3589
0.47	.1808	.3192	.6808	.3616	.6384	.3572
0.48	.1844	.3156	.6844	.3688	.6312	.3555
0.49	.1879	.3121	.6879	.3758	.6242	.3538
0.50	.1915	.3085	.6915	.3830	.6170	.3521
0.51	.1950	.3050	.6950	.3900	.6100	.3503
0.52	.1985	.3015	.6985	.3970	.6030	.3485
0.53	.2019	.2981	.7019	.4038	.5962	.3467
0.54	.2054	.2946	.7054	.4108	.5892	.3448
0.55	.2088	.2912	.7088	.4176	.5824	.3429
0.56	.2123	.2877	.7123	.4246	.5754	.3410
0.57	.2157	.2843	.7157	.4314	.5686	.3391
0.58	.2190	.2810	.7190	.4380	.5620	.3372
0.59	.2224	.2776	.7224	.4448	.5552	.3352
0.60	.2257	.2743	.7257	.4514	.5486	.3332
0.61	.2291	.2709	.7291	.4582	.5418	.3312
0.62	.2324	.2676	.7324	.4648	5352	.3292
0.63	.2357	.2643	.7357	.4714	.5286	.3271
0.64	.2389	.2611	.7389	.4778	.5222	.3251
0.65	.2422	.2578	.7422	.4844	.5156	.3230
0.66	.2454	.2546	.7454	.4908	.5092	.3209
0.67	.2486	.2514	.7486	.4972	.5028	.3187
0.68	.2517	.2483	.7517	.5034	.4966	.3166
0.69	.2549	.2451	.7549	.5098	.4902	.3144
0.70	.2580	.2420	.7580	.5160	.4840	.3123
0.71	.2611	.2389	.7611	.5222	.4778	.3101
0.72	.2642	.2358	.7642	.5284	.4716	.3079
0.73	.2673	.2327	.7673	.5346	.4654	.3056
0.74	.2704	.2296	.7704	.5408	.4592	.3034
0.75	.2734	.2266	.7734	.5468	.4532	.3011
0.76	.2764	.2236	.7764	.5528	.4472	.2989
0.77	.2794	.2206	.7794	.5588	.4412	.2966
0.78	.2823	.2177	.7823	.5646	.4354	.2943
0.79	.2852	.2148	.7852	.5704	.4296	.2920
0.80	.2881	.2119	.7881	.5762	.4238	.2897
0.81	.2910	.2090	.7910	.5820	.4180	.2874
0.82	.2939	.2061	.7939	.5878	.4122	.2850
0.83	.2967	.2033	.7967	.5934	.4066	.2827
0.84	.2995	.2005	.7995	.5990	.4010	.2803

table B-2 (continued)

Z	$A(Z)$	$B(Z)$	$C(Z)$	$D(Z)$	$E(Z)$	$F(Z)$
0.85	.3023	.1977	.8023	.6046	.3954	.2780
0.86	.3051	.1949	.8051	.6102	.3898	.2756
0.87	.3078	.1922	.8078	.6156	.3844	.2732
0.88	.3106	.1894	.8106	.6212	.3788	.2709
0.89	.3133	.1867	.8133	.6266	.3734	.2685
0.90	.3159	.1841	.8159	.6318	.3682	.2661
0.91	.3186	.1814	.8186	.6372	.3628	.2637
0.92	.3212	.1788	.8212	.6424	.3576	2613
0.93	.3238	.1762	.8238	.6476	.3524	.2589
0.94	.3264	.1736	.8264	.6528	.3472	.2565
0.95	.3289	.1711	.8289	.6578	.3422	.2541
0.96	.3315	.1685	.8315	.6630	.3370	.2516
0.97	.3340	.1660	.8340	.6680	.3320	.2492
0.98	.3365	.1635	.8365	.6730	.3270	.2468
0.99	.3389	.1611	.8389	.6778	.3222	.2444
1.00	.3413	.1587	.8413	.6826	.3174	.2420
1.01	.3438	.1562	.8438	.6876	.3124	.2396
1.02	.3461	.1539	.8461	.6922	.3078	.2371
1.03	.3485	.1515	8485	.6970	.3030	.2347
1.04	.3508	.1492	.8508	.7016	.2984	.2323
1.05	.3531	.1469	.8531	.7062	.2938	.2299
1.06	3554	.1446	.8554	.7108	.2892	.2275
1.07	.3577	.1423	.8577	.7154	.2846	.2251
1.08	.3599	.1401	.8599	.7198	.2802	.2227
1.09	.3621	.1379	.8621	.7242	.2758	.2203
1.10	.3643	.1357	.8643	.7286	.2714	.2179
1.11	.3665	.1335	.8665	.7330	.2670	.2155
1.12	.3686	.1314	.8686	.7372	.2628	.2131
1.13	.3708	.1292	.8708	.7416	.2584	.2107
1.14	.3729	.1271	.8729	.7458	.2542	.2083
1.15	.3749	.1251	.8749	.7498	.2502	.2059
1.16	.3770	.1230	.8770	.7540	.2460	.2036
1.17	.3790	.1210	.8790	.7580	.2420	.2012
1.18	.3810	.1190	.8810	.7620	.2380	.1989
1.19	.3830	.1170	.8830	.7660	.2340	.1965
1.20	.3849	.1151	.8849	.7698	.2302	.1942
1.21	.3869	.1131	.8869	.7738	.2262	.1919
1.22	.3888	.1112	.8888	.7776	.2224	.1895
1.23	.3907	.1093	.8907	.7814	.2186	.1872
1.24	.3925	.1075	.8925	.7850	.2150	.1849
1.25	.3944	.1056	.8944	.7888	.2112	.1826
1.26	.3962	.1038	.8962	.7924	.2076	.1804
1.27	.3980	.1020	.8980	.7960	.2040	.1781
1.28	.3997	.1003	.8997	.7994	.2006	.1758
1.29	.4015	.0985	.9015	.8030	.1970	1736

table B-2 (continued)

Z	$A(Z)$	$B(Z)$	$C(Z)$	$D(Z)$	$E(Z)$	$F(Z)$
1.30	.4032	.0968	.9032	.8064	.1936	.1714
1.31	.4049	.0951	.9049	.8098	.1902	.1691
1.32	.4066	.0934	.9066	.8132	.1868	.1669
1.33	.4082	.0918	9082	.8164	.1836	.1647
1.34	.4099	.0901	.9099	.8198	.1802	.1626
1.35	.4115	.0885	.9115	.8230	.1770	.1604
1.36	.4131	.0869	.9131	.8262	.1738	.1582
1.37	.4147	.0853	.9147	.8294	.1706	.1561
1.38	.4162	.0838	.9162	.8324	.1676	.1539
1.39	.4177	.0823	.9177	.8354	.1646	.1518
1.40	4192	.0808	.9192	.8384	.1616	.1497
1.41	.4207	.0793	.9207	.8414	.1586	.1476
1.42	.4222	.0778	.9222	.8444	.1556	.1456
1.43	.4236	.0764	.9236	.8472	.1528	.1435
1.44	.4251	.0749	.9251	.8502	.1498	.1415
1.45	.4265	.0735	.9265	.8530	.1470	.1394
1.46	.4279	.0721	.9279	.8558	.1442	.1374
1.47	.4292	.0708	.9292	.8584	.1416	.1354
1.48	.4306	.0694	.9306	.8612	.1388	.1334
1.49	.4319	.0681	.9319	.8638	.1362	.1315
1.50	.4332	.0668	.9332	.8664	.1336	.1295
1.51	.4345	.0655	.9345	.8690	.1310	.1276
1.52	.4357	.0643	.9357	.8714	.1286	.1257
1.53	.4370	.0630	.9370	.8740	.1260	.1238
1.54	.4382	.0618	.9382	.8764	.1236	.1219
1.55	.4394	.0606	.9394	.8788	.1212	.1200
1.56	.4406	.0594	.9406	.8812	.1188	.1182
1.57	.4418	.0582	.9418	.8836	.1164	.1163
1.58	.4429	.0571	.9429	.8858	.1142	.1145
1.59	.4441	.0559	.9441	.8882	.1118	.1127
1.60	.4452	.0548	.9452	.8904	.1096	.1109
1.61	.4463	.0537	.9463	.8926	.1074	.1092
1.62	.4474	.0526	.9474	.8948	.1052	.1074
1.63	.4484	.0516	.9484	.8968	.1032	.1057
1.64	.4495	.0505	.9495	.8990	.1010	.1040
1.65	.4505	.0495	.9505	.9010	.0990	.1023
1.66	.4515	.0485	.9515	.9030	.0970	.1006
1.67	.4525	.0475	.9525	.9050	.0950	.0989
1.68	.4535	.0465	.9535	.9070	.0930	.0973
1.69	.4545	.0455	.9545	.9090	.0910	.0957
1.70	.4554	.0446	.9554	.9108	.0892	.0940
1.71	.4564	.0436	.9564	.9128	.0872	.0925
1.72	.4573	.0427	.9573	.9146	.0854	.0909
1.73	.4582	.0418	.9582	.9164	.0836	.0893
1.74	.4591	.0409	.9591	.9182	.0818	.0878

table B-2 (continued)

Z	$A(Z)$	$B(Z)$	$C(Z)$	$D(Z)$	$E(Z)$	$F(Z)$
1.75	.4599	.0401	.9599	.9198	.0802	.0863
1.76	.4608	0392	.9608	.9216	.0784	.0848
1.77	.4616	.0384	.9616	.9232	.0768	.0833
1.78	.4625	.0375	.9625	.9250	.0750	.0818
1.79	.4633	.0367	.9633	.9266	.0734	.0804
1.80	.4641	.0359	.9641	.9282	.0718	.0790
1.81	.4649	.0351	.9649	.9298	.0702	.0775
1.82	.4656	.0344	.9656	.9312	.0688	.0761
1.83	.4664	.0336	.9664	.9328	.0672	.0748
1.84	.4671	.0329	.9671	.9342	.0658	.0734
1.85	.4678	.0322	.9678	.9356	.0644	.0721
1.86	.4686	.0314	.9686	.9372	.0628	.0707
1.87	.4693	.0307	.9693	.9386	.0614	.0694
1.88	.4699	.0301	.9699	.9398	.0602	.0681
1.89	.4706	.0294	.9706	.9412	.0588	.0669
1.90	.4713	.0287	.9713	.9426	.0574	.0656
1.91	.4719	.0281	.9719	.9438	.0562	.0644
1.92	.4726	.0274	.9726	.9452	.0548	.0632
1.93	.4732	.0268	.9732	.9464	.0536	.0620
1.94	.4738	.0262	.9738	.9476	.0524	.0608
1.95	.4744	.0256	.9744	.9488	.0512	.0596
1.96	.4750	.0250	.9750	.9500	.0500	.0584
1.97	.4756	.0244	.9756	.9512	.0488	0573
1.98	.4761	.0239	.9761	.9522	0478	.0562
1.99	.4767	.0233	.9767	.9534	.0466	.0551
2.00	.4772	.0228	.9772	.9544	.0456	.0540
2.01	.4778	.0222	.9778	.9556	.0444	.0529
2.02	.4783	.0217	.9783	.9566	.0434	.0519
2.03	.4788	.0212	.9788	.9576	.0424	.0508
2.04	.4793	.0207	.9793	.9586	.0414	.0498
2.05	.4798	.0202	.9798	.9596	.0404	.0488
2.06	.4803	.0197	.9803	.9606	.0394	.0478
2.07	.4808	.0192	.9808	.9616	.0384	.0468
2.08	.4812	.0188	.9812	.9624	.0376	.0459
2.09	.4817	.0183	.9817	.9634	.0366	.0449
2.10	.4821	.0179	.9821	.9642	.0358	.0440
2.11	.4826	.0174	.9826	.9652	.0348	.0431
2.12	.4830	.0170	.9830	.9660	.0340	.0422
2.13	.4834	.0166	.9834	.9668	.0332	.0413
2.14	.4838	.0162	.9838	.9676	.0324	.0404
2.15	.4842	.0158	.9842	.9684	.0316	.0396
2.16	.4846	.0154	.9846	.9692	.0308	.0387
2.17	.4850	.0150	.9850	.9700	.0300	.0379
2.18	.4854	.0146	.9854	.9708	.0292	.0371
2.19	.4857	.0143	.9857	.9714	.0286	.0363

table B-2 (continued)

Z	$A(Z)$	$B(Z)$	$C(Z)$	$D(Z)$	$E(Z)$	$F(Z)$
2.20	.4861	.0139	.9861	.9722	.0278	.0355
2.21	.4864	.0136	.9864	.9728	.0272	.0347
2.22	.4868	.0132	.9868	.9736	.0264	.0339
2.23	.4871	.0129	.9871	.9742	.0258	.0332
2.24	.4875	.0125	.9875	.9750	.0250	.0325
2.25	.4878	.0122	.9878	.9756	.0244	.0317
2.26	.4881	.0119	.9881	.9762	.0238	.0310
2.27	.4884	.0116	.9884	.9768	.0232	.0303
2.28	.4887	.0113	.9887	.9774	.0226	.0297
2.29	.4890	.0110	.9890	.9780	.0220	.0290
2.30	.4893	.0107	.9893	.9786	.0214	.0283
2.31	.4896	.0104	.9896	.9792	.0208	.0277
2.32	.4898	.0102	.9898	.9796	.0204	.0270
2.33	.4901	.0099	.9901	.9802	.0198	.0264
2.34	.4904	.0096	.9904	.9808	.0192	.0258
2.35	.4906	.0094	.9906	.9812	.0188	.0252
2.36	.4909	.0091	.9909	.9818	.0182	.0246
2.37	.4911	.0089	.9911	.9822	.0178	.0241
2.38	.4913	.0087	.9913	.9826	.0174	.0235
2.39	.4916	.0084	.9916	.9832	.0168	.0229
2.40	.4918	.0082	.9918	.9836	.0164	.0224
2.41	.4920	.0080	.9920	.9840	.0160	.0219
2.42	.4922	.0078	.9922	.9844	.0156	.0213
2.43	.4925	.0075	.9925	.9850	.0150	.0208
2.44	.4927	.0073	.9927	.9854	.0146	.0203
2.45	.4929	.0071	.9929	.9858	.0142	.0198
2.46	.4931	.0069	.9931	.9862	.0138	.0194
2.47	.4932	.0068	.9932	.9864	.0136	.0189
2.48	.4934	.0066	.9934	.9868	.0132	.0184
2.49	.4936	.0064	.9936	.9872	.0128	.0180
2.50	.4938	.0062	.9938	.9876	.0124	.0175
2.51	.4940	.0060	.9940	.9880	.0120	.0171
2.52	.4941	.0059	.9941	.9882	.0118	.0167
2.53	.4943	.0057	.9943	.9886	.0114	.0163
2.54	.4945	.0055	9945	.9890	.0110	.0158
2.55	.4946	.0054	.9946	.9892	.0108	.0154
2.56	.4948	0052	.9948	.9896	.0104	.0151
2.57	.4949	.0051	.9949	.9898	.0102	.0147
2.58	.4951	.0049	.9951	.9902	.0098	.0143
2.59	.4952	.0048	.9952	.9904	.0096	.0139
2.60	.4953	.0047	.9953	.9906	.0094	.0136
2.61	.4955	.0045	.9955	.9910	.0090	.0132
2.62	.4956	.0044	.9956	.9912	.0088	.0129
2.63	.4957	.0043	.9957	.9914	.0086	.0126
2.64	.4959	.0041	.9959	.9918	.0082	.0122

table B-2 (continued)

Z	$A(Z)$	$B(Z)$	$C(Z)$	$D(Z)$	$E(Z)$	$F(Z)$
2.65	.4960	.0040	.9960	.9920	.0080	.0119
2.66	.4961	.0039	.9961	.9922	.0078	.0116
2.67	.4962	.0038	.9962	.9924	.0076	.0113
2.68	.4963	.0037	.9963	.9926	.0074	.0110
2.69	.4964	.0036	.9964	.9928	.0072	.0107
2.70	.4965	.0035	.9965	.9930	.0070	.0104
2.71	.4966	.0034	.9966	.9932	.0068	.0101
2.72	.4967	.0033	.9967	.9934	.0066	.0099
2.73	.4968	.0032	.9968	.9936	.0064	.0096
2.74	.4969	.0031	.9969	.9938	.0062	.0093
2.75	.4970	.0030	.9970	.9940	.0060	.0091
2.76	.4971	.0029	.9971	.9942	.0058	.0088
2.77	.4972	.0028	.9972	.9944	.0056	.0086
2.78	.4973	.0027	.9973	.9946	.0054	.0084
2.79	.4974	.0026	.9974	.9948	.0052	.0081
2.80	.4974	.0026	.9974	.9948	.0052	.0079
2.81	.4975	.0025	.9975	.9950	.0050	.0077
2.82	.4976	.0024	.9976	.9952	.0048	.0075
2.83	.4977	.0023	.9977	.9954	.0046	.0073
2.84	.4977	.0023	.9977	.9954	.0046	.0071
2.85	.4978	.0022	.9978	.9956	.0044	.0069
2.86	.4979	.0021	.9979	.9958	.0042	.0067
2.87	.4979	.0021	.9979	9958	.0042	.0065
2.88	.4980	.0020	.9980	.9960	.0040	.0063
2.89	.4981	.0019	.9981	.9962	.0038	.0061
2.90	.4981	.0019	.9981	.9962	.0038	.0060
2.91	.4982	.0018	.9982	.9964	.0036	.0058
2.92	.4982	.0018	.9982	.9964	.0036	.0056
2.93	.4983	.0017	.9983	.9966	.0034	.0055
2.94	.4984	.0016	.9984	.9968	.0032	.0053
2.95	.4984	.0016	.9984	.9968	.0032	.0051
2.96	.4985	.0015	.9985	.9970	.0030	.0050
2.97	.4985	.0015	.9985	.9970	.0030	.0048
2.98	.4986	.0014	.9986	.9972	.0028	.0047
2.99	.4986	0014	.9986	.9972	.0028	.0046
3.00	.4987	.0013	.9987	.9974	.0026	.0044
3.01	.4987	.0013	.9987	.9974	.0026	.0043
3.02	.4987	.0013	.9987	.9974	.0026	.0042
3.03	.4988	.0012	.9988	.9976	.0024	.0040
3.04	.4988	.0012	.9988	.9976	.0024	.0039
3.05	.4989	.0011	.9989	.9978	.0022	.0038
3.06	.4989	.0011	.9989	.9978	.0022	.0037
3.07	.4989	.0011	.9989	.9978	.0022	.0036
3.08	.4990	.0010	.9990	.9980	.0020	.0035
3.09	.4990	.0010	.9990	.9980	.0020	.0034

table B-2 (continued)

Z	$A(Z)$	$B(Z)$	$C(Z)$	$D(Z)$	$E(Z)$	$F(Z)$
3.10	.4990	.0010	.9990	.9980	.0020	.0033
3.11	.4991	.0009	.9991	.9982	.0018	.0032
3.12	.4991	.0009	.9991	9982	.0018	.0031
3.13	.4991	.0009	.9991	.9982	.0018	.0030
3.14	.4992	.0008	.9992	.9984	.0016	.0029
3.15	.4992	.0008	.9992	.9984	.0016	.0028
3.16	.4992	.0008	.9992	.9984	.0016	.0027
3.17	.4992	.0008	.9992	.9984	.0016	.0026
3.18	.4993	.0007	.9993	.9986	.0014	.0025
3.19	.4993	.0007	.9993	.9986	.0014	.0025
3.20	.4993	.0007	.9993	.9986	.0014	.0024

table B-3 *Percentiles of t distributions*

df	.005	.01	.025	.05	.95	.975	.99	.995
				PERCENTILE				
1	−63.657	−31.821	−12.706	−6.314	6.314	12.706	31.821	63.657
2	−9.925	−6.965	−4.303	−2.920	2.920	4.303	6.965	9.925
3	−5.841	−4.541	−3.182	−2.353	2.353	3.182	4.541	5.841
4	−4.604	−3.747	−2.776	−2.132	2.132	2.776	3.747	4.604
5	−4.032	−3.365	−2.571	−2.015	2.015	2.571	3.365	4.032
6	−3.707	−3.143	−2.447	−1.943	1.943	2.447	3.143	3.707
7	−3.499	−2.998	−2.365	−1.895	1.895	2.365	2.998	3.499
8	−3.355	−2.896	−2.306	−1.860	1.860	2.306	2.896	3.355
9	−3.250	−2.821	−2.262	−1.833	1.833	2.262	2.821	3.250
10	−3.169	−2.764	−2.228	−1.812	1.812	2.228	2.764	3.169
11	−3.106	−2.718	−2.201	−1.796	1.796	2.201	2.718	3.106
12	−3.055	−2.681	−2.179	−1.782	1.782	2.179	2.681	3.055
13	−3.012	−2.650	−2.160	−1.771	1.771	2.160	2.650	3.012
14	−2.977	−2.624	−2.145	−1.761	1.761	2.145	2.624	2.977
15	−2.947	−2.602	−2.131	−1.753	1.753	2.131	2.602	2.947
16	−2.921	−2.583	−2.120	−1.746	1.746	2.120	2.583	2.921
17	−2.898	−2.567	−2.110	−1.740	1.740	2.110	2.567	2.898
18	−2.878	−2.552	−2.101	−1.734	1.734	2.101	2.552	2.878
19	−2.861	−2.539	−2.093	−1.729	1.729	2.093	2.539	2.861
20	−2.845	−2.528	−2.086	−1.725	1.725	2.086	2.528	2.845
21	−2.831	−2.518	−2.080	−1.721	1.721	2.080	2.518	2.831
22	−2.819	−2.508	−2.074	−1.717	1.717	2.074	2.508	2.819
23	−2.807	−2.500	−2.069	−1.714	1.714	2.069	2.500	2.807
24	−2.797	−2.492	−2.064	−1.711	1.711	2.064	2.492	2.797
25	−2.787	−2.485	−2.060	−1.708	1.708	2.060	2.485	2.787
26	−2.779	−2.479	−2.056	−1.706	1.706	2.056	2.479	2.779
27	−2.771	−2.473	−2.052	−1.703	1.703	2.052	2.473	2.771
28	−2.763	−2.467	−2.048	−1.701	1.701	2.048	2.467	2.763
29	−2.756	−2.462	−2.045	−1.699	1.699	2.045	2.462	2.756
30	−2.750	−2.457	−2.042	−1.697	1 697	2.042	2.457	2.750
40	−2.704	−2.423	−2.021	−1.684	1.684	2.021	2.423	2.704
50	−2.678	−2.403	−2.009	−1.676	1.676	2.009	2.403	2.678
60	−2.660	−2.390	−2.000	−1.671	1.671	2.000	2.390	2.660
80	−2.639	−2.374	−1.990	−1.664	1.664	1.990	2.374	2.639
100	−2.626	−2.365	−1.984	−1.660	1.660	1.984	2.365	2.626
120	−2.617	−2.358	−1.980	−1.658	1.658	1.980	2.358	2.617
200	−2.601	−2.345	−1.972	−1.653	1.653	1.972	2.345	2.601
500	−2.586	−2.334	−1.965	−1.648	1.648	1.965	2.334	2.586
∞	−2.576	−2.326	−1.960	−1.645	1.645	1.960	2.326	2.576

table B-4 *Percentiles of chi-square distributions*

	PERCENTILE							
df	.005	.01	.025	.05	.95	.975	.99	.995
1	.00004	.00016	.00098	.00393	3.84	5.02	6.63	7.88
2	.0100	.0201	.0506	.103	5.99	7.38	9.21	10.60
3	.072	.115	.216	.352	7.81	9.35	11.34	12.84
4	.207	.297	.484	.711	9.49	11.14	13.28	14.86
5	.412	.554	.831	1.145	11.07	12.83	15.09	16.75
6	.676	.872	1.24	1.64	12.59	14.45	16.81	18.55
7	.989	1.24	1.69	2.17	14.07	16.01	18.48	20.28
8	1.34	1.65	2.18	2.73	15.51	17.53	20.09	21.96
9	1.73	2.09	2.70	3.33	16.92	19.02	21.67	23.59
10	2.16	2.56	3.25	3.94	18.31	20.48	23.21	25.19
11	2.60	3.05	3.82	4.57	19.68	21.92	24.72	26.76
12	3.07	3.57	4.40	5.23	21.03	23.34	26.22	28.30
13	3.57	4.11	5.01	5.89	22.36	24.74	27.69	29.82
14	4.07	4.66	5.63	6.57	23.68	26.12	29.14	31.32
15	4.60	5.23	6.26	7.26	25.00	27.49	30.58	32.80
16	5.14	5.81	6.91	7.96	26.30	28.85	32.00	34.27
17	5.70	6.41	7.56	8.67	27.59	30.19	33.41	35.72
18	6.26	7.01	8.23	9.39	28.87	31.53	34.81	37.16
19	6.84	7.63	8.91	10.12	30.14	32.85	36.19	38.58
20	7.43	8.26	9.59	10.85	31.41	34.17	37.57	40.00
21	8.03	8.90	10.28	11.59	32.67	35.48	38.93	41.40
22	8.64	9.54	10.98	12.34	33.92	36.78	40.29	42.80
23	9.26	10.20	11.69	13.09	35.17	38.08	41.64	44.18
24	9.89	10.86	12.40	13.85	36.42	39.36	42.98	45.56
25	10.52	11.52	13.12	14.61	37.65	40.65	44.31	46.93
26	11.16	12.20	13.84	15.38	38.89	41.92	45.64	48.29
27	11.81	12.88	14.57	16.15	40.11	43.19	46.96	49.64
28	12.46	13.56	15.31	16.93	41.34	44.46	48.28	50.99
29	13.21	14.26	16.05	17.71	42.56	45.72	49.59	53.34
30	13.79	14.95	16.79	18.49	43.77	46.98	50.89	53.67
40	20.71	22.16	24.43	26.51	55.76	59.34	63.69	66.77
50	27.99	29.71	32.36	34.76	67.50	71.42	76.15	79.49
60	35.53	37.48	40.48	43.19	79.08	83.30	88.38	91.95
70	43.28	45.44	48.76	51.74	90.53	95.02	100.4	104.2
80	51.17	53.54	57.15	60.39	101.9	106.6	112.3	116.3
90	59.20	61.75	65.65	69.13	113.1	118.1	124.1	128.3
100	67.33	70.06	74.22	77.93	124.3	129.6	135.8	140.2
120	83.85	86.92	91.58	95.70	146.6	152.2	159.0	163.6

table B-5 *Percentiles of F distributions*

df		PERCENTILE							
numerator	denominator	.005	.01	.025	.05	.95	.975	.99	.995
1	1	.0001	.0002	.0015	.0062	161.45	647.79	4,052.2	16,211.
	2	.0000	.0002	.0012	.0050	18.513	38.506	98.503	198.50
	3	.0000	.0002	.0012	.0046	10.128	17.443	34.116	55.552
	4	.0000	.0002	.0011	.0044	7.7086	12.218	21.198	31.333
	5	.0000	.0002	.0011	.0043	6.6079	10.007	16.258	22.785
	6	.0000	.0002	.0011	.0043	5.9874	8.8131	13.745	18.635
	7	.0000	.0002	.0010	.0042	5.5914	8.0727	12.246	16.236
	8	.0000	.0002	.0010	.0042	5.3177	7.5709	11.259	14.688
	9	.0000	.0002	.0010	.0042	5.1174	7.2093	10.561	13.614
	10	.0000	.0002	.0010	.0041	4.9646	6.9367	10.044	12.826
	12	.0000	.0002	.0010	.0041	4.7472	6.5538	9.3302	11.754
	15	.0000	.0002	.0010	.0041	4.5431	6.1995	8.6831	10.798
	20	.0000	.0002	.0010	.0040	4.3513	5.8715	8.0960	9.9439
	24	.0000	.0002	.0010	.0040	4.2597	5.7167	7.8229	9.5513
	30	.0000	.0002	.0010	.0040	4.1709	5.5675	7.5625	9.1797
	40	.0000	.0002	.0010	.0040	4.0848	5.4239	7.3141	8.8278
	60	.0000	.0002	.0010	.0040	4.0012	5.2857	7.0771	8.4946
	120	.0000	.0002	.0010	.0039	3.9201	5.1524	6.8510	8.1790
	∞	.0000	.0002	.0010	.0039	3.8415	5.0239	6.6349	7.8794
2	1	.0050	.0101	.0260	.0540	199.50	799.50	4,999.5	20,000.
	2	.0050	.0101	.0256	.0526	19.000	39.000	99.000	199.00
	3	.0050	.0101	.0255	.0522	9.5521	16.044	30.817	49.799
	4	.0050	.0101	.0255	.0520	6.9443	10.649	18.000	26.284
	5	.0050	.0101	.0254	.0518	5.7861	8.4336	13.274	18.314
	6	.0050	.0101	.0254	.0517	5.1433	7.2598	10.925	14.544
	7	.0050	.0101	.0254	.0517	4.7374	6.5415	9.5466	12.404
	8	.0050	.0101	.0254	.0516	4.4590	6.0595	8.6491	11.042
	9	.0050	.0101	.0254	.0516	4.2565	5.7147	8.0215	10.107
	10	.0050	.0101	.0254	.0516	4.1028	5.4564	7.5594	9.4270
	12	.0050	.0101	.0254	.0515	3.8853	5.0959	6.9266	8.5096
	15	.0050	.0101	.0254	.0515	3.6823	4.7650	6.3589	7.7008
	20	.0050	.0101	.0253	.0514	3.4928	4.4613	5.8489	6.9865
	24	.0050	.0100	.0253	.0514	3.4028	4.3187	5.6136	6.6610
	30	.0050	.0100	.0253	.0514	3.3158	4.1821	5.3904	6.3547
	40	.0050	.0100	.0253	.0514	3.2317	4.0510	5.1785	6.0664
	60	.0050	.0100	.0253	.0513	3.1504	3.9253	4.9774	5.7950
	120	.0050	.0100	.0253	.0513	3.0718	3.8046	4.7865	5.5398
	∞	.0050	.0100	.0253	.0513	2.9957	3.6889	4.6052	5.2983

table B-5 (continued)

df		PERCENTILE							
numerator	denominator	.005	.01	.025	.05	.95	.975	.99	.995
3	1	.0180	.0293	.0573	.0987	215.71	864.16	5,403.3	21,615.
	2	.0201	.0324	.0623	.1047	19.164	39.165	99.166	199.17
	3	.0211	.0339	.0648	.1078	9.2766	15.439	29.457	47.467
	4	.0216	.0348	.0662	.1097	6.5914	9.9792	16.694	24.259
	5	.0220	.0354	.0672	.1109	5.4095	7.7636	12.060	16.530
	6	.0223	.0358	.0679	.1118	4.7571	6.5988	9.7795	12.917
	7	.0225	.0361	.0684	.1125	4.3468	5.8898	8.4513	10.882
	8	.0227	.0364	.0688	.1131	4.0662	5.4160	7.5910	9.5965
	9	0228	.0366	.0691	.1135	3.8626	5.0781	6.9919	8.7171
	10	.0229	.0367	.0693	.1138	3.7083	4.8256	6.5523	8.0807
	12	.0230	.0370	.0697	.1144	3.4903	4.4742	5.9526	7.2258
	15	.0232	.0372	.0702	.1149	3.2874	4.1528	5.4170	6.4760
	20	.0234	.0375	.0706	.1155	3.0984	3.8587	4 9382	5.8177
	24	.0235	.0376	.0708	.1158	3.0088	3.7211	4.7181	5.5190
	30	.0235	.0377	.0710	.1161	2.9223	3.5894	4.5097	5.2388
	40	.0236	.0379	.0712	.1163	2.8387	3.4633	4.3126	4.9759
	60	.0237	.0380	.0715	.1167	2.7581	3.3425	4.1259	4.7290
	120	.0238	.0381	.0717	.1170	2.6802	3.2270	3.9493	4.4973
	∞	.0239	.0383	.0719	.1173	2.6049	3.1161	3.7816	4.2794
4	1	.0319	.0472	.0818	.1297	224.58	899.58	5,624.6	22,500.
	2	.0380	.0556	.0939	.1440	19.247	39.248	99.249	199.25
	3	.0412	.0599	.1002	.1517	9.1172	15.101	28.710	46.195
	4	.0432	.0626	.1041	.1565	6.3883	9.6045	15.977	23.155
	5	.0445	.0644	.1068	.1598	5.1922	7.3879	11.392	15.556
	6	.0455	.0658	.1087	.1623	4.5337	6.2272	9.1483	12.028
	7	.0462	.0668	.1102	.1641	4.1203	5.5226	7.8467	10.050
	8	.0468	.0676	.1114	.1655	3.8378	5.0526	7.0060	8.8051
	9	.0473	.0682	.1123	.1667	3.6331	4.7181	6.4221	7.9559
	10	.0477	.0687	.1131	.1677	3.4780	4.4683	5.9943	7.3428
	12	.0483	.0696	.1143	.1692	3.2592	4.1212	5.4119	6.5211
	15	.0489	.0704	.1155	.1707	3.0556	3.8043	4.8932	5.8029
	20	.0496	.0713	.1168	.1723	2.8661	3.5147	4.4307	5.1743
	24	.0499	.0718	.1175	.1732	2.7763	3.3794	4.2184	4.8898
	30	.0503	.0723	.1182	.1740	2.6896	3.2499	4.0179	4.6233
	40	.0506	.0727	.1189	.1749	2.6060	3.1261	3.8283	4.3738
	60	.0510	.0732	.1196	.1758	2.5252	3.0077	3.6491	4.1399
	120	.0514	.0738	.1203	.1767	2.4472	2.8943	3.4796	3.9207
	∞	.0517	.0743	.1211	.1777	2.3719	2.7858	3.3192	3.7151

table B-5 (continued)

df		PERCENTILE							
numerator	denominator	.005	.01	.025	.05	.95	.975	.99	.995
5	1	.0439	.0615	.0999	.1513	230.16	921.85	5,763.7	23,056.
	2	.0546	.0753	.1186	.1728	19.296	39.298	99.299	199.30
	3	.0605	.0829	.1288	.1849	9.0135	14.885	28.237	45.392
	4	.0643	.0878	.1354	.1926	6.2560	9.3645	15.522	22.456
	5	.0669	.0912	.1399	.1980	5.0503	7.1464	10.967	14.940
	6	.0689	.0937	.1433	.2020	4.3874	5.9876	8.7459	11.464
	7	.0704	.0956	.1459	.2051	3.9715	5.2852	7.4604	9.5221
	8	.0716	.0972	.1480	.2075	3.6875	4.8173	6.6318	8.3018
	9	.0726	.0984	.1497	.2095	3.4817	4.4844	6.0569	7.4711
	10	.0734	.0995	.1511	.2112	3.3258	4.2361	5.6363	6.8723
	12	.0747	.1011	1533	.2138	3.1059	3.8911	5.0643	6.0711
	15	.0761	.1029	.1556	.2165	2.9013	3.5764	4.5556	5.3721
	20	.0775	.1047	.1580	.2194	2.7109	3.2891	4.1027	4.7616
	24	.0782	.1056	.1593	.2209	2.6207	3.1548	3.8951	4.4857
	30	.0790	.1066	.1606	.2224	2.5336	3.0265	3.6990	4.2276
	40	.0798	.1076	.1619	.2240	2.4495	2.9037	3.5138	3.9860
	60	.0806	.1087	.1633	.2257	2.3683	2.7863	3.3389	3.7600
	120	.0815	.1097	.1648	.2274	2.2900	2.6740	3.1735	3.5482
	∞	.0823	.1109	.1662	.2291	2.2141	2.5665	3.0173	3.3499
6	1	.0537	.0727	.1135	.1670	233.99	937.11	5,859.0	23,437.
	2	.0688	.0915	.1377	.1944	19.330	39.331	99.332	199.33
	3	.0774	.1022	.1515	.2102	8.9406	14.735	27.911	44.838
	4	.0831	.1093	.1606	.2206	6.1631	9.1973	15.207	21.975
	5	.0872	.1143	.1670	.2279	4.9503	6.9777	10.672	14.513
	6	.0903	.1181	.1718	.2334	4.2839	5.8197	8.4661	11.073
	7	.0927	.1211	.1756	.2377	3.8660	5.1186	7.1914	9.1554
	8	.0946	.1234	.1786	.2411	3.5806	4.6517	6.3707	7.9520
	9	.0962	.1254	.1810	.2440	3.3738	4.3197	5.8018	7.1338
	10	.0976	.1270	.1831	.2463	3.2172	4.0721	5.3858	6.5446
	12	.0997	.1296	.1863	.2500	2.9961	3.7283	4.8206	5.7570
	15	.1019	.1323	.1898	.2539	2.7905	3.4147	4.3183	5.0708
	20	.1043	.1352	.1935	.2581	2.5990	3.1283	3.8714	4.4721
	24	.1055	.1367	.1954	.2603	2.5082	2.9946	3.6667	4.2019
	30	.1069	.1383	.1974	.2626	2.4205	2.8667	3.4735	3.9492
	40	.1082	.1400	.1995	.2649	2.3359	2.7444	3.2910	3.7129
	60	.1096	.1417	.2017	.2674	2.2540	2.6274	3.1187	3.4918
	120	.1111	.1435	.2039	.2699	2.1750	2.5154	2.9559	3.2849
	∞	.1126	.1453	.2062	.2726	2.0986	2.4082	2.8020	3.0913

table B-5 (continued)

df numerator	*df* denominator	.005	.01	.025	.05	.95	.975	.99	.995
7	1	.0616	.0817	.1239	.1788	236.77	948.22	5,928.3	23,715.
	2	.0806	.1047	.1529	.2111	19.353	39.355	99.356	199.36
	3	.0919	.1183	.1698	.2300	8.8868	14.624	27.672	44.434
	4	.0995	.1274	.1811	.2427	6.0942	9.0741	14.976	21.622
	5	.1050	.1340	.1892	.2518	4.8759	6.8531	10.456	14.200
	6	.1092	.1390	.1954	.2587	4.2066	5.6955	8.2600	10.786
	7	.1125	.1430	.2002	.2641	3.7870	4.9949	6.9928	8.8854
	8	.1152	.1462	.2041	.2684	3.5005	4.5286	6.1776	7.6942
	9	.1175	.1488	.2073	.2720	3.2927	4.1971	5.6129	6.8849
	10	.1193	.1511	.2100	.2750	3.1355	3.9498	5.2001	6.3025
	12	.1223	.1546	.2143	.2797	2.9134	3.6065	4.6395	5.5245
	15	.1255	.1584	.2189	.2848	2.7066	3.2934	4.1415	4.8473
	20	.1290	.1625	.2239	.2903	2.5140	3.0074	3.6987	4.2569
	24	.1308	.1646	.2265	.2932	2.4226	2.8738	3.4959	3.9905
	30	.1327	.1669	.2292	.2962	2.3343	2.7460	3.3045	3.7416
	40	.1347	.1692	.2321	.2994	2.2490	2.6238	3.1238	3.5088
	60	.1368	.1717	.2351	.3026	2.1665	2.5068	3.9530	3.2911
	120	.1390	.1743	.2382	.3060	2.0867	2.3948	3.7918	3.0874
	∞	.1413	.1770	.2414	.3096	2.0096	2.2875	3.6393	2.8968
8	1	.0681	.0888	.1321	.1880	238.88	956.66	5,981.6	23,925.
	2	.0906	.1156	.1650	.2243	19.371	39.373	99.374	199.37
	3	.1042	.1317	.1846	.2459	8.8452	14.540	27.489	44.126
	4	.1136	.1427	.1979	.2606	6.0410	8.9796	14.799	21.352
	5	.1205	.1508	.2076	.2712	4.8183	6.7572	10.289	13.961
	6	.1257	.1570	.2150	.2793	4.1468	5.5996	8.1016	10.566
	7	.1300	.1619	.2208	.2857	3.7257	4.8994	6.8401	8.6781
	8	.1334	.1659	.2256	.2909	3.4381	4.4332	6.0289	7.4960
	9	.1363	.1692	.2295	.2951	3.2296	4.1020	5.4671	6.6933
	10	.1387	.1720	.2328	.2988	3.0717	3.8549	5.0567	6.1159
	12	.1425	.1765	.2381	.3045	2.8486	3.5118	4.4994	5.3451
	15	.1467	.1813	.2438	.3107	2.6408	3.1987	4.0045	4.6743
	20	.1513	.1866	.2500	.3174	2.4471	2.9128	3.5644	4.0900
	24	.1538	.1894	.2533	.3210	2.3551	2.7791	3.3629	3.8264
	30	.1563	.1924	.2568	.3247	2.2662	2.6513	3.1726	3.5801
	40	.1590	.1955	.2604	.3286	2.1802	2.5289	2.9930	3.3498
	60	.1619	.1987	.2642	.3328	2.0970	2.4117	2.8233	3.1344
	120	.1649	.2022	.2682	.3370	2.0164	2.2994	2.6629	2.9330
	∞	.1680	.2058	.2725	.3416	1.9384	2.1918	2.5113	2.7444

table B-5 (continued)

df numerator	df denominator	.005	.01	.025	.05	.95	.975	.99	.995
9	1	.0734	.0947	.1387	.1954	240.54	963.28	6,022.5	24,091.
	2	.0989	.1247	.1750	.2349	19.385	39.387	99.388	199.39
	3	.1147	.1430	.1969	.2589	8.8123	14.473	27.345	43.882
	4	.1257	.1557	.2119	.2752	5.9988	8.9047	14.659	21.139
	5	.1339	.1651	.2230	.2872	4.7725	6.6810	10.158	13.772
	6	.1402	.1724	.2315	.2964	4.0990	5.5234	7.9761	10.391
	7	.1453	.1782	.2383	.3037	3.6767	4.8232	6.7188	8.5138
	8	.1494	.1829	.2438	.3096	3.3881	4.3572	5.9106	7.3386
	9	.1529	.1869	.2484	.3146	3.1789	4.0260	5.3511	6.5411
	10	.1558	.1902	.2523	.3188	3.0204	3.7790	4.9424	5.9676
	12	.1606	.1956	.2585	.3254	2.7964	3.4358	4.3875	5.2021
	15	.1658	.2015	.2653	.3327	2.5876	3.1227	3.8948	4.5364
	20	.1715	.2080	.2727	.3405	2.3928	2.8365	3.4567	3.9564
	24	.1745	.2115	.2767	.3448	2.3002	2.7027	3.2560	3.6949
	30	.1778	.2151	.2809	.3492	2.2107	2.5746	3.0665	3.4505
	40	.1812	.2190	.2853	.3539	2.1240	2.4519	2.8876	3.2220
	60	.1843	.2231	.2899	.3588	2.0401	2.3344	2.7185	3.0083
	120	.1887	.2274	.2948	.3640	1.9588	2.2217	2.5586	2.8083
	∞	.1928	.2320	.3000	.3694	1.8799	2.1136	2.4073	2.6210
10	1	.0780	.0996	.1442	.2014	241.88	968.63	6,055.8	24,224.
	2	.1061	.1323	.1833	.2437	19.396	39.398	99.399	199.40
	3	.1237	.1526	.2072	.2697	8.7855	14.419	27.229	43.686
	4	.1362	.1668	.2238	.2875	5.9644	8.8439	14.546	20.967
	5	.1455	.1774	.2361	.3007	4.7351	6.6192	10.051	13.618
	6	.1528	.1857	.2456	.3108	4.0600	5.4613	7.8741	10.250
	7	.1587	.1923	.2532	.3189	3.6365	4.7611	6.6201	8.3803
	8	.1635	.1978	.2594	.3255	3.3472	4.2951	5.8143	7.2107
	9	.1676	.2023	.2646	.3311	3.1373	3.9639	5.2565	6.4171
	10	.1710	.2062	.2690	.3358	2.9782	3.7168	4.8492	5.8467
	12	.1766	.2125	.2762	.3433	2.7534	3.3736	4.2961	5.0855
	15	.1828	.2194	.2840	.3515	2.5437	3.0602	3.8049	4.4236
	20	.1896	.2270	.2925	.3605	2.3479	2.7737	3.3682	3.8470
	24	.1933	.2311	.2971	.3653	2.2547	2.6396	3.1681	3.5870
	30	.1972	.2355	.3020	.3704	2.1646	2.5112	2.9791	3.3440
	40	.2014	.2401	.3072	.3758	2.0772	2.3882	2.8005	3.1167
	60	.2058	.2450	.3127	.3815	1.9926	2.2702	2.6318	2.9042
	120	.2105	.2502	.3185	.3876	1.9105	2.1570	2.4721	2.7052
	∞	.2156	.2558	.3247	.3940	1.8307	2.0483	2.3209	2.5188

table B-5 (continued)

df		PERCENTILE							
numerator	denominator	.005	.01	.025	.05	.95	.975	.99	.995
12	1	.0851	.1072	.1526	.2106	243.91	976.71	6,106.3	24,426.
	2	.1175	.1444	.1962	.2574	19.413	39.415	99.416	199.42
	3	.1384	.1680	.2235	.2865	8.7446	14.337	27.052	43.387
	4	.1533	.1848	.2426	.3068	5.9117	8.7512	14.374	20.705
	5	.1647	.1975	.2570	.3220	4.6777	6.5246	9.8883	13.384
	6	.1737	.2074	.2682	.3338	3.9999	5.3662	7.7183	10.034
	7	.1810	.2155	.2773	.3432	3.5747	4.6658	6.4691	8.1764
	8	.1871	.2222	.2847	.3510	3.2840	4.1997	5.6668	7.0149
	9	.1922	.2279	.2910	.3576	3.0729	3.8682	5.1114	6.2274
	10	.1966	.2328	.2964	.3632	2.9130	3.6209	4.7059	5.6613
	12	.2038	.2407	.3051	.3722	2.6866	3.2773	4.1553	4.9063
	15	.2118	.2494	.3147	.3821	2.4753	2.9633	3.6662	4.2498
	20	.2208	.2592	.3254	.3931	2.2776	2.6758	3.2311	3.6779
	24	.2257	.2645	.3313	.3991	2.1834	2.5412	3.0316	3.4199
	30	.2309	.2702	.3375	.4055	2.0921	2.4120	2.8431	3.1787
	40	.2365	.2763	.3441	.4122	2.0035	2.2882	2.6648	2.9531
	60	.2426	.2828	.3511	.4194	1.9174	2.1692	2.4961	2.7419
	120	.2491	.2899	.3588	.4272	1.8337	2.0548	2.3363	2.5439
	∞	.2561	.2976	.3670	.4355	1.7522	1.9447	2.1848	2.3583
15	1	.0926	.1152	.1613	.2201	245.95	984.87	6,157.3	24,630.
	2	.1299	.1573	.2099	.2716	19.429	39.431	99.432	99.43
	3	.1544	.1846	.2408	.3042	8.7029	14.253	26.872	43.085
	4	.1723	.2044	.2629	.3273	5.8578	8.6565	14.198	20.438
	5	.1861	.2195	.2796	.3447	4.6188	6.4277	9.7222	13.146
	6	.1972	.2316	.2928	.3584	3.9381	5.2687	7.5590	9.8140
	7	.2063	.2415	.3036	.3695	3.5108	4.5678	6.3143	7.9678
	8	.2139	.2497	.3126	.3787	3.2184	4.1012	5.5151	6.8143
	9	.2204	.2567	.3202	.3865	3.0061	3.7694	4.9621	6.0325
	10	.2261	.2628	.3268	.3931	2.8450	3.5217	4.5582	5.4707
	12	.2353	.2728	.3375	.4040	2.6169	3.1772	4.0096	4.7214
	15	.2457	.2839	.3494	.4161	2.4035	2.8621	3.5222	4.0698
	20	.2576	.2966	.3629	.4297	2.2033	2.5731	3.0880	3.5020
	24	.2641	.3036	.3703	.4371	2.1077	2.4374	2.8887	3.2456
	30	.2712	.3111	.3783	.4451	2.0148	2.3072	2.7002	3.0057
	40	.2789	.3193	.3868	.4537	1.9245	2.1819	2.5216	2.7811
	60	.2873	.3282	.3962	.4629	1.8364	2.0613	2.3523	2.5705
	120	.2965	.3379	.4063	.4730	1.7505	1.9450	2.1915	2.3727
	∞	.3067	.3486	.4175	.4841	1.6664	1.8326	2.0385	2.1868

table B-5 (continued)

df numer-ator	*df* denom-inator	.005	.01	.025	.05	.95	.975	.99	.995
20	1	.1006	.1235	.1703	.2298	248.01	993.10	6,208.7	24,836.
	2	.1431	.1710	.2241	.2863	19.446	39.448	99.449	199.45
	3	.1719	.2025	.2591	.3227	8.6602	14.167	26.690	42.778
	4	.1933	.2257	.2845	.3489	5.8025	8.5599	14.020	20.167
	5	.2100	.2437	.3040	.3689	4.5581	6.3285	9.5527	12.903
	6	.2236	.2583	.3197	.3848	3.8742	5.1684	7.3958	9.5888
	7	.2349	.2704	.3325	.3978	3.4445	4.4667	6.1554	7.7540
	8	.2445	.2805	.3433	.4086	3.1503	3.9995	5.3591	6.6082
	9	.2528	.2893	.3525	.4179	2.9365	3.6669	4.8080	5.8318
	10	.2599	.2969	.3605	.4259	2.7740	3.4186	4.4054	5.2740
	12	.2719	.3095	.3737	.4391	2.5436	3.0728	3.8584	4.5299
	15	.2855	.3238	.3886	.4539	2.3275	2.7559	3.3719	3.8826
	20	.3014	.3404	.4058	.4708	2.1242	2.4645	2.9377	3.3178
	24	.3104	.3497	.4153	.4802	2.0267	2.3273	2.7380	3.0624
	30	.3202	.3599	.4258	.4904	1.9317	2.1952	2.5487	2.8230
	40	.3310	.3711	.4372	.5015	1.8389	2.0677	2.3689	2.5984
	60	.3430	.3835	.4498	.5138	1.7480	1.9445	2.1978	2.3872
	120	.3564	.3973	.4638	.5273	1.6587	1.8249	2.0346	2.1881
	∞	.3717	.4130	.4795	.5425	1.5705	1.7085	1.8783	1.9998
24	1	.1047	.1278	.1749	.2348	249.05	997.25	6,234.6	24,940.
	2	.1501	.1781	.2316	.2939	19.454	39.456	99.458	199.46
	3	.1812	.2119	.2687	.3324	8.6385	14.124	26.598	42.622
	4	.2045	.2371	.2959	.3602	5.7744	8.5109	13.929	20.030
	5	.2229	.2567	.3170	.3816	4.5272	6.2780	9.4665	12.780
	6	.2380	.2727	.3339	.3987	3.8415	5.1172	7.3127	9.4741
	7	.2506	.2860	.3480	.4128	3.4105	4.4150	6.0743	7.6450
	8	.2613	.2974	.3598	.4246	3.1152	3.9472	5.2793	6.5029
	9	.2706	.3071	.3700	.4347	2.9005	3.6142	4.7290	5.7292
	10	.2788	.3157	.3788	.4435	2.7372	3.3654	4.3269	5.1732
	12	.2924	.3299	.3935	.4580	2.5055	3.0187	3.7805	4.4315
	15	.3081	.3462	.4103	.4744	2.2878	2.7006	3.2940	3.7859
	20	.3265	.3652	.4297	.4934	2.0825	2.4076	2.8594	3.2220
	24	.3371	.3761	.4407	.5041	1.9838	2.2693	2.6591	2.9667
	30	.3487	.3880	.4527	.5157	1.8874	2.1359	2.4689	2.7272
	40	.3616	.4012	.4660	.5285	1.7929	2.0069	2.2880	2.5020
	60	.3761	.4161	.4808	.5428	1.7001	1.8817	2.1154	2.2898
	120	.3927	.4329	.4975	.5588	1.6084	1.7597	1.9500	2.0890
	∞	.4119	.4523	.5167	.5770	1.5173	1.6402	1.7908	1.8983

table B-5 (continued)

df		PERCENTILE							
numerator	denominator	.005	.01	.025	.05	.95	.975	.99	.995
30	1	.1089	.1322	.1796	.2398	250.09	1,001.4	6,260.7	25,044.
	2	.1574	.1855	.2391	.3016	19.462	39.465	99.466	199.47
	3	.1909	.2217	.2786	.3422	8.6166	14.081	26.505	42.466
	4	.2163	.2489	.3077	.3718	5.7459	8.4613	13.838	19.892
	5	.2365	.2703	.3304	.3947	4.4957	6.2269	9.3793	12.656
	6	.2532	.2879	.3488	.4131	3.8082	5.0652	7.2285	9.3583
	7	.2673	.3026	.3642	.4284	3.3758	4.3624	5.9921	7.5345
	8	.2793	.3152	.3772	.4413	3.0794	3.8940	5.1981	6.3961
	9	.2898	.3261	.3884	.4523	2.8637	3.5604	4.6486	5.6248
	10	.2990	.3357	.3982	.4620	2.6996	3.3110	4.2469	5.0705
	12	.3146	.3517	.4146	.4780	2.4663	2.9633	3.7008	4.3309
	15	.3327	.3703	.4334	.4963	2.2468	2.6437	3.2141	3.6867
	20	.3542	.3924	.4555	.5177	2.0391	2.3486	2.7785	3.1234
	24	.3667	.4050	.4682	.5298	1.9390	2.2090	2.5773	2.8679
	30	.3805	.4191	.4822	.5432	1.8409	2.0739	2.3860	2.6278
	40	.3962	.4349	.4978	.5581	1.7444	1.9429	2.2034	2.4015
	60	.4141	.4529	.5155	.5748	1.6491	1.8152	2.0285	2.1874
	120	.4348	.4738	.5358	.5940	1.5543	1.6899	1.8600	1.9839
	∞	.4596	.4984	.5597	.6164	1.4591	1.5660	1.6964	1.7891
40	1	.1133	.1367	.1844	.2448	251.14	1,005.6	6,286.3	25,148.
	2	.1648	.1931	.2468	.3094	19.471	39.473	99.474	199.47
	3	.2010	.2319	.2887	.3523	8.5944	14.037	26.411	42.308
	4	.2286	.2612	.3199	.3837	5.7170	8.4111	13.745	19.752
	5	.2509	.2846	.3444	.4082	4.4638	6.1751	9.2912	12.530
	6	.2693	.3039	.3644	.4281	3.7743	5.0125	7.1432	9.2408
	7	.2850	.3201	.3811	.4446	3.3404	4.3089	5.9084	7.4225
	8	.2985	.3341	.3954	.4587	3.0428	3.8398	5.1156	6.2875
	9	.3104	.3463	.4078	.4708	2.8259	3.5055	4.5667	5.5186
	10	.3208	.3571	.4187	.4814	2.6609	3.2554	4.1653	4.9659
	12	.3386	.3753	.4370	.4991	2.4259	2.9063	3.6192	4.2282
	15	.3596	.3966	.4583	.5196	2.2043	2.5850	3.1319	3.5850
	20	.3848	.4221	.4836	.5438	1.9938	2.2873	2.6947	3.0215
	24	.3997	.4371	.4983	.5578	1.8920	2.1460	2.4923	2.7654
	30	.4164	.4538	.5147	.5733	1.7918	2.0089	2.2992	2.5241
	40	.4356	.4730	.5333	.5907	1.6928	1.8752	2.1142	2.2958
	60	.4579	.4952	.5547	.6108	1.5943	1.7440	1.9360	2.0789
	120	.4846	.5216	.5800	.6343	1.4952	1.6141	1.7628	1.8709
	∞	.5177	.5541	.6108	.6627	1.3940	1.4835	1.5923	1.6691

table B-5 (continued)

df numerator	df denominator	.005	.01	.025	.05	.95	.975	.99	.995
60	1	.1177	.1413	.1892	.2499	252.20	1,009.8	6,313.0	25,253.
	2	.1726	.2009	.2548	.3174	19.479	39.481	99.483	199.48
	3	.2115	.2424	.2992	.3626	8.5720	13.992	26.316	42.149
	4	.2415	.2740	.3325	.3960	5.6878	8.3604	13.652	19.611
	5	.2660	.2995	.3589	.4222	4.4314	6.1225	9.2020	12.402
	6	.2864	.3206	.3806	.4437	3.7398	4.9589	7.0568	9.1219
	7	.3038	.2530	.3989	.4616	3.3043	4.2544	5.8236	7.3088
	8	.3190	.3542	.4147	.4769	3.0053	3.7844	5.0316	6.1772
	9	.3324	.3678	.4284	.4902	2.7872	3.4493	4.4831	5.4104
	10	.3443	.3800	.4405	.5019	2.6211	3.1984	4.0819	4.8592
	12	.3647	.4006	.4610	.5215	2.3842	2.8478	3.5355	4.1229
	15	.3890	.4251	.4851	.5445	2.1601	2.5242	3.0471	3.4803
	20	.4189	.4550	.5143	.5721	1.9464	2.2234	2.6077	2.9159
	24	.4367	.4727	.5314	.5882	1.8424	2.0799	2.4035	2.6585
	30	.4572	.4930	.5509	.6064	1.7396	1.9400	2.2079	2.4151
	40	.4810	.5165	.5734	.6272	1.6373	1.8028	2.0194	2.1838
	60	.5096	.5446	.6000	.6518	1.5343	1.6668	1.8363	1.9622
	120	.5452	.5793	.6325	.6815	1.4290	1.5299	1.6557	1.7469
	∞	.5922	.6248	.6747	.7198	1.3180	1.3883	1.4730	1.5325
120	1	.1223	.1460	.1941	.2551	253.25	1,014.0	6,339.4	25,359.
	2	.1805	.2089	.2628	.3255	19.487	39.490	99.491	199.49
	3	.2224	.2532	.3099	.3731	8.5494	13.947	26.221	41.989
	4	.2551	.2874	.3455	.4086	5.6581	8.3092	13.558	19.468
	5	.2818	.3151	.3740	.4367	4.3984	6.0693	9.1118	12.274
	6	.3044	.3383	.3976	.4598	3.7047	4.9045	6.9690	9.0015
	7	.3239	.2637	.4176	.4792	3.2674	4.1989	5.7372	7.1933
	8	.3409	.3755	.4349	.4959	2.9669	3.7279	4.9460	6.0649
	9	.3561	.3908	.4501	.5105	2.7475	3.3918	4.3978	5.3001
	10	.3697	.4045	.4636	.5234	2.5801	3.1399	3.9965	4.7501
	12	.3931	.4280	.4867	.5454	2.3410	2.7874	3.4494	4.0149
	15	.4215	.4563	.5141	.5713	2.1141	2.4611	2.9595	3.3722
	20	.4570	.4915	.5480	.6029	1.8963	2.1562	2.5168	2.8058
	24	.4787	.5128	.5683	.6217	1.7897	2.0099	2.3099	2.5463
	30	.5041	.5376	.5918	.6434	1.6835	1.8664	2.1107	2.2997
	40	.5345	.5673	.6195	.6688	1.5766	1.7242	1.9172	2.0635
	60	.5724	.6040	.6536	.6998	1.4673	1.5810	1.7263	1.8341
	120	.6229	.6523	.6980	.7397	1.3519	1.4327	1.5330	1.6055
	∞	.6988	.7244	.7631	.7975	1.2214	1.2684	1.3246	1.3637

table B-5 (continued)

df numer-ator	*df* denom-inator	PERCENTILE .005	.01	.025	.05	.95	.975	.99	.995
∞	1	.1269	.1507	.1990	.2603	254.32	1,018.3	6,366.0	25,465.
	2	.1887	.2171	.2711	.3338	19.496	39.498	99.501	199.51
	3	.2337	.2644	.3209	.3839	8.5265	13.902	26.125	41.829
	4	.2692	.3013	.3590	.4216	5.6281	8.2573	13.463	19.325
	5	.2985	.3314	.3896	.4517	4.3650	6.0153	9.0204	12.144
	6	.3235	.3569	.4152	.4765	3.6688	4.8491	6.8801	8.8793
	7	.3452	.2748	.4372	.4976	3.2298	4.1423	5.6495	7.0760
	8	.3644	.3982	.4562	.5159	2.9276	3.6702	4.8588	5.9505
	9	.3815	.4154	.4731	.5319	2.7067	3.3329	4.3105	5.1875
	10	.3970	.4309	.4882	.5462	2.5379	3.0798	3.9090	4.6385
	12	.4240	.4577	.5142	.5707	2.2962	2.7249	3.3608	3.9039
	15	.4573	.4906	.5457	.6001	2.0658	2.3953	2.8684	3.2602
	20	.5001	.5324	.5853	.6367	1.8432	2.0853	2.4212	2.6904
	24	.5268	.5584	.6097	.6591	1.7331	1.9353	2.2107	2.4276
	30	.5589	.5895	.6386	.6854	1.6223	1.7867	2.0062	2.1760
	40	.5991	.6280	.6741	.7174	1.5089	1.6371	1.8047	1.9318
	60	.6525	.6789	.7203	.7587	1.3893	1.4822	1.6006	1.6885
	120	.7333	.7549	.7884	.8187	1.2539	1.3104	1.3805	1.4311
	∞	1.0000	1.0000	1.0000	1.0000	1.0000	1.0000	1.0000	1.0000

table B-6 *Percentage points of the ratio* s^2_{max}/s^2_{min}†

ν‡	k‡ = 2	3	4	5	6	7	8	9	10	11	12
					UPPER .05 POINTS						
2	39.0	87.5	142	202	266	333	403	475	550	626	704
3	15.4	27.8	39.2	50.7	62.0	72.9	83.5	93.9	104	114	124
4	9.60	15.5	20.6	25.2	29.5	33.6	37.5	41.1	44.6	48.0	51.4
5	7.15	10.8	13.7	16.3	18.7	20.8	22.9	24.7	26.5	28.2	29.9
6	5.82	8.38	10.4	12.1	13.7	15.0	16.3	17.5	18.6	19.7	20.7
7	4.99	6.94	8.44	9.70	10.8	11.8	12.7	13.5	14.3	15.1	15.8
8	4.43	6.00	7.18	8.12	9.03	9.78	10.5	11.1	11.7	12.2	12.7
9	4.03	5.34	6.31	7.11	7.80	8.41	8.95	9.45	9.91	10.3	10.7
10	3.72	4.85	5.67	6.34	6.92	7.42	7.87	8.28	8.66	9.01	9.34
12	3.28	4.16	4.79	5.30	5.72	6.09	6.42	6.72	7.00	7.25	7.48
15	2.86	3.54	4.01	4.37	4.68	4.95	5.19	5.40	5.59	5.77	5.93
20	2.46	2.95	3.29	3.54	3.76	3.94	4.10	4.24	4.37	4.49	4.59
30	2.07	2.40	2.61	2.78	2.91	3.02	3.12	3.21	3.29	3.36	3.39
60	1.67	1.85	1.96	2.04	2.11	2.17	2.22	2.26	2.30	2.33	2.36
∞	1.00	1.00	1.00	1.00	1.00	1.00	1.00	1.00	1.00	1.00	1.00

table B-6 (continued)

ν‡	k‡ = 2	3	4	5	6	7	8	9	10	11	12
					UPPER .01 POINTS						
2	199	448	729	1,036	1,362	1,705	2,063	2,432	2,813	3,204	3,605
3	47.5	85	120	151	184	21(6)	24(9)	28(1)	31(0)	33(7)	36(1)
4	23.2	37	49	59	69	79	89	97	106	113	120
5	14.9	22	28	33	38	42	46	50	54	57	60
6	11.1	15.5	19.1	22	25	27	30	32	34	36	37
7	8.89	12.1	14.5	16.5	18.4	20	22	23	24	26	27
8	7.50	9.9	11.7	13.2	14.5	15.8	16.9	17.9	18.9	19.8	21
9	6.54	8.5	9.9	11.1	12.1	13.1	13.9	14.7	15.3	16.0	16.6
10	5.85	7.4	8.6	9.6	10.4	11.1	11.8	12.4	12.9	13.4	13.9
12	4.91	6.1	6.9	7.6	8.2	8.7	9.1	9.5	9.9	10.2	10.6
15	4.07	4.9	5.5	6.0	6.4	6.7	7.1	7.3	7.5	7.8	8.0
20	3.32	3.8	4.3	4.6	4.9	5.1	5.3	5.5	5.6	5.8	5.9
30	2.63	3.0	3.3	3.4	3.6	3.7	3.8	3.9	4.0	4.1	4.2
60	1.96	2.2	2.3	2.4	2.4	2.5	2.5	2.6	2.6	2.7	2.7
∞	1.00	1.0	1.0	1.0	1.0	1.0	1.0	1.0	1.0	1.0	1.0

†s^2_{max} is the largest and s^2_{min} the smallest in a set of k independent mean squares, each based on ν degrees of freedom.

‡Values in the column $k = 2$ and in the rows $\nu = 2$ and ∞ are exact. Elsewhere the third digit may be in error by a few units for the .05 points and several units for the .01 points. The third-digit figures in brackets for $\nu = 3$ are the most uncertain.

Source: Reprinted from Pearson and Hartley (1966) by kind permission of E. S. Pearson for the Biometrika Trustees.

table B-7 *Percentage points of the studentized range statistic*

UPPER .10 POINTS

n‡	m† = 2	3	4	5	6	7	8	9	10
1	8.93	13.44	16.36	18.49	20.15	21.51	22.64	23.62	24.48
2	4.13	5.73	6.77	7.54	8.14	8.63	9.05	9.41	9.72
3	3.33	4.47	5.20	5.74	6.16	6.51	6.81	7.06	7.29
4	3.01	3.98	4.59	5.03	5.39	5.68	5.93	6.14	6.33
5	2.85	3.72	4.26	4.66	4.98	5.24	5.46	5.65	5.82
6	2.75	3.56	4.07	4.44	4.73	4.97	5.17	5.34	5.50
7	2.68	3.45	3.93	4.28	4.55	4.78	4.97	5.14	5.28
8	2.63	3.37	3.83	4.17	4.43	4.65	4.83	4.99	5.13
9	2.59	3.32	3.76	4.08	4.34	4.54	4.72	4.87	5.01
10	2.56	3.27	3.70	4.02	4.26	4.47	4.64	4.78	4.91
11	2.54	3.23	3.66	3.96	4.20	4.40	4.57	4.71	4.84
12	2.52	3.20	3.62	3.92	4.16	4.35	4.51	4.65	4.78
13	2.50	3.18	3.59	3.88	4.12	4.30	4.46	4.60	4.72
14	2.49	3.16	3.56	3.85	4.08	4.27	4.42	4.56	4.68
15	2.48	3.14	3.54	3.83	4.05	4.23	4.39	4.52	4.64
16	2.47	3.12	3.52	3.80	4.03	4.21	4.36	4.49	4.61
17	2.46	3.11	3.50	3.78	4.00	4.18	4.33	4.46	4.58
18	2.45	3.10	3.49	3.77	3.98	4.16	4.31	4.44	4.55
19	2.45	3.09	3.47	3.75	3.97	4.14	4.29	4.42	4.53
20	2.44	3.08	3.46	3.74	3.95	4.12	4.27	4.40	4.51
24	2.42	3.05	3.42	3.69	3.90	4.07	4.21	4.34	4.44
30	2.40	3.02	3.39	3.65	3.85	4.02	4.16	4.28	4.38
40	2.38	2.99	3.35	3.60	3.80	3.96	4.10	4.21	4.32
60	2.36	2.96	3.31	3.56	3.75	3.91	4.04	4.16	4.25
120	2.34	2.93	3.28	3.52	3.71	3.86	3.99	4.10	4.19
∞	2.33	2.90	3.24	3.48	3.66	3.81	3.93	4.04	4.13

†Number of ranks spanned by means.
‡Degrees of freedom per mean.

UPPER .10 POINTS										
n	$m = 11$	12	13	14	15	16	17	18	19	20
1	25.24	25.92	26.54	27.10	27.62	28.10	28.54	28.96	29.35	29.71
2	10.01	10.26	10.49	10.70	10.89	11.07	11.24	11.39	11.54	11.68
3	7.49	7.67	7.83	7.98	8.12	8.25	8.37	8.48	8.58	8.68
4	6.49	6.65	6.78	6.91	7.02	7.13	7.23	7.33	7.41	7.50
5	5.97	6.10	6.22	6.34	6.44	6.54	6.63	6.71	6.79	6.86
6	5.64	5.76	5.87	5.98	6.07	6.16	6.25	6.32	6.40	6.47
7	5.41	5.53	5.64	5.74	5.83	5.91	5.99	6.06	6.13	6.19
8	5.25	5.36	5.46	5.56	5.64	5.72	5.80	5.87	5.93	6.00
9	5.13	5.23	5.33	5.42	5.51	5.58	5.66	5.72	5.79	5.85
10	5.03	5.13	5.23	5.32	5.40	5.47	5.54	5.61	5.67	5.73
11	4.95	5.05	5.15	5.23	5.31	5.38	5.45	5.51	5.57	5.63
12	4.89	4.99	5.08	5.16	5.24	5.31	5.37	5.44	5.49	5.55
13	4.83	4.93	5.02	5.10	5.18	5.25	5.31	5.37	5.43	5.48
14	4.79	4.88	4.97	5.05	5.12	5.19	5.26	5.32	5.37	5.43
15	4.75	4.84	4.93	5.01	5.08	5.15	5.21	5.27	5.32	5.38
16	4.71	4.81	4.89	4.97	5.04	5.11	5.17	5.23	5.28	5.33
17	4.68	4.77	4.86	4.93	5.01	5.07	5.13	5.19	5.24	5.30
18	4.65	4.75	4.83	4.90	4.98	5.04	5.10	5.16	5.21	5.26
19	4.63	4.72	4.80	4.88	4.95	5.01	5.07	5.13	5.18	5.23
20	4.61	4.70	4.78	4.85	4.92	4.99	5.05	5.10	5.16	5.20
24	4.54	4.63	4.71	4.78	4.85	4.91	4.97	5.02	5.07	5.12
30	4.47	4.56	4.64	4.71	4.77	4.83	4.89	4.94	4.99	5.03
40	4.41	4.49	4.56	4.63	4.69	4.75	4.81	4.86	4.90	4.95
60	4.34	4.42	4.49	4.56	4.62	4.67	4.73	4.78	4.82	4.86
120	4.28	4.35	4.42	4.48	4.54	4.60	4.65	4.69	4.74	4.78
∞	4.21	4.28	4.35	4.41	4.47	4.52	4.57	4.61	4.65	4.69

table B-7 (continued)

				UPPER .05 POINTS					
n	$m = 2$	3	4	5	6	7	8	9	10
1	17.97	26.98	32.82	37.08	40.41	43.12	45.40	47.36	49.07
2	6.08	8.33	9.80	10.88	11.74	12.44	13.03	13.54	13.99
3	4.50	5.91	6.82	7.50	8.04	8.48	8.85	9.18	9.46
4	3.93	5.04	5.76	6.29	6.71	7.05	7.35	7.60	7.83
5	3.64	4.60	5.22	5.67	6.03	6.33	6.58	6.80	6.99
6	3.46	4.34	4.90	5.30	5.63	5.90	6.12	6.32	6.49
7	3.34	4.16	4.68	5.06	5.36	5.61	5.82	6.00	6.16
8	3.26	4.04	4.53	4.89	5.17	5.40	5.60	5.77	5.92
9	3.20	3.95	4.41	4.76	5.02	5.24	5.43	5.59	5.74
10	3.15	3.88	4.33	4.65	4.91	5.12	5.30	5.46	5.60
11	3.11	3.82	4.26	4.57	4.82	5.03	5.20	5.35	5.49
12	3.08	3.77	4.20	4.51	4.75	4.95	5.12	5.27	5.39
13	3.06	3.73	4.15	4.45	4.69	4.88	5.05	5.19	5.32
14	3.03	3.70	4.11	4.41	4.64	4.83	4.99	5.13	5.25
15	3.01	3.67	4.08	4.37	4.59	4.78	4.94	5.08	5.20
16	3.00	3.65	4.05	4.33	4.56	4.74	4.90	5.03	5.15
17	2.98	3.63	4.02	4.30	4.52	4.70	4.86	4.99	5.11
18	2.97	3.61	4.00	4.28	4.49	4.67	4.82	4.96	5.07
19	2.96	3.59	3.98	4.25	4.47	4.65	4.79	4.92	5.04
20	2.95	3.58	3.96	4.23	4.45	4.62	4.77	4.90	5.01
24	2.92	3.53	3.90	4.17	4.37	4.54	4.68	4.81	4.92
30	2.89	3.49	3.85	4.10	4.30	4.46	4.60	4.72	4.82
40	2.86	3.44	3.79	4.04	4.23	4.39	4.52	4.63	4.73
60	2.83	3.40	3.74	3.98	4.16	4.31	4.44	4.55	4.65
120	2.80	3.36	3.68	3.92	4.10	4.24	4.36	4.47	4.56
∞	2.77	3.31	3.63	3.86	4.03	4.17	4.29	4.39	4.47

UPPER .05 POINTS										
n	$m = 11$	12	13	14	15	16	17	18	19	20
1	50.59	51.96	53.20	54.33	55.36	56.32	57.22	58.04	58.83	59.56
2	14.39	14.75	15.08	15.38	15.65	15.91	16.14	16.37	16.57	16.77
3	9.72	9.95	10.15	10.35	10.52	10.69	10.84	10.98	11.11	11.24
4	8.03	8.21	8.37	8.52	8.66	8.79	8.91	9.03	9.13	9.23
5	7.17	7.32	7.47	7.60	7.72	7.83	7.93	8.03	8.12	8.21
6	6.65	6.79	6.92	7.03	7.14	7.24	7.34	7.43	7.51	7.59
7	6.30	6.43	6.55	6.66	6.76	6.85	6.94	7.02	7.10	7.17
8	6.05	6.18	6.29	6.39	6.48	6.57	6.65	6.73	6.80	6.87
9	5.87	5.98	6.09	6.19	6.28	6.36	6.44	6.51	6.58	6.64
10	5.72	5.83	5.93	6.03	6.11	6.19	6.27	6.34	6.40	6.47
11	5.61	5.71	5.81	5.90	5.98	6.06	6.13	6.20	6.27	6.33
12	5.51	5.61	5.71	5.80	5.88	5.95	6.02	6.09	6.15	6.21
13	5.43	5.53	5.63	5.71	5.79	5.86	5.93	5.99	6.05	6.11
14	5.36	5.46	5.55	5.64	5.71	5.79	5.85	5.91	5.97	6.03
15	5.31	5.40	5.49	5.57	5.65	5.72	5.78	5.85	5.90	5.96
16	5.26	5.35	5.44	5.52	5.59	5.66	5.73	5.79	5.84	5.90
17	5.21	5.31	5.39	5.47	5.54	5.61	5.67	5.73	5.79	5.84
18	5.17	5.27	5.35	5.43	5.50	5.57	5.63	5.69	5.74	5.79
19	5.14	5.23	5.31	5.39	5.46	5.53	5.59	5.65	5.70	5.75
20	5.11	5.20	5.28	5.36	5.43	5.49	5.55	5.61	5.66	5.71
24	5.01	5.10	5.18	5.25	5.32	5.38	5.44	5.49	5.55	5.59
30	4.92	5.00	5.08	5.15	5.21	5.27	5.33	5.38	5.43	5.47
40	4.82	4.90	4.98	5.04	5.11	5.16	5.22	5.27	5.31	5.36
60	4.73	4.81	4.88	4.94	5.00	5.06	5.11	5.15	5.20	5.24
120	4.64	4.71	4.78	4.84	4.90	4.95	5.00	5.04	5.09	5.13
∞	4.55	4.62	4.68	4.74	4.80	4.85	4.89	4.93	4.97	5.01

table B-7 (continued)

	UPPER .01 POINTS								
n	$m = 2$	3	4	5	6	7	8	9	10
1	90.03	135.0	164.3	185.6	202.2	215.8	227.2	237.0	245.6
2	14.04	19.02	22.29	24.72	26.63	28.20	29.53	30.68	31.69
3	8.26	10.62	12.17	13.33	14.24	15.00	15.64	16.20	16.69
4	6.51	8.12	9.17	9.96	10.58	11.10	11.55	11.93	12.27
5	5.70	6.98	7.80	8.42	8.91	9.32	9.67	9.97	10.24
6	5.24	6.33	7.03	7.56	7.97	8.32	8.61	8.87	9.10
7	4.95	5.92	6.54	7.01	7.37	7.68	7.94	8.17	8.37
8	4.75	5.64	6.20	6.62	6.96	7.24	7.47	7.68	7.86
9	4.60	5.43	5.96	6.35	6.66	6.91	7.13	7.33	7.49
10	4.48	5.27	5.77	6.14	6.43	6.67	6.87	7.05	7.21
11	4.39	5.15	5.62	5.97	6.25	6.48	6.67	6.84	6.99
12	4.32	5.05	5.50	5.84	6.10	6.32	6.51	6.67	6.81
13	4.26	4.96	5.40	5.73	5.98	6.19	6.37	6.53	6.67
14	4.21	4.89	5.32	5.63	5.88	6.08	6.26	6.41	6.54
15	4.17	4.84	5.25	5.56	5.80	5.99	6.16	6.31	6.44
16	4.13	4.79	5.19	5.49	5.72	5.92	6.08	6.22	6.35
17	4.10	4.74	5.14	5.43	5.66	5.85	6.01	6.15	6.27
18	4.07	4.70	5.09	5.38	5.60	5.79	5.94	6.08	6.20
19	4.05	4.67	5.05	5.33	5.55	5.73	5.89	6.02	6.14
20	4.02	4.64	5.02	5.29	5.51	5.69	5.84	5.97	6.09
24	3.96	4.55	4.91	5.17	5.37	5.54	5.69	5.81	5.92
30	3.89	4.45	4.80	5.05	5.24	5.40	5.54	5.65	5.76
40	3.82	4.37	4.70	4.93	5.11	5.26	5.39	5.50	5.60
60	3.76	4.28	4.59	4.82	4.99	5.13	5.25	5.36	5.45
120	3.70	4.20	4.50	4.71	4.87	5.01	5.12	5.21	5.30
∞	3.64	4.12	4.40	4.60	4.76	4.88	4.99	5.08	5.16

					UPPER .01 POINTS					
n	$m = 11$	12	13	14	15	16	17	18	19	20
1	253.2	260.0	266.2	271.8	277.0	281.8	286.3	290.4	294.3	298.0
2	32.59	33.40	34.13	34.81	35.43	36.00	36.53	37.03	37.50	37.95
3	17.13	17.53	17.89	18.22	18.52	18.81	19.07	19.32	19.55	19.77
4	12.57	12.84	13.09	13.32	13.53	13.73	13.91	14.08	14.24	14.40
5	10.48	10.70	10.89	11.08	11.24	11.40	11.55	11.68	11.81	11.93
6	9.30	9.48	9.65	9.81	9.95	10.08	10.21	10.32	10.43	10.54
7	8.55	8.71	8.86	9.00	9.12	9.24	9.35	9.46	9.55	9.65
8	8.03	8.18	8.31	8.44	8.55	8.66	8.76	8.85	8.94	9.03
9	7.65	7.78	7.91	8.03	8.13	8.23	8.33	8.41	8.49	8.57
10	7.36	7.49	7.60	7.71	7.81	7.91	7.99	8.08	8.15	8.23
11	7.13	7.25	7.36	7.46	7.56	7.65	7.73	7.81	7.88	7.95
12	6.94	7.06	7.17	7.26	7.36	7.44	7.52	7.59	7.66	7.73
13	6.79	6.90	7.01	7.10	7.19	7.27	7.35	7.42	7.48	7.55
14	6.66	6.77	6.87	6.96	7.05	7.13	7.20	7.27	7.33	7.39
15	6.55	6.66	6.76	6.84	6.93	7.00	7.07	7.14	7.20	7.26
16	6.46	6.56	6.66	6.74	6.82	6.90	6.97	7.03	7.09	7.15
17	6.38	6.48	6.57	6.66	6.73	6.81	6.87	6.94	7.00	7.05
18	6.31	6.41	6.50	6.58	6.65	6.73	6.79	6.85	6.91	6.97
19	6.25	6.34	6.43	6.51	6.58	6.65	6.72	6.78	6.84	6.89
20	6.19	6.28	6.37	6.45	6.52	6.59	6.65	6.71	6.77	6.82
24	6.02	6.11	6.19	6.26	6.33	6.39	6.45	6.51	6.56	6.61
30	5.85	5.93	6.01	6.08	6.14	6.20	6.26	6.31	6.36	6.41
40	5.69	5.76	5.83	5.90	5.96	6.02	6.07	6.12	6.16	6.21
60	5.53	5.60	5.67	5.73	5.78	5.84	5.89	5.93	5.97	6.01
120	5.37	5.44	5.50	5.56	5.61	5.66	5.71	5.75	5.79	5.83
∞	5.23	5.29	5.35	5.40	5.45	5.49	5.54	5.57	5.61	5.65

Source: Reprinted from Pearson and Hartley (1966) by permission of E. S. Pearson for the Biometrika Trustees.

table B-8 *Percentage points for Duncan's multiple range test*†

	$\alpha = .10$																	
n§	m‡ = 2	3	4	5	6	7	8	9	10	11	12	13	14	15	16	17	18	19
2	4.130																	
3	3.328	3.330																
4	3.015	3.074	3.081															
5	2.850	2.934	2.964	2.970														
6	2.748	2.846	2.890	2.908	2.911													
7	2.680	2.785	2.838	2.864	2.876	2.878												
8	2.630	2.742	2.800	2.832	2.849	2.857	2.858											
9	2.592	2.708	2.771	2.808	2.829	2.840	2.845	2.847										
10	2.563	2.682	2.748	2.788	2.813	2.827	2.835	2.839	2.839									
11	2.540	2.660	2.730	2.772	2.799	2.817	2.827	2.833	2.835	2.835								
12	2.521	2.643	2.714	2.759	2.789	2.808	2.821	2.828	2.832	2.833	2.833							
13	2.505	2.628	2.701	2.748	2.779	2.800	2.815	2.824	2.829	2.832	2.832	2.832						
14	2.491	2.616	2.690	2.739	2.771	2.794	2.810	2.820	2.827	2.831	2.832	2.833	2.833					
15	2.479	2.605	2.681	2.731	2.765	2.789	2.805	2.817	2.825	2.830	2.833	2.834	2.834	2.834				
16	2.469	2.596	2.673	2.723	2.759	2.784	2.802	2.815	2.824	2.829	2.833	2.835	2.836	2.836	2.836			
17	2.460	2.588	2.665	2.717	2.753	2.780	2.798	2.812	2.822	2.829	2.834	2.836	2.838	2.838	2.838	2.838		
18	2.452	2.580	2.659	2.712	2.749	2.776	2.796	2.810	2.821	2.828	2.834	2.838	2.840	2.840	2.840	2.840	2.840	
19	2.445	2.574	2.653	2.707	2.745	2.773	2.793	2.808	2.820	2.828	2.834	2.839	2.841	2.842	2.843	2.843	2.843	2.843
20	2.439	2.568	2.648	2.702	2.741	2.770	2.791	2.807	2.819	2.828	2.834	2.839	2.843	2.845	2.845	2.845	2.845	2.845
24	2.420	2.550	2.632	2.688	2.729	2.760	2.783	2.801	2.816	2.827	2.835	2.842	2.848	2.851	2.854	2.856	2.857	2.857
30	2.400	2.532	2.615	2.674	2.717	2.750	2.776	2.796	2.813	2.826	2.837	2.846	2.853	2.859	2.863	2.867	2.869	2.871
40	2.381	2.514	2.600	2.660	2.705	2.741	2.769	2.791	2.810	2.825	2.838	2.849	2.858	2.866	2.873	2.878	2.883	2.887
60	2.363	2.497	2.584	2.646	2.694	2.731	2.761	2.786	2.807	2.825	2.839	2.853	2.864	2.874	2.883	2.890	2.897	2.903
120	2.344	2.479	2.568	2.632	2.682	2.722	2.754	2.781	2.804	2.824	2.842	2.857	2.871	2.883	2.893	2.903	2.912	2.920
∞	2.326	2.462	2.552	2.619	2.670	2.712	2.746	2.776	2.801	2.824	2.844	2.861	2.877	2.892	2.905	2.918	2.929	2.939

†The entries in this table were tabulated and made available by H. Leon Harter.
‡Number of ranks spanned by means.
§Degrees of freedom per mean.

table B-8 (continued)

	$\alpha = .05$																	
n	$m = 2$	3	4	5	6	7	8	9	10	11	12	13	14	15	16	17	18	19
2	6.085																	
3	4.501	4.516																
4	3.927	4.013	4.033															
5	3.635	3.749	3.797	3.814														
6	3.461	3.587	3.649	3.680	3.694													
7	3.344	3.477	3.548	3.588	3.611	3.622												
8	3.261	3.399	3.475	3.521	3.549	3.566	3.575											
9	3.199	3.339	3.420	3.470	3.502	3.523	3.536	3.544										
10	3.151	3.293	3.376	3.430	3.465	3.489	3.505	3.516	3.522									
11	3.113	3.256	3.342	3.397	3.435	3.462	3.480	3.493	3.501	3.506								
12	3.082	3.225	3.313	3.370	3.410	3.439	3.459	3.474	3.484	3.491	3.496							
13	3.055	3.200	3.289	3.348	3.389	3.419	3.442	3.458	3.470	3.478	3.484	3.488						
14	3.033	3.178	3.268	3.329	3.372	3.403	3.426	3.444	3.457	3.467	3.474	3.479	3.482					
15	3.014	3.160	3.250	3.312	3.356	3.389	3.413	3.432	3.446	3.457	3.465	3.471	3.476	3.478				
16	2.998	3.144	3.235	3.298	3.343	3.376	3.402	3.422	3.437	3.449	3.458	3.465	3.470	3.473	3.477			
17	2.984	3.130	3.222	3.285	3.331	3.366	3.392	3.412	3.429	3.441	3.451	3.459	3.465	3.469	3.473	3.475		
18	2.971	3.118	3.210	3.274	3.321	3.356	3.383	3.405	3.421	3.435	3.445	3.454	3.460	3.465	3.470	3.472	3.474	
19	2.960	3.107	3.199	3.264	3.311	3.347	3.375	3.397	3.415	3.429	3.440	3.449	3.456	3.462	3.467	3.470	3.472	3.473
20	2.950	3.097	3.190	3.255	3.303	3.339	3.368	3.391	3.409	3.424	3.436	3.445	3.453	3.459	3.464	3.467	3.470	3.472
24	2.919	3.066	3.160	3.226	3.276	3.315	3.345	3.370	3.390	3.406	3.420	3.432	3.441	3.449	3.456	3.461	3.465	3.469
30	2.888	3.035	3.131	3.199	3.250	3.290	3.322	3.349	3.371	3.389	3.405	3.418	3.430	3.439	3.447	3.454	3.460	3.466
40	2.858	3.006	3.102	3.171	3.224	3.266	3.300	3.328	3.352	3.373	3.390	3.405	3.418	3.429	3.439	3.448	3.456	3.463
60	2.829	2.976	3.073	3.143	3.198	3.241	3.277	3.307	3.333	3.355	3.374	3.391	3.406	3.419	3.431	3.442	3.451	3.460
120	2.800	2.947	3.045	3.116	3.172	3.217	3.254	3.287	3.314	3.337	3.359	3.377	3.394	3.409	3.423	3.435	3.446	3.457
∞	2.772	2.918	3.017	3.089	3.146	3.193	3.232	3.265	3.294	3.320	3.343	3.363	3.382	3.399	3.414	3.428	3.442	3.454

table B-8 (continued)

$\alpha = .01$

n	$m = 2$	3	4	5	6	7	8	9	10	11	12	13	14	15	16	17	18	19
2	14.04																	
3	8.261	8.321																
4	6.512	6.677	6.740															
5	5.702	5.893	5.989	6.040														
6	5.243	5.439	5.549	5.614	5.655													
7	4.949	5.145	5.260	5.334	5.383	5.416												
8	4.746	4.939	5.057	5.135	5.189	5.227	5.256											
9	4.596	4.787	4.906	4.986	5.043	5.086	5.118	5.142										
10	4.482	4.671	4.790	4.871	4.931	4.975	5.010	5.037	5.058									
11	4.392	4.579	4.697	4.780	4.841	4.887	4.924	4.952	4.975	4.994								
12	4.320	4.504	4.622	4.706	4.767	4.815	4.852	4.883	4.907	4.927	4.944							
13	4.260	4.442	4.560	4.644	4.706	4.755	4.793	4.824	4.850	4.872	4.889	4.904						
14	4.210	4.391	4.508	4.591	4.654	4.704	4.743	4.775	4.802	4.824	4.843	4.859	4.872					
15	4.168	4.347	4.463	4.547	4.610	4.660	4.700	4.733	4.760	4.783	4.803	4.820	4.834	4.846				
16	4.131	4.309	4.425	4.509	4.572	4.622	4.663	4.696	4.724	4.748	4.768	4.786	4.800	4.813	4.825			
17	4.099	4.275	4.391	4.475	4.539	4.589	4.630	4.664	4.693	4.717	4.738	4.756	4.771	4.785	4.797	4.807		
18	4.071	4.246	4.362	4.445	4.509	4.560	4.601	4.635	4.664	4.689	4.711	4.729	4.745	4.759	4.772	4.783	4.792	
19	4.046	4.220	4.335	4.419	4.483	4.534	4.575	4.610	4.639	4.665	4.686	4.705	4.722	4.736	4.749	4.761	4.771	4.780
20	4.024	4.197	4.312	4.395	4.459	4.510	4.552	4.587	4.617	4.642	4.664	4.684	4.701	4.716	4.729	4.741	4.751	4.761
24	3.956	4.126	4.239	4.322	4.386	4.437	4.480	4.516	4.546	4.573	4.596	4.616	4.634	4.651	4.665	4.678	4.690	4.700
30	3.889	4.056	4.168	4.250	4.314	4.366	4.409	4.445	4.477	4.504	4.528	4.550	4.569	4.586	4.601	4.615	4.628	4.640
40	3.825	3.988	4.098	4.180	4.244	4.296	4.339	4.376	4.408	4.436	4.461	4.483	4.503	4.521	4.537	4.553	4.566	4.579
60	3.762	3.922	4.031	4.111	4.174	4.226	4.270	4.307	4.340	4.368	4.394	4.417	4.438	4.456	4.474	4.490	4.504	4.518
120	3.702	3.858	3.965	4.044	4.107	4.158	4.202	4.239	4.272	4.301	4.327	4.351	4.372	4.392	4.410	4.426	4.442	4.456
∞	3.643	3.796	3.900	3.978	4.040	4.091	4.135	4.172	4.205	4.235	4.261	4.285	4.307	4.327	4.345	4.363	4.379	4.394

table B-8 (continued)

									$\alpha = .005$									
n	$m = 2$	3	4	5	6	7	8	9	10	11	12	13	14	15	16	17	18	19
2	19.93																	
3	10.55	10.63																
4	7.916	8.126	8.210															
5	6.751	6.980	7.100	7.167														
6	6.105	6.334	6.466	6.547	6.600													
7	5.699	5.922	6.057	6.145	6.207	6.250												
8	5.420	5.638	5.773	5.864	5.930	5.978	6.014											
9	5.218	5.430	5.565	5.657	5.725	5.776	5.815	5.846										
10	5.065	5.273	5.405	5.498	5.567	5.620	5.662	5.695	5.722									
11	4.945	5.149	5.280	5.372	5.442	5.496	5.539	5.574	5.603	5.626								
12	4.849	5.048	5.178	5.270	5.341	5.396	5.439	5.475	5.505	5.531	5.552							
13	4.770	4.966	5.094	5.186	5.256	5.312	5.356	5.393	5.424	5.450	5.472	5.492						
14	4.704	4.897	5.023	5.116	5.185	5.241	5.286	5.324	5.355	5.382	5.405	5.425	5.442					
15	4.647	4.838	4.964	5.055	5.125	5.181	5.226	5.264	5.297	5.324	5.348	5.368	5.386	5.402				
16	4.599	4.787	4.912	5.003	5.073	5.129	5.175	5.213	5.245	5.273	5.298	5.319	5.338	5.354	5.368			
17	4.557	4.744	4.867	4.958	5.027	5.084	5.130	5.168	5.201	5.229	5.254	5.275	5.295	5.311	5.327	5.340		
18	4.521	4.705	4.828	4.918	4.987	5.043	5.090	5.129	5.162	5.190	5.215	5.237	5.256	5.274	5.289	5.303	5.316	
19	4.488	4.671	4.793	4.883	4.952	5.008	5.054	5.093	5.127	5.156	5.181	5.203	5.222	5.240	5.256	5.270	5.283	5.295
20	4.460	4.641	4.762	4.851	4.920	4.976	5.022	5.061	5.095	5.124	5.150	5.172	5.193	5.210	5.226	5.241	5.254	5.266
24	4.371	4.547	4.666	4.753	4.822	4.877	4.924	4.963	4.997	5.027	5.053	5.076	5.097	5.116	5.133	5.148	5.162	5.175
30	4.285	4.456	4.572	4.658	4.726	4.781	4.827	4.867	4.901	4.931	4.958	4.981	5.003	5.022	5.040	5.056	5.071	5.085
40	4.202	4.369	4.482	4.566	4.632	4.687	4.733	4.772	4.806	4.837	4.864	4.888	4.910	4.930	4.948	4.965	4.980	4.995
60	4.122	4.284	4.394	4.476	4.541	4.595	4.640	4.679	4.713	4.744	4.771	4.796	4.818	4.838	4.857	4.874	4.890	4.905
120	4.045	4.201	4.308	4.388	4.452	4.505	4.550	4.588	4.622	4.652	4.679	4.704	4.726	4.747	4.766	4.784	4.800	4.815
∞	3.970	4.121	4.225	4.303	4.365	4.417	4.461	4.499	4.532	4.562	4.589	4.614	4.636	4.657	4.676	4.694	4.710	4.726

table B-8 (continued)

n	$m = 2$	3	4	5	6	7	8	9	10	11	12	13	14	15	16	17	18	19
									$\alpha = .001$									
2	44.69																	
3	18.28	18.45																
4	12.18	12.52	12.67															
5	9.714	10.05	10.24	10.35														
6	8.427	8.743	8.932	9.055	9.139													
7	7.648	7.943	8.127	8.252	8.342	8.409												
8	7.130	7.407	7.584	7.708	7.799	7.869	7.924											
9	6.762	7.024	7.195	7.316	7.407	7.478	7.535	7.582										
10	6.487	6.738	6.902	7.021	7.111	7.182	7.240	7.287	7.327									
11	6.275	6.516	6.676	6.791	6.880	6.950	7.008	7.056	7.097	7.132								
12	6.106	6.340	6.494	6.607	6.695	6.765	6.822	6.870	6.911	6.947	6.978							
13	5.970	6.195	6.346	6.457	6.543	6.612	6.670	6.718	6.759	6.795	6.826	6.854						
14	5.856	6.075	6.223	6.332	6.416	6.485	6.542	6.590	6.631	6.667	6.699	6.727	6.752					
15	5.760	5.974	6.119	6.225	6.309	6.377	6.433	6.481	6.522	6.558	6.590	6.619	6.644	6.666				
16	5.678	5.888	6.030	6.135	6.217	6.284	6.340	6.388	6.429	6.465	6.497	6.525	6.551	6.574	6.595			
17	5.608	5.813	5.953	6.056	6.138	6.204	6.260	6.307	6.348	6.384	6.416	6.444	6.470	6.493	6.514	6.533		
18	5.546	5.748	5.886	5.988	6.068	6.134	6.189	6.236	6.277	6.313	6.345	6.373	6.399	6.422	6.443	6.462	6.480	
19	5.492	5.691	5.826	5.927	6.007	6.072	6.127	6.174	6.214	6.250	6.281	6.310	6.336	6.359	6.380	6.400	6.418	6.434
20	5.444	5.640	5.774	5.873	5.952	6.017	6.071	6.117	6.158	6.193	6.225	6.254	6.279	6.303	6.324	6.344	6.362	6.379
24	5.297	5.484	5.612	5.708	5.784	5.846	5.899	5.945	5.984	6.020	6.051	6.079	6.105	6.129	6.150	6.170	6.188	6.205
30	5.156	5.335	5.457	5.549	5.622	5.682	5.734	5.778	5.817	5.851	5.882	5.910	5.935	5.958	5.980	6.000	6.018	6.036
40	5.022	5.191	5.308	5.396	5.466	5.524	5.574	5.617	5.654	5.688	5.718	5.745	5.770	5.793	5.814	5.834	5.852	5.869
60	4.894	5.055	5.166	5.249	5.317	5.372	5.420	5.461	5.498	5.530	5.559	5.586	5.610	5.632	5.653	5.672	5.690	5.707
120	4.771	4.924	5.029	5.109	5.173	5.226	5.271	5.311	5.346	5.377	5.405	5.431	5.454	5.476	5.496	5.515	5.532	5.549
∞	4.654	4.798	4.898	4.974	5.034	5.085	5.128	5.166	5.199	5.229	5.256	5.280	5.303	5.324	5.343	5.361	5.378	5.394

Source: Reprinted from Edwards (1960) by permission of the publisher.

table B-9 *Percentage points of Dunnett's t statistic*

	$\alpha = .05$, ONE-TAILED TEST								
df	$p† = 1$	2	3	4	5	6	7	8	9
5	2.02	2.44	2.68	2.85	2.98	3.08	3.16	3.24	3.30
6	1.94	2.34	2.56	2.71	2.83	2.92	3.00	3.07	3.12
7	1.89	2.27	2.48	2.62	2.73	2.82	2.89	2.95	3.01
8	1.86	2.22	2.42	2.55	2.66	2.74	2.81	2.87	2.92
9	1.83	2.18	2.37	2.50	2.60	2.68	2.75	2.81	2.86
10	1.81	2.15	2.34	2.47	2.56	2.64	2.70	2.76	2.81
11	1.80	2.13	2.31	2.44	2.53	2.60	2.67	2.72	2.77
12	1.78	2.11	2.29	2.41	2.50	2.58	2.64	2.69	2.74
13	1.77	2.09	2.27	2.39	2.48	2.55	2.61	2.66	2.71
14	1.76	2.08	2.25	2.37	2.46	2.53	2.59	2.64	2.69
15	1.75	2.07	2.24	2.36	2.44	2.51	2.57	2.62	2.67
16	1.75	2.06	2.23	2.34	2.43	2.50	2.56	2.61	2.65
17	1.74	2.05	2.22	2.33	2.42	2.49	2.54	2.59	2.64
18	1.73	2.04	2.21	2.32	2.41	2.48	2.53	2.58	2.62
19	1.73	2.03	2.20	2.31	2.40	2.47	2.52	2.57	2.61
20	1.72	2.03	2.19	2.30	2.39	2.46	2.51	2.56	2.60
24	1.71	2.01	2.17	2.28	2.36	2.43	2.48	2.53	2.57
30	1.70	1.99	2.15	2.25	2.33	2.40	2.45	2.50	2.54
40	1.68	1.97	2.13	2.23	2.31	2.37	2.42	2.47	2.51
60	1.67	1.95	2.10	2.21	2.28	2.35	2.39	2.44	2.48
120	1.66	1.93	2.08	2.18	2.26	2.32	2.37	2.41	2.45
∞	1.64	1.92	2.06	2.16	2.23	2.29	2.34	2.38	2.42

†Number of treatment means (excluding the control).

table B-9 (continued)

				α = .01, ONE-TAILED TEST					
df	*p* = 1	2	3	4	5	6	7	8	9
5	3.37	3.90	4.21	4.43	4.60	4.73	4.85	4.94	5.03
6	3.14	3.61	3.88	4.07	4.21	4.33	4.43	4.51	4.59
7	3.00	3.42	3.66	3.83	3.96	4.07	4.15	4.23	4.30
8	2.90	3.29	3.51	3.67	3.79	3.88	3.96	4.03	4.09
9	2.82	3.19	3.40	3.55	3.66	3.75	3.82	3.89	3.94
10	2.76	3.11	3.31	3.45	3.56	3.64	3.71	3.78	3.83
11	2.72	3.06	3.25	3.38	3.48	3.56	3.63	3.69	3.74
12	2.68	3.01	3.19	3.32	3.42	3.50	3.56	3.62	3.67
13	2.65	2.97	3.15	3.27	3.37	3.44	3.51	3.56	3.61
14	2.62	2.94	3.11	3.23	3.32	3.40	3.46	3.51	3.56
15	2.60	2.91	3.08	3.20	3.29	3.36	3.42	3.47	3.52
16	2.58	2.88	3.05	3.17	3.26	3.33	3.39	3.44	3.48
17	2.57	2.86	3.03	3.14	3.23	3.30	3.36	3.41	3.45
18	2.55	2.84	3.01	3.12	3.21	3.27	3.33	3.38	3.42
19	2.54	2.83	2.99	3.10	3.18	3.25	3.31	3.36	3.40
20	2.53	2.81	2.97	3.08	3.17	3.23	3.29	3.34	3.38
24	2.49	2.77	2.92	3.03	3.11	3.17	3.22	3.27	3.31
30	2.46	2.72	2.87	2.97	3.05	3.11	3.16	3.21	3.24
40	2.42	2.68	2.82	2.92	2.99	3.05	3.10	3.14	3.18
60	2.39	2.64	2.78	2.87	2.94	3.00	3.04	3.08	3.12
120	2.36	2.60	2.73	2.82	2.89	2.94	2.99	3.03	3.06
∞	2.33	2.56	2.68	2.77	2.84	2.89	2.93	2.97	3.00

Source: Reprinted from Dunnett (1955) by permission of C. W. Dunnett and the American Statistical Association.

table B-9 (continued)

						$\alpha = .05$, TWO-TAILED TEST‡								
df	p† = 1	2	3	4	5	6	7	8	9	10	11	12	15	20
5	2.57	$3.03^{2.3}$	$3.29^{3.6}$	$3.48^{4.6}$	$3.62^{5.4}$	$3.73^{5.9}$	$3.82^{6.4}$	$3.90^{6.8}$	$3.97^{7.2}$	$4.03^{7.5}$	$4.09^{7.8}$	$4.14^{8.0}$	$4.26^{8.7}$	$4.42^{9.4}$
6	2.45	$2.86^{2.1}$	$3.10^{3.4}$	$3.26^{4.3}$	$3.39^{5.0}$	$3.49^{5.8}$	$3.57^{6.0}$	$3.64^{6.4}$	$3.71^{6.8}$	$3.76^{7.1}$	$3.81^{7.4}$	$3.86^{7.6}$	$3.97^{8.2}$	$4.11^{9.0}$
7	2.36	$2.75^{2.0}$	$2.97^{3.2}$	$3.12^{4.1}$	$3.24^{4.8}$	$3.33^{5.3}$	$3.41^{5.7}$	$3.47^{6.1}$	$3.53^{6.5}$	$3.58^{6.7}$	$3.63^{7.0}$	$3.67^{7.2}$	$3.78^{7.8}$	$3.91^{8.6}$
8	2.31	$2.67^{2.0}$	$2.88^{3.1}$	$3.02^{3.9}$	$3.13^{4.5}$	$3.22^{5.1}$	$3.29^{5.5}$	$3.35^{5.9}$	$3.41^{6.2}$	$3.46^{6.5}$	$3.50^{6.7}$	$3.54^{6.9}$	$3.64^{7.5}$	$3.76^{8.2}$
9	2.26	$2.61^{1.9}$	$2.81^{3.0}$	$2.95^{3.8}$	$3.05^{4.4}$	$3.14^{4.9}$	$3.20^{5.3}$	$3.26^{5.6}$	$3.32^{5.9}$	$3.36^{6.2}$	$3.40^{6.5}$	$3.44^{6.7}$	$3.53^{7.2}$	$3.65^{7.9}$
10	2.23	$2.57^{1.8}$	$2.76^{2.9}$	$2.89^{3.6}$	$2.99^{4.2}$	$3.07^{4.7}$	$3.14^{5.1}$	$3.19^{5.4}$	$3.24^{5.7}$	$3.29^{6.0}$	$3.33^{6.2}$	$3.36^{6.5}$	$3.45^{7.0}$	$3.57^{7.7}$
11	2.20	$2.53^{1.8}$	$2.72^{2.8}$	$2.84^{3.5}$	$2.94^{4.1}$	$3.02^{4.6}$	$3.08^{4.9}$	$3.14^{5.3}$	$3.19^{5.6}$	$3.23^{5.8}$	$3.27^{6.1}$	$3.30^{6.3}$	$3.39^{6.8}$	$3.50^{7.5}$
12	2.18	$2.50^{1.7}$	$2.68^{2.7}$	$2.81^{3.4}$	$2.90^{4.0}$	$2.98^{4.4}$	$3.04^{4.8}$	$3.09^{5.1}$	$3.14^{5.4}$	$3.18^{5.7}$	$3.22^{5.9}$	$3.25^{6.1}$	$3.34^{6.6}$	$3.45^{7.3}$
13	2.16	$2.48^{1.7}$	$2.65^{2.7}$	$2.78^{3.4}$	$2.87^{3.9}$	$2.94^{4.3}$	$3.00^{4.7}$	$3.06^{5.0}$	$3.10^{5.3}$	$3.14^{5.5}$	$3.18^{5.8}$	$3.21^{6.0}$	$3.29^{6.5}$	$3.40^{7.1}$
14	2.14	$2.46^{1.7}$	$2.63^{2.6}$	$2.75^{3.3}$	$2.84^{3.8}$	$2.91^{4.2}$	$2.97^{4.6}$	$3.02^{4.9}$	$3.07^{5.2}$	$3.11^{5.4}$	$3.14^{5.6}$	$3.18^{5.8}$	$3.26^{6.3}$	$3.36^{7.0}$
15	2.13	$2.44^{1.7}$	$2.61^{2.6}$	$2.73^{3.2}$	$2.82^{3.8}$	$2.89^{4.2}$	$2.95^{4.5}$	$3.00^{4.8}$	$3.04^{5.1}$	$3.08^{5.3}$	$3.12^{5.5}$	$3.15^{5.7}$	$3.23^{6.2}$	$3.33^{6.8}$
16	2.12	$2.42^{1.6}$	$2.59^{2.5}$	$2.71^{3.2}$	$2.80^{3.7}$	$2.87^{4.1}$	$2.92^{4.4}$	$2.97^{4.7}$	$3.02^{5.0}$	$3.06^{5.2}$	$3.09^{5.4}$	$3.12^{5.6}$	$3.20^{6.1}$	$3.30^{6.7}$
17	2.11	$2.41^{1.6}$	$2.58^{2.5}$	$2.69^{3.1}$	$2.78^{3.6}$	$2.85^{4.0}$	$2.90^{4.4}$	$2.95^{4.7}$	$3.00^{4.9}$	$3.03^{5.1}$	$3.07^{5.3}$	$3.10^{5.5}$	$3.18^{6.0}$	$3.27^{6.4}$
18	2.10	$2.40^{1.6}$	$2.56^{2.5}$	$2.68^{3.1}$	$2.76^{3.6}$	$2.83^{4.0}$	$2.89^{4.3}$	$2.94^{4.6}$	$2.98^{4.8}$	$3.01^{5.1}$	$3.05^{5.3}$	$3.08^{5.4}$	$3.16^{5.9}$	$3.25^{6.5}$
19	2.09	$2.39^{1.6}$	$2.55^{2.5}$	$2.66^{3.1}$	$2.75^{3.5}$	$2.81^{3.9}$	$2.87^{4.2}$	$2.92^{4.5}$	$2.96^{4.8}$	$3.00^{5.0}$	$3.03^{5.2}$	$3.06^{5.4}$	$3.14^{5.8}$	$3.23^{6.4}$
20	2.09	$2.38^{1.6}$	$2.54^{2.4}$	$2.65^{3.0}$	$2.73^{3.5}$	$2.80^{3.9}$	$2.86^{4.2}$	$2.90^{4.5}$	$2.95^{4.7}$	$2.98^{4.9}$	$3.02^{5.1}$	$3.05^{5.3}$	$3.12^{5.7}$	$3.22^{6.3}$
24	2.06	$2.35^{1.5}$	$2.51^{2.3}$	$2.61^{2.9}$	$2.70^{3.4}$	$2.76^{3.7}$	$2.81^{4.0}$	$2.86^{4.3}$	$2.90^{4.5}$	$2.94^{4.7}$	$2.97^{4.9}$	$3.00^{5.1}$	$3.07^{5.5}$	$3.16^{6.0}$
30	2.04	$2.32^{1.5}$	$2.47^{2.3}$	$2.58^{2.8}$	$2.66^{3.2}$	$2.72^{3.6}$	$2.77^{3.9}$	$2.82^{4.1}$	$2.86^{4.3}$	$2.89^{4.5}$	$2.92^{4.7}$	$2.95^{4.8}$	$3.02^{5.2}$	$3.11^{5.8}$
40	2.02	$2.29^{1.4}$	$2.44^{2.2}$	$2.54^{2.7}$	$2.62^{3.1}$	$2.68^{3.4}$	$2.73^{3.7}$	$2.77^{3.9}$	$2.81^{4.1}$	$2.85^{4.3}$	$2.87^{4.5}$	$2.90^{4.6}$	$2.97^{5.0}$	$3.06^{5.5}$
60	2.00	$2.27^{1.4}$	$2.41^{2.1}$	$2.51^{2.6}$	$2.58^{3.0}$	$2.64^{3.3}$	$2.69^{3.5}$	$2.73^{3.7}$	$2.77^{3.9}$	$2.80^{4.1}$	$2.83^{4.2}$	$2.86^{4.4}$	$2.92^{4.7}$	$3.00^{5.1}$
120	1.98	$2.24^{1.3}$	$2.38^{2.0}$	$2.47^{2.5}$	$2.55^{2.8}$	$2.60^{3.1}$	$2.65^{3.3}$	$2.69^{3.5}$	$2.73^{3.7}$	$2.76^{3.8}$	$2.79^{4.0}$	$2.81^{4.1}$	$2.87^{4.4}$	$2.95^{4.8}$
∞	1.96	$2.21^{1.3}$	$2.35^{1.9}$	$2.44^{2.3}$	$2.51^{2.7}$	$2.57^{2.9}$	$2.61^{3.1}$	$2.65^{3.3}$	$2.69^{3.5}$	$2.72^{3.6}$	$2.74^{3.7}$	$2.77^{3.8}$	$2.83^{4.1}$	$2.91^{4.5}$

†Number of treatment means (excluding the control).
‡The tabular value is the critical value of t appropriate when $\rho = 0.5$ or $n_c/n_t = 1$. The value shown as a superscript, when multiplied by $(1 - 2\rho)/(1 - \rho)$ or $1 - n_t/n_c$, gives the percentage increase required in the critical value of t valid for $.125 < \rho < .5$ or $n_c/n_t > 1$.

table B-9 (continued)

						$\alpha = .01$, TWO-TAILED TEST							
df	$p = 1$	2	3	4	5	6	7	8	9	10	11	12	15	20
5	4.03	4.63$^{1.8}$	4.98$^{3.0}$	5.22$^{3.9}$	5.41$^{4.6}$	5.56$^{5.2}$	5.69$^{5.7}$	5.80$^{6.1}$	5.89$^{6.5}$	5.98$^{6.9}$	6.05$^{7.2}$	6.12$^{7.4}$	6.30$^{8.1}$	6.52$^{9.0}$
6	3.71	4.21$^{1.6}$	4.51$^{2.7}$	4.71$^{3.5}$	4.87$^{4.1}$	5.00$^{4.6}$	5.10$^{5.1}$	5.20$^{5.5}$	5.28$^{5.8}$	5.35$^{6.1}$	5.41$^{6.4}$	5.47$^{6.7}$	5.62$^{7.3}$	5.81$^{8.1}$
7	3.50	3.95$^{1.5}$	4.21$^{2.4}$	4.39$^{3.1}$	4.53$^{3.7}$	4.64$^{4.2}$	4.74$^{4.6}$	4.82$^{5.0}$	4.89$^{5.3}$	4.95$^{5.6}$	5.01$^{5.8}$	5.06$^{6.1}$	5.19$^{6.7}$	5.36$^{7.4}$
8	3.36	3.77$^{1.3}$	4.00$^{2.2}$	4.17$^{2.9}$	4.29$^{3.4}$	4.40$^{3.9}$	4.48$^{4.2}$	4.56$^{4.6}$	4.62$^{4.9}$	4.68$^{5.1}$	4.73$^{5.4}$	4.78$^{5.6}$	4.90$^{6.1}$	5.05$^{6.9}$
9	3.25	3.63$^{1.2}$	3.85$^{2.1}$	4.01$^{2.7}$	4.12$^{3.2}$	4.22$^{3.6}$	4.30$^{3.9}$	4.37$^{4.2}$	4.43$^{4.5}$	4.48$^{4.8}$	4.53$^{5.0}$	4.57$^{5.2}$	4.68$^{5.7}$	4.82$^{6.4}$
10	3.17	3.53$^{1.2}$	3.74$^{1.9}$	3.88$^{2.5}$	3.99$^{3.0}$	4.08$^{3.4}$	4.16$^{3.7}$	4.22$^{4.0}$	4.28$^{4.2}$	4.33$^{4.5}$	4.37$^{4.7}$	4.42$^{4.9}$	4.52$^{5.4}$	4.65$^{6.0}$
11	3.11	3.45$^{1.1}$	3.65$^{1.8}$	3.79$^{2.4}$	3.89$^{2.8}$	3.98$^{3.2}$	4.05$^{3.5}$	4.11$^{3.8}$	4.16$^{4.0}$	4.21$^{4.2}$	4.25$^{4.4}$	4.29$^{4.6}$	4.39$^{5.1}$	4.52$^{5.7}$
12	3.05	3.39$^{1.1}$	3.58$^{1.7}$	3.71$^{2.3}$	3.81$^{2.7}$	3.89$^{3.0}$	3.96$^{3.3}$	4.02$^{3.6}$	4.07$^{3.8}$	4.12$^{4.0}$	4.16$^{4.2}$	4.19$^{4.4}$	4.29$^{4.8}$	4.41$^{5.4}$
13	3.01	3.33$^{1.0}$	3.52$^{1.7}$	3.65$^{2.2}$	3.74$^{2.6}$	3.82$^{2.9}$	3.89$^{3.2}$	3.94$^{3.4}$	3.99$^{3.6}$	4.04$^{3.8}$	4.08$^{4.0}$	4.11$^{4.2}$	4.20$^{4.6}$	4.32$^{5.2}$
14	2.98	3.29$^{1.0}$	3.47$^{1.6}$	3.59$^{2.1}$	3.69$^{2.5}$	3.76$^{2.8}$	3.83$^{3.0}$	3.88$^{3.3}$	3.93$^{3.5}$	3.97$^{3.7}$	4.01$^{3.9}$	4.05$^{4.0}$	4.13$^{4.4}$	4.24$^{5.0}$
15	2.95	3.25$^{0.9}$	3.43$^{1.5}$	3.55$^{2.0}$	3.64$^{2.4}$	3.71$^{2.7}$	3.78$^{2.9}$	3.83$^{3.2}$	3.88$^{3.4}$	3.92$^{3.6}$	3.95$^{3.7}$	3.99$^{3.9}$	4.07$^{4.3}$	4.18$^{4.8}$
16	2.92	3.22$^{0.9}$	3.39$^{1.5}$	3.51$^{1.9}$	3.60$^{2.3}$	3.67$^{2.6}$	3.73$^{2.8}$	3.78$^{3.1}$	3.83$^{3.3}$	3.87$^{3.4}$	3.91$^{3.6}$	3.94$^{3.8}$	4.02$^{4.1}$	4.13$^{4.6}$
17	2.90	3.19$^{0.9}$	3.36$^{1.5}$	3.47$^{1.9}$	3.56$^{2.2}$	3.63$^{2.5}$	3.69$^{2.7}$	3.74$^{3.0}$	3.79$^{3.2}$	3.83$^{3.3}$	3.86$^{3.5}$	3.90$^{3.6}$	3.98$^{4.0}$	4.08$^{4.5}$
18	2.88	3.17$^{0.9}$	3.33$^{1.4}$	3.44$^{1.8}$	3.53$^{2.2}$	3.60$^{2.4}$	3.66$^{2.7}$	3.71$^{2.9}$	3.75$^{3.1}$	3.79$^{3.2}$	3.83$^{3.4}$	3.86$^{3.5}$	3.94$^{3.9}$	4.04$^{4.4}$
19	2.86	3.15$^{0.9}$	3.31$^{1.4}$	3.42$^{1.8}$	3.50$^{2.1}$	3.57$^{2.4}$	3.63$^{2.6}$	3.68$^{2.8}$	3.72$^{3.0}$	3.76$^{3.2}$	3.79$^{3.3}$	3.83$^{3.4}$	3.90$^{3.8}$	4.00$^{4.3}$
20	2.85	3.13$^{0.8}$	3.29$^{1.4}$	3.40$^{1.7}$	3.48$^{2.1}$	3.55$^{2.3}$	3.60$^{2.5}$	3.65$^{2.7}$	3.69$^{2.9}$	3.73$^{3.1}$	3.77$^{3.2}$	3.80$^{3.4}$	3.87$^{3.7}$	3.97$^{4.2}$
24	2.80	3.07$^{0.8}$	3.22$^{1.3}$	3.32$^{1.6}$	3.40$^{1.9}$	3.47$^{2.1}$	3.52$^{2.4}$	3.57$^{2.5}$	3.61$^{2.7}$	3.64$^{2.8}$	3.68$^{3.0}$	3.70$^{3.1}$	3.78$^{3.4}$	3.87$^{3.8}$
30	2.75	3.01$^{0.7}$	3.15$^{1.2}$	3.25$^{1.5}$	3.33$^{1.8}$	3.39$^{2.0}$	3.44$^{2.2}$	3.49$^{2.3}$	3.52$^{2.5}$	3.56$^{2.6}$	3.59$^{2.7}$	3.62$^{2.8}$	3.69$^{3.1}$	3.78$^{3.5}$
40	2.70	2.95$^{0.7}$	3.09$^{1.1}$	3.19$^{1.4}$	3.26$^{1.6}$	3.32$^{1.8}$	3.37$^{2.0}$	3.41$^{2.1}$	3.44$^{2.3}$	3.48$^{2.4}$	3.51$^{2.5}$	3.53$^{2.6}$	3.60$^{2.8}$	3.68$^{3.2}$
60	2.66	2.90$^{0.6}$	3.03$^{1.0}$	3.12$^{1.3}$	3.19$^{1.5}$	3.25$^{1.6}$	3.29$^{1.8}$	3.33$^{1.9}$	3.37$^{2.0}$	3.40$^{2.1}$	3.42$^{2.2}$	3.45$^{2.3}$	3.51$^{2.5}$	3.59$^{2.8}$
120	2.62	2.85$^{0.6}$	2.97$^{0.9}$	3.06$^{1.1}$	3.12$^{1.3}$	3.18$^{1.5}$	3.22$^{1.6}$	3.26$^{1.7}$	3.29$^{1.8}$	3.32$^{1.9}$	3.35$^{2.0}$	3.37$^{2.1}$	3.43$^{2.2}$	3.51$^{2.5}$
∞	2.58	2.79$^{0.5}$	2.92$^{0.8}$	3.00$^{1.0}$	3.06$^{1.2}$	3.11$^{1.3}$	3.15$^{1.4}$	3.19$^{1.5}$	3.22$^{1.6}$	3.25$^{1.7}$	3.27$^{1.7}$	3.29$^{1.8}$	3.35$^{1.9}$	3.42$^{2.2}$

Source: Reprinted from Dunnett (1964) by permission of C. W. Dunnett.

table B-10 *Critical values of L*†

number of rankings or replications m	n = 3	4	5	6	7	8	9	10
	...	...	109	178	269	388	544	726
2	...	60	106	173	261	376	520	696
	28	58	103	166	252	362	500	670
	...	89	160	260	394	567	790	1,056
3	42	87	155	252	382	549	761	1,019
	41	84	150	244	370	532	736	987
	56	117	210	341	516	743	1,032	1,382
4	55	114	204	331	501	722	999	1,339
	54	111	197	321	487	701	971	1,301
	70	145	259	420	637	917	1,273	1,704
5	68	141	251	409	620	893	1,236	1,656
	66	137	244	397	603	869	1,204	1,614
	83	172	307	499	757	1,090	1,512	2,025
6	81	167	299	486	737	1,063	1,472	1,972
	79	163	291	474	719	1,037	1,436	1,927
	96	198	355	577	876	1,262	1,750	2,344
7	93	193	346	563	855	1,232	1,706	2,288
	91	189	338	550	835	1,204	1,668	2,238
	109	225	403	655	994	1,433	1,987	2,662
8	106	220	393	640	972	1,401	1,940	2,602
	104	214	384	625	950	1,371	1,900	2,549
	121	252	451	733	1,113	1,603	2,223	2,980
9	119	246	441	717	1,088	1,569	2,174	2,915
	116	240	431	701	1,065	1,537	2,131	2,859
	134	278	499	811	1,230	1,773	2,459	3,296
10	131	272	487	793	1,205	1,736	2,407	3,228
	128	266	477	777	1,180	1,703	2,361	3,169
	147	305	546	888	1,348	1,943	2,694	3,612
11	144	298	534	869	1,321	1,905	2,639	3,541
	141	292	523	852	1,295	1,868	2,592	3,478

†For given m and n, numbers represent L values at or beyond which the null hypothesis may be rejected in favor of the ordered alternative. For the .001 level use the upper figure in each cell; for .01 the middle figure; for .05 the lower figure, all probabilities one-tailed. Those values enclosed within double lines are based upon exact distributions; those for larger m and n are based upon normal-deviate approximations.

table B-10 (continued)

number of rankings or replications m	number of treatments or objects ranked n							
	3	4	5	6	7	8	9	10
	160	331	593	965	1,465	2,112	2,929	3,927
12	156	324	581	946	1,437	2,072	2,872	3,852
	153	317	570	928	1,410	2,035	2,822	3,788
	172	358	642	1,044	1,585	2,285	3,163	4,241
13	169	350	628	1,022	1,553	2,240	3,104	4,164
	165	343	615	1,003	1,525	2,201	3,052	4,097
	185	384	689	1,121	1,702	2,453	3,397	4,556
14	181	376	674	1,098	1,668	2,407	3,335	4,475
	178	368	661	1,078	1,639	2,367	3,281	4,405
	197	410	736	1,197	1,818	2,622	3,631	4,869
15	194	402	721	1,174	1,784	2,574	3,567	4,786
	190	394	707	1,153	1,754	2,532	3,511	4,714
	210	436	783	1,274	1,935	2,790	3,864	5,183
16	206	427	767	1,249	1,899	2,740	3,798	5,096
	202	420	754	1.228	1,868	2,697	3,741	5,022
	223	463	830	1,350	2,051	2,958	4,098	5,496
17	218	453	814	1,325	2,014	2,907	4,029	5,407
	215	445	800	1,303	1,982	2,862	3,970	5,330
	235	489	876	1,427	2,167	3,126	4,330	5,808
18	231	479	860	1,401	2,130	3,073	4,260	5,717
	227	471	846	1,378	2,097	3,028	4,199	5,638
	248	515	923	1,503	2,283	3,294	4,563	6,121
19	243	505	906	1,476	2,245	3,240	4,491	6,027
	239	496	891	1,453	2,217	3,193	4,428	5,946
	260	541	970	1,579	2,399	3,461	4,796	6,433
20	256	531	953	1,552	2,360	3,406	4,722	6,337
	251	522	937	1,528	2,325	3,358	4,657	6,253
	273	567	1,017	1,656	2,515	3,629	5,028	6,745
21	268	556	999	1,628	2,475	3,572	4,952	6,647
	263	547	983	1,603	2,439	3,523	4,886	6,561
	285	593	1,063	1,732	2,631	3,796	5,260	7,057
22	280	582	1,045	1,703	2,589	3,738	5,182	6,956
	275	573	1,029	1,678	2,553	3,687	5,115	6,868
	298	619	1,110	1,808	2,747	3,963	5,492	7,368
23	292	608	1,091	1,778	2,704	3,904	5,413	7,265
	288	598	1,075	1,753	2,667	3,852	5,343	7,176
	310	644	1,157	1,884	2,863	4,130	5,724	7,679
24	305	633	1,138	1,854	2,819	4,070	5,643	7,574
	300	624	1,121	1,828	2,781	4,017	5,572	7,483

table B-10 (continued)

number of rankings or replications m	number of treatments or objects ranked n							
	3	4	5	6	7	8	9	10
	322	670	1,203	1,960	2,978	4,297	5,955	7,991
25	317	659	1,184	1,929	2,934	4,235	5,873	7,883
	312	649	1,167	1,903	2,895	4,181	5,801	7,790
	335	696	1,250	2,036	3,094	4,464	6,187	8,302
26	329	685	1,230	2,004	3,048	4,401	6,103	8,192
	324	675	1,213	1,977	3,009	4,346	6,029	8,097
	347	722	1,296	2,111	3,209	4,631	6,418	8,612
27	342	710	1,276	2,080	3,163	4,567	6,332	8,501
	337	700	1,258	2,052	3,123	4,511	6,257	8,404
	360	748	1,343	2,187	3,325	4,798	6,650	8,923
28	354	736	1,322	2,155	3,277	4,732	6,562	8,810
	349	726	1,304	2,127	3,236	4,675	6,486	8,711
	372	774	1,389	2,263	3,440	4,964	6,881	9,234
29	366	762	1,368	2,230	3,392	4,898	6,792	9,118
	361	751	1,350	2,202	3,350	4,840	6,714	9,017
	385	800	1,436	2,339	3,555	5,131	7,112	9,544
30	379	787	1,414	2,305	3,506	5,063	7,021	9,426
	373	777	1,396	2,276	3,464	5,004	6,942	9,324
	397	825	1,482	2,415	3,670	5,297	7,343	9,854
31	391	813	1,460	2,381	3,621	5,228	7,251	9,735
	385	802	1,441	2,351	3,578	5,168	7,170	9,631
	409	851	1,528	2,490	3,786	5,464	7,574	10,164
32	403	839	1,506	2,456	3,735	5,394	7,480	10,043
	398	827	1,487	2,426	3,691	5,333	7,398	9,937
	422	877	1,575	2,566	3,901	5,630	7,804	10,474
33	415	864	1,552	2,531	3,849	5,559	7,709	10,351
	410	853	1,533	2,500	3,805	5,497	7,627	10,244
	434	903	1,621	2,641	4,016	5,796	8,035	10,784
34	428	890	1,598	2,606	3,964	5,724	7,939	10,659
	422	878	1,579	2,575	3,918	5,661	7,855	10,550
	447	929	1,667	2,717	4,131	5,963	8,266	11,094
35	440	915	1,644	2,681	4,078	5,889	8,168	10,967
	434	904	1,624	2,649	4,032	5,825	8,083	10,856
	459	954	1,714	2,793	4,246	6,129	8,496	11,404
36	452	941	1,690	2,756	4,192	6,054	8,397	11,275
	447	929	1,670	2,724	4,146	5,990	8,311	11,163
	471	980	1,760	2,868	4,361	6,295	8,727	11,713
37	465	966	1,736	2,831	4,307	6,219	8,626	11,583
	459	954	1,716	2,799	4,259	6,154	8,538	11,469

table B-10 (continued)

number of rankings or replications m	number of treatments or objects ranked n							
	3	4	5	6	7	8	9	10
	484	1,006	1,806	2,944	4,476	6,461	8,957	12,023
38	477	992	1,782	2,906	4,421	6,385	8,855	11,890
	471	980	1,761	2,873	4,373	6,318	8,766	11,775
	496	1,032	1,853	3,019	4,591	6,627	9,188	12,332
39	489	1,018	1,828	2,981	4,535	6,549	9,084	12,198
	483	1,005	1,807	2,948	4,486	6,482	8,994	12,081
	508	1,057	1,899	3,094	4,705	6,793	9,418	12,641
40	501	1,043	1,874	3,056	4,649	6,714	9,313	12,506
	495	1,031	1,853	3,022	4,600	6,646	9,222	12,387
	521	1,083	1,945	3,170	4,820	6,959	9,648	12,951
41	514	1,069	1,920	3,131	4,763	6,879	9,542	12,813
	507	1,056	1,898	3,097	4,713	6,810	9,450	12,694
	533	1,109	1,991	3,245	4,935	7,124	9,878	13,260
42	526	1,094	1,966	3,206	4,877	7,044	9,771	13,121
	520	1,081	1,944	3,171	4,827	6,974	9,677	13,000
	545	1,134	2,037	3,321	5,050	7,290	10,108	13,569
43	538	1,120	2,012	3,281	4,991	7,209	10,000	13,428
	532	1,107	1,990	3,246	4,940	7,138	9,905	13,306
	558	1,160	2,084	3,396	5,164	7,456	10,338	13,878
44	550	1,145	2,058	3,355	5,105	7,374	10,228	13,736
	544	1,132	2,035	3,320	5,054	7,302	10,133	13,611
	570	1,186	2,130	3,471	5,279	7,622	10,568	14,187
45	563	1,171	2,104	3,430	5,219	7,539	10,457	14,043
	556	1,157	2,081	3,395	5,167	7,466	10,360	13,917
	582	1,211	2,176	3,547	5,394	7,787	10,798	14,496
46	575	1,196	2,150	3,505	5,333	7,703	10,686	14,350
	568	1,183	2,126	3,469	5,280	7,630	10,588	14,223
	595	1,237	2,222	3,622	5,508	7,953	11,028	14,804
47	587	1,222	2,195	3,580	5,447	7,868	10,914	14,657
	580	1,208	2,172	3,544	5,394	7,794	10,815	14,529
	607	1,263	2,268	3,697	5,623	8,119	11,258	15,113
48	599	1,247	2,241	3,655	5,561	8,033	11,143	14,964
	593	1,234	2,218	3,618	5,507	7,958	11,043	14,835
	619	1,288	2,314	3,772	5,737	8,284	11,487	15,422
49	612	1,273	2,287	3,730	5,675	8,197	11,371	15,271
	605	1,259	2,263	3,692	5,621	8,122	11,271	15,141
	632	1,314	2,360	3,848	5,852	8,450	11,717	15,730
50	624	1,298	2,333	3,804	5,789	8,362	11,600	15,579
	617	1,284	2,309	3,767	5,734	8,286	11,498	15,446

Source: Reprinted from Page (1963) by permission of E. B. Page and the American Statistical Association.

table B-11 *Orthogonal-polynomial coefficients*

$\sum c_{ij}^2$	number of treatment levels		1	2	3	4	5	6	7	8	9	10	11	12	13	14	15	16
2	3	Linear	−1	0	1													
6		Quadratic	1	−2	1													
20	4	Linear	−3	−1	1	3												
4		Quadratic	1	−1	−1	1												
20		Cubic	−1	3	−3	1												
10	5	Linear	−2	−1	0	1	2											
14		Quadratic	2	−1	−2	−1	2											
10		Cubic	−1	2	0	−2	1											
70	6	Linear	−5	−3	−1	1	3	5										
84		Quadratic	5	−1	−4	−4	−1	5										
180		Cubic	−5	7	4	−4	−7	5										
28	7	Linear	−3	−2	−1	0	1	2	3									
84		Quadratic	5	0	−3	−4	−3	0	5									
6		Cubic	−1	1	1	0	−1	−1	1									
168	8	Linear	−7	−5	−3	−1	1	3	5	7								
168		Quadratic	7	1	−3	−5	−5	−3	1	7								
264		Cubic	−7	5	7	3	−3	−7	−5	7								
60	9	Linear	−4	−3	−2	−1	0	1	2	3	4							
2,772		Quadratic	28	7	−8	−17	−20	−17	−8	7	28							
990		Cubic	−14	7	13	9	0	−9	−13	−7	14							
330	10	Linear	−9	−7	−5	−3	−1	1	3	5	7	9						
132		Quadratic	6	2	−1	−3	−4	−4	−3	−1	2	6						
8,580		Cubic	−42	14	35	31	12	−12	−31	−35	−14	42						
110	11	Linear	−5	−4	−3	−2	−1	0	1	2	3	4	5					
858		Quadratic	15	6	−1	−6	−9	−10	−9	−6	−1	6	15					
4,290		Cubic	−30	6	22	23	14	0	−14	−23	−22	−6	30					
572	12	Linear	−11	−9	−7	−5	−3	−1	1	3	5	7	9	11				
12,012		Quadratic	55	25	1	−17	−29	−35	−35	−29	−17	1	25	55				
5,148		Cubic	−33	3	21	25	19	7	−7	−19	−25	−21	−3	33				

table B-11 (continued)

Σc_{ij}^2	number of treatment levels		1	2	3	4	5	6	7	8	9	10	11	12	13	14	15	16
182	13	Linear	−6	−5	−4	−3	−2	−1	0	1	2	3	4	5	6			
2,002		Quadratic	22	11	2	−5	−10	−13	−14	−13	−10	−5	2	11	22			
572		Cubic	−11	0	6	8	7	4	0	−4	−7	−8	−6	0	11			
910	14	Linear	−13	−11	−9	−7	−5	−3	−1	1	3	5	7	9	11	13		
728		Quadratic	13	7	2	−2	−5	−7	−8	−8	−7	−5	−2	2	7	13		
97,240		Cubic	−143	−11	66	98	95	67	24	−24	−67	−95	−98	−66	11	143		
280	15	Linear	−7	−6	−5	−4	−3	−2	−1	0	1	2	3	4	5	6	7	
37,128		Quadratic	91	52	19	−8	−29	−44	−53	−56	−53	−44	−29	−8	19	52	91	
39,780		Cubic	−91	−13	35	58	61	49	27	0	−27	−49	−61	−58	−35	13	91	
1,360	16	Linear	−15	−13	−11	−9	−7	−5	−3	−1	1	3	5	7	9	11	13	15
5,712		Quadratic	35	21	9	−1	−9	−15	−19	−21	−21	−19	−15	−9	1	9	21	35
1,007,760		Cubic	−455	−91	143	267	301	265	179	63	−63	−179	−265	−301	−267	−143	91	455

table B-12 *Inverse sine transformations†*

proportion (X)	$2 \sin^{-1} \sqrt{X}$	proportion (X)	$2 \sin^{-1} \sqrt{X}$	proportion (X)	$2 \sin^{-1} \sqrt{X}$
.01	.2003	.36	1.2870	.71	2.0042
.02	.2838	.37	1.3078	.72	2.0264
.03	.3482	.38	1.3284	.73	2.0488
.04	.4027	.39	1.3490	.74	2.0715
.05	.4510	.40	1.3694	.75	2.0944
.06	.4949	.41	1.3898	.76	2.1176
.07	.5355	.42	1.4101	.77	2.1412
.08	.5735	.43	1.4303	.78	2.1652
.09	.6094	.44	1.4505	.79	2.1895
.10	.6435	.45	1.4706	.80	2.2143
.11	.6761	.46	1.4907	.81	2.2395
.12	.7075	.47	1.5108	.82	2.2653
.13	.7377	.48	1.5308	.83	2.2916
.14	.7670	.49	1.5508	.84	2.3186
.15	.7954	.50	1.5708	.85	2.3462
.16	.8230	.51	1.5908	.86	2.3746
.17	.8500	.52	1.6108	.87	2.4039
.18	.8763	.53	1.6308	.88	2.4341
.19	.9021	.54	1.6509	.89	2.4655
.20	.9273	.55	1.6710	.90	2.4981
.21	.9521	.56	1.6911	.91	2.5322
.22	.9764	.57	1.7113	.92	2.5681
.23	1.0004	.58	1.7315	.93	2.6061
.24	1.0239	.59	1.7518	.94	2.6467
.25	1.0472	.60	1.7722	.95	2.6906
.26	1.0701	.61	1.7926	.96	2.7389
.27	1.0928	.62	1.8132	.97	2.7934
.28	1.1152	.63	1.8338	.98	2.8578
.29	1.1374	.64	1.8546	.99	2.9413
.30	1.1593	.65	1.8755		
.31	1.1810	.66	1.8965		
.32	1.2025	.67	1.9177		
.33	1.2239	.68	1.9391		
.34	1.2451	.69	1.9606		
.35	1.2661	.70	1.9823		

†Angles are reported in radians.

references

Allen, D. W., & McDonald, F. J. The effects of self-selection on learning in programmed instruction. *Amer. Educ. Res. J.*, 1966, **3,** 1–6.

Bartlett, M. S. A note on the multiplying of factors for various chi-squared approximations. *J. Royal Statist. Soc.*, Series B, 1954, **16,** 296–298.

Boersma, F. J., DeJonge, J. J., & Stellwagen, W. R. A power comparison of the F and L tests. *Psychol. Bull.*, 1964, **71,** 505–513.

Boneau, C. A. The effects of violations of assumptions underlying the t test. *Psychol. Bull.*, 1960, **57,** 49–64.

Box, G. E. P. Problems in the analysis of growth and wear curves. *Biometrics*, 1950, **6,** 362–389.

Box, G. E. P. Non-normality and tests on variance. *Biometrika*, 1953, **40,** 318–335.

Box, G. E. P. Some theorems on quadratic forms applied in the study of analysis of variance problems. *Ann. Math. Statist.*, 1954, **25,** 290–302.

Campbell, D. T., & Stanley, J. C. *Experimental and quasi-experimental designs for research.* Chicago: Rand McNally, 1963.

Cochran, W. G., & Cox, G. M. *Experimental designs.* New York: Wiley, 1957.

Cooley, W. W., & Lohnes, P. R. *Multivariate procedures for the behavioral sciences.* New York: Wiley, 1962.

Cornfield, J., & Tukey, J. W. Average values of mean squares in factorials. *Ann. Math. Statist.*, 1956, **27,** 907–949.

Duncan, D. B. Multiple range and multiple *F* tests. *Biometrics*, 1955, **11,** 1–42.

Dunnett, C. W. A multiple comparison procedure for comparing several treatments with a control. *J. Amer. Statist. Assoc.*, 1955, **50,** 1096–1121.

Dunnett, C. W. New tables for multiple comparisons with a control. *Biometrics*, 1964, **20,** 482–491.

Edwards, A. L. *Experimental design in psychological research.* (Rev. ed.) New York: Holt, 1960.

Edwards, A. L. *Expected values of discrete random variables and elementary statistics.* New York: Wiley, 1964.

Evans, S. H., & Anastasio, E. J. Misuse of analysis of covariance when treatment effect and covariate are confounded. *Psychol. Bull.*, 1968, **69,** 225–234.

Faddeeva, V. N. *Computational methods of linear algebra.* New York: Dover, 1959.

Faust, G. W., & Anderson, R. C. Effects of incidental material in a programmed Russian vocabulary lesson. *J. Educ. Psychol.*, 1967, **58,** 3–10.

Federer, W. T. *Experimental design, theory and application.* New York: Macmillan, 1955.

Feldt, L. S. A comparison of the precision of three experimental designs employing a concomitant variable. *Psychometrika*, 1958, **23,** 335–354.

Fisher, R. A. *The design of experiments.* New York: Hafner, 1960.

Glass, G. V. Testing homogeneity of variance. *Amer. Educ. Res. J.*, 1966, **3,** 187–190.

Goldhaber, G. M. Listener comprehension of compressed speech when the difficulty level of the content and the sex of the listener are varied. Unpublished M. A. thesis, University of Maryland, 1967.

Herman, W. L., Potterfield, J. E., Dayton, C. M., & Amershek, K. G. The relationship of teacher-centered activities and pupil-centered activities to pupil achievement in 18 fifth-grade social studies classes. *Amer. Educ. Res. J.*, 1969, **6,** 227–239.

Hopkins, K. D., & Chadbourn, R. A. A schema for proper utilization of multiple comparisons in research and a case study. *Amer. Educ. Res. J.*, 1967, **4,** 407–412.

Horst, P. *Matrix algebra for social scientists.* New York: Holt, 1963.

Jaech, J. L. An alternate approach to missing value estimation. *Amer. Statist.*, 1966, **20,** 27–29.

Jester, R. E., & Travers, R. M. W. The effect of various presentation patterns on the comprehension of speeded speech. *Amer. Educ. Res. J.*, 1967, **4,** 353–360.

Johnson, R. B. The effects of prompting, practice and feedback in programmed videotape. *Amer. Educ. Res. J.*, 1968, **5,** 73–79.

Kendall, M. G., & Stuart, A. *The advanced theory of statistics.* Vol. 2. London: Griffin, 1961.

Lang, G., & Hochman, I. Teacher aide service as a means of enriching a sophomore course in educational psychology. *Amer. Educ. Res. J.*, 1966, **3,** 113–124.

Lindquist, E. F. *Design and analysis of experiments in psychology and education.* Boston: Houghton Mifflin, 1953.

Mann, H. B. *Analysis and design of experiments.* New York: Dover, 1949.

Merrill, M. D., & Stolurow, L. M. Hierarchical preview vs. problem oriented review

in learning an imaginary science. *Amer. Educ. Res. J.*, 1966, **3,** 251–261.
Morrison, D. F. *Multivariate statistical methods.* New York: McGraw-Hill, 1967.
Myers, J. L. *Fundamentals of experimental design.* Boston: Allyn & Bacon, 1966.
Page, E. B. Ordered hypotheses for multiple comparisons: a significance test for linear ranks. *J. Amer. Statist. Assoc.*, 1963, **58,** 216–230.
Pearson, E. S., & Hartley, H. O. *Biometrika tables for statisticians.* (3d ed.) Vol. 1. New York: Cambridge, 1966.
Peng, K. C. *The design and analysis of scientific experiments.* Reading, Mass.: Addison-Wesley, 1967.
Potts, M. The effect of second-language instruction on the reading proficiency and general school achievement of primary-grade children. *Amer. Educ. Res. J.*, 1967, **4,** 367–373.
Rao, C. R. *Advanced statistical methods in biometric research.* New York: Wiley, 1952.
Ray, W. S. *An introduction to experimental design.* New York: Macmillan, 1960.
Scannell, D. P., & Marshall, J. C. The effect of selected composition errors on grades assigned to essay examinations. *Amer. Educ. Res. J.*, 1966, **3,** 125–130.
Scheffé, H. *The analysis of variance.* New York: Wiley, 1959.
Schurdak, J. J. An approach to the use of computers in the instructional process and an evaluation. *Amer. Educ. Res. J.*, 1967, **4,** 59–73.
Stevens, S. S. (Ed.) *Handbook of experimental psychology,* New York: Wiley, 1951, pp. 1–49.
Stevens, S. S. Measurement, statistics, and the schemapiric view. *Science,* 1968, **161,** 849–856.
Tagatz, G. E. Effects of strategy, sex, and age on conceptual behavior of elementary school children. *J. Educ. Psychol.*, 1967, **58,** 103–109.
Williams, J. P., & Levy, E. I. Retention of introductory and review programs as a function of response mode. *Amer. Educ. Res. J.*, 1964, **1,** 211–218.
Winer, B. J. *Statistical principles in experimental design.* New York: McGraw-Hill, 1962.

index

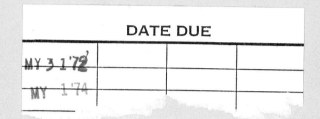